THE NEW DIGITAL ERA GOVERNANCE

THE NEW DIGITAL ERA GOVERNANCE

HOW NEW DIGITAL TECHNOLOGIES ARE SHAPING PUBLIC GOVERNANCE

edited by:

Evrim Tan and Joep Crompvoets

Wageningen Academic
Publishers

EAN: 9789086863785
e-EAN: 9789086869305
ISBN: 978-90-8686-378-5
eISBN: 978-90-8686-930-5
DOI: 10.3920/978-90-8686-930-5

First published, 2022

Wageningen Academic Publishers,
P.O. Box 220,
NL-6700 AE Wageningen,
The Netherlands.
www.WageningenAcademic.com
copyright@WageningenAcademic.com

Table of contents

Acknowledgements

We would like to acknowledge the support of all those who, in various ways, have contributed to the successful completion of the book. We thank first and foremost our respective families for their unconditional love and support over the years.

The most serious expression of gratitude is reserved for those authors who contributed to this book. Our collaboration through the DIGI4FED project has been a blessing for us. Without their expertise and support, this book would not have been possible. We also deeply appreciate the participation and feedback of colleagues in the follow-up and steering committees of DIGI4FED. Their guidance and suggestions helped us significantly improve our research. We also extend our gratitude to our colleagues at KU Leuven, UNamur, ULiège, and UAntwerpen, who did not contribute to the chapters directly, but in various ways, their support guided us throughout the completion of individual chapters.

Last but not the least, we would like to thank Belgian Federal Science Policy Office (BELSPO) for their financial support, and for providing us with the means to conduct the research that led up to this volume.

Evrim Tan

Joep Crompvoets

Endorsements

Public sector digitalization receives more attention than ever. Clients expect their public organizations to confidently shape digital environments and to be prepared for the challenges of our and coming times. The book 'The new digital era governance' identifies pathways to apply the most recent technologies creating a reliable public sector of the future.

Jörg Becker,
University of Münster, Professor Chair for Information Systems and Information Management

Shifting to digital societies includes not only digital public governance but requires a consolidated public governance of this digital era. This book is about unpacking this digital governance agenda. Tan and Crompvoets show very clearly how to upgrade a traditional management and governance agenda to the next decades.

Geert Bouckaert,
KU Leuven and University College London, Professor at the Public Governance Institute

Digital era governance has emerged as an increasingly important, perhaps now even the dominant, 'quasi-paradigm' for analysing key changes in contemporary public administration. In the current third wave of digital-era changes (accelerated by the Covid-19 pandemic) manifold issues have opened up around the use of artificial intelligence, blockchain, and automated decision-making in liberal democratic governance. This edited collection provides many insights into this rapidly changing picture and will be of wide interest for scholars of modern public management

Patrick Dunleavy,
London Schools of Economics, Emeritus Professor of Political Science and Public Policy

'The new digital era governance' outlines the opportunities and threats that come with adopting big data, blockchain technology, artificial intelligence, and other technical innovations in government. The book notes the important role government employees, politicians, data scientists, and legal experts play in working together to achieve effective public governance – often by applying strategies of learning while doing through policy experiments and living labs. The book offers a useful, up-to-date assessment on the contemporary transition from e-government to digital governance.

Mary K. Feeney,
University of Arizona, Professor and Lincoln Professor of Ethics in Public Affairs

New digital technologies are transforming the playing field. For those looking into technologies for the digital era this is a perfect primer to understand what is going on. The works will help you to question your organization's choices and provide starting points for realizing the benefits of the technologies. This comprehensive book will help you to gain new insights and practical recommendations which are based on the Belgium situation to move forward.

Marijn Janssen,
Technical University Delft, Full Professor and the Antoni van Leeuwenhoek chair in ICT and Governance

Evrim Tan and Joep Crompvets have – together with an impressive list of scholars – put together an important corner-stone volume on the issues of digital transformation of our public space. This includes their interdisciplinary analysis of how emerging technologies such as blockchain or artificial intelligence will impact our future interaction and collaboration with the state. As such it is a must read for scholars and practitioners alike!

Robert Krimmer,
Tartu University, ERA Chair professor of e-governance

This book answers many issues happening currently in Digital Government. First of all, this book addresses a wide coverage of areas, and the challenges and issues this book is addressing are common to many countries and public services. There is not enough space to detail the relevance and importance of every chapter but it is important to note that the first question to reflect on is public value creation. This question is fundamental for the design, development and delivery of any public service not only to make public services more efficient but also as a way to ensure trust. The book addresses the importance of the transformation brought by emerging technologies on digital public governance and also the governance of these emerging technologies, including the importance of interoperability (based on the European Interoperability Framework). Finally, the book outlines the importance of citizen-driven innovation through emerging technologies. For all the reasons above, I think this book constitutes a milestone in public administration research, will certainly provide many references to future research, and serve as an inspiration to implementers.

Georges Lobo,
European Commission, Program manager at DIGIT

This book provides a framework on how we can think about the new digital era governance. Specifically, it helps readers understand how to conceptualize data-driven service delivery through technologies that automate and at the same time dehumanize public service delivery. The authors provide important insights on how new digital era governance will contribute to public value creation and ultimately trust in government operations.

Ines Mergel,
University of Konstanz, Professor of Public Administration – Digital Governance

For three reasons I highly recommend this book for those interested in the policy and practice of digital governance and e-government. Firstly, it creates clarity in complexity, elucidating the complex interplay of public administration and digital technologies. Secondly, it is based on a sound combination of theory and practice. Thirdly, it presents the reader a solid understanding of the impact of two foundational digital technologies, artificial intelligence and blockchain, on public administrations and the governance challenges of these technologies.

Paul H. Timmers,
University of Oxford, Research Associate and former director at the European Commission for Digital Society, Trust & Cybersecurity

Chapter 1

A new era of digital governance

E. Tan* and J. Crompvoets

KU Leuven, Public Governance Institute, Parkstraat 45 – bus 3609, 3000 Leuven, Belgium; evrim.tan@kuleuven.be

Abstract

This chapter introduces the 'New Digital Era Governance' as a new paradigm in public administration scholarship by compounding on the changing roles of digital technologies in public governance. Particularly, the chapter highlights the challenges inflicted upon public administration and society with the advancements of new digital technologies, such as big data, artificial intelligence, and blockchain. By comparing the changing role of digital technologies in government functions, public value creation, human resources, and governance, the chapter asserts that the theoretical and practical implications of 'New Digital Era Governance' differ from Dunleavy *et al.*'s 'Digital Era Governance'.

Keywords: new digital technologies, Digital Era Governance, data, public governance, government functions

1.1 Introduction

> As Gregor Samsa awoke one morning from uneasy dreams, he found himself transformed in his bed into a monstrous vermin. He was lying on his hard, as it were armour-plated, back and when he lifted his head a little he could see his domelike brown belly divided into stiff arched segments on top of which the bed quilt could hardly keep in position and was about to slide off completely. His numerous legs, which were pitifully thin compared to the rest of his bulk, waved helplessly before his eyes. 'What has happened to me?' he thought. It was no dream. (The Metamorphosis, F. Kafka)

Kafka's The Metamorphosis is often interpreted as an allegory of human suffering propagated by the alienation to self-image and its surroundings at the advent of modernity. In the novella, Gregor Samsa tries to adjust himself in his new form but finds himself in growing pain to fit into his social reality and eventually chooses to starve himself to death. Public administration finds itself today in a similar predicament. The surge of 'disruptive technologies' such as artificial intelligence, blockchain, Internet of Things, augmented and virtual reality, robotics, and so forth, are forcing public administration to engage in a rapid transformation of traditional modes of operation in search of new business models, products and services. Flourishing new and diverse data sources to gain insight from, rapid advancements in data processing technologies

DOI 10.3920/978-90-8686-930-5_1, © Wageningen Academic Publishers 2022

and capabilities, and changing societal expectations in the way that citizens, organisations, and governments should interact herald a new digital-era governance. In a short span of time in the 2010s, new governance ideas and regulatory techniques (e.g. e-governance, agile governance), new organisational forms (e.g. open government, government by algorithm, government as a platform), and new means of engagement between citizens and governments (e.g. smart cities, digital platforms) have emerged to cultivate public value using digital technologies. Today, digital technologies are not only recognised as strategic drivers for improving public sector efficiency or bringing issues back under government control, but they are at the heart of government transformation into open, transparent, innovative, participatory, and trustworthy governments (OECD, 2014). However, the multiplication of technological options, automation of public policy processes, and the replacement of humans with digital agents are challenging the skills and capabilities existing in public administration and risking the alienation of public servants from their traditional role, responsibilities, and behaviours. Being subjected to brisk techno-social changes, could the gloomy story of Gregor Samsa be what awaits public administration in this new digital era?

This book focuses on this transformation experienced by public administration by expounding on the changes in administrative and business processes, legislative and regulatory actions, and perceptions and reactions inside and outside administration. Particularly, the promise of the book is to provide a better understanding of the challenges inflicted upon public administration and society with the advancements of new digital technologies such as big data, artificial intelligence, and blockchain and to advance our theoretical foundations for the widescale adoption of these technologies in public governance.

1.2 From e-government to a new era of digital governance

Technological innovation has shaped and transformed society again and again in the past, and the radical transformation of public administration through the adoption of digital technologies is not necessarily a new phenomenon. So, why would we think that the impact of these new types of digital technologies would be any different from the past experiences with technological innovation? Indeed, in the mid-2000s, Dunleavy and his colleagues had already been declaring the replacement of the new public management (NPM) era – the dominant public administration ideology of the 1990s – with an emerging 'Digital-Era Governance', whereby the government changes at its core by focussing on the reintegration of services, and holistic, and joined-up approaches to policymaking through the extensive digitalisation of administrative operations (Dunleavy *et al.*, 2006: 7). During the same period, many governments announced their national strategies for the digital transformation of public services using information and communication technologies (ICT) to improve public service performance. Through the conceptualisation of e-government and digital government, paper-based processes and services are gradually being transformed into digital services. The transition from paper-based to digital-based systems is accompanied by infrastructural, administrative, and cultural shifts in public administration. Digitalisation was not exclusive to the public sector, and specifically, with the proliferation of smartphones and social media, societies experienced a rapid transition in the use of digital technologies for business and public services. International organisations, such as OECD, UN, and the EU developed multiple metrics and indexes to gauge the advancement in the digital capabilities of public administration and society. The findings show a global trend in growing practices and capabilities in the use of ICT in government services. For instance, the UN e-government development index (EGDI) surveys from 2008 to 2020 shows that global e-government capacities have been improving gradually both at a *vertical* level, i.e. increasing level of complexity in e-government services, and at a *horizontal* level, i.e. increase in the uptake of e-government services. As of 2020, 65% of 193 countries are categorised in the high or very high EGDI group, and more than 84% of countries now offer at least one online transactional service (e.g. application for birth certificate, paying for public utilities) (UN, 2020). In a similar vein, as of 2021, 51.4% of the global population uses the Internet because of the widespread increase in fixed and mobile broadband connections.[1] Notwithstanding the positive correlation between EGDI ranking and the income level of a country, a strong political will, strategic leadership, and the commitment to expanding the provision of digital services are found to be key factors for a country to achieve a higher EGDI ranking (UN, 2020).

These past experiences suggest that through investment in infrastructure and human capital and with the right managerial approaches, it is possible to exploit the benefits of digital innovation for governments and improve the quality and extent of public services. Hence, why should we be wary of the impact of the new digital technologies and rethink our assumptions and theories about the impact of digital technologies in public governance? To answer this question, we need to look at the changing role of digital technologies in government functions, public value creation, human resources, and governance.

1.3 Changing roles of digital technologies in government functions

According to Hood (1983), the key administrative function of governments is to control society. To this end, Hoods assigns four types of resources to government, namely *nodality, authority, treasure,* and *organisation,* from which government can draw its tools for *detecting* the information in society and *effecting* upon that information to make an impact on society. Nodality infers that government holds a central position at the junction of the information channels in a society, and government can act upon this position to collect and disseminate information. Authority denotes the legal and official power to demand, forbid, guarantee, and adjudicate. Treasure refers to the financial means or fungible assets held by governments that can be used as a means of trying to influence outsiders or as a way of purchasing information and talents. Organisation is the physical ability of government to act directly using its own human resources. The idea is that by combining these four properties with detectors and effectors, government uses eight different kinds of tools to steer society according to its will.

[1] The data for this research is accessible at https://www.statista.com/study/12322/global-internet-usage-statista-dossier/.

Table 1.1. Tools of government in the first era of digital governance.

Type of government tool	Detectors	Effectors
Nodality	Online tools to acquire information from citizens and media (e.g. e-mails, media-monitoring systems)	Online tools to disseminate information to citizens and media (e.g. government websites, information databases)
Authority	Electronic databases to store information on citizenship records, rights, benefits and liabilities (e.g. taxation, ID systems, etc.)	Online systems to post information about legal and regulatory issues (e.g. penalty systems, price and service information systems used in government regulation of financial or product markets)
Treasure	Computerised forms to collect information from citizens and businesses about financial issues (e.g. tax, income, profits, and property registries)	Online and/or digital payment systems for citizens and businesses (e.g. welfare state benefit payment systems, automated traffic-charging systems for cities)
Organisation[1]	Management information and monitoring systems inside and outside the administration	Grants and information transfer systems for sustaining public sector delivery chains (e.g. medical record systems in public health care systems)

[1] Dunleavy *et al.* (2006) add one additional category of 'expertise' by splitting the category of organisation. They use 'expertise' to refer to the accumulation of knowledge and professional expertise on how to handle social problems in government agencies. In our categorisation, we treat such tools under the organisation category, and do not include them as a separate category.

Building upon this analytical framework, Hood and Margetts (2006) and Dunleavy *et al.* (2006) assign different tools for governments in the digital age. Table 1.1 illustrates the type of tools that are assigned for each category of tools in the first era of digital governance.[2]

What are the mutual characteristics of these tools? The innovations in the first era of digital governance are primarily built upon the advancements in ICT. The digital innovation in governments was about the establishment of the ICT infrastructure and its integration into the external telecommunication networks. Through the utilisation of computerisation and the Internet at private and public domains, governments attained new capabilities and tools to collect and disseminate information. Dunleavy *et al.* (2006) underline three immediate impacts of these digital tools on government functions. One imminent impact is on the reintegration of

[2] 'First era of digital governance' or 'first digital-era governance' is used interchangeably in the remainder of the chapter to denote the ICT-centric digital governance.

government services and reclamation of government services that had been dispersed previously through the phases of agentification and privatisation in the New Public Management (NPM) era. Digital technologies brought a new set of toolkits to facilitate information exchange across various departments and agencies and led to the creation of centralised back offices for data storage and processing. Although it is debatable to what extent the rollback of agentification in the post-NPM area can be attributed solely to digital technologies, it is plausible to assume that the creation of centralised services to better coordinate information management systems contributed to the reinstating central processes both inside and across public sector organisations. The second imminent impact of digital technologies appears in the needs-based holism denoting a shift towards a citizen-based, services-based, or needs-based foundation for organisation rather than the operational efficiencies stressed by NPM. Creation of one-stop shops, ask-one principle, end-to-end service re-engineering, and interactive information-giving and seeking characterise some modes of digital government associated with needs-based holism. The third imminent impact is the organisational, behavioural, and cultural changes inside the government sector brought about by the digitisation changes in public administration that are often encapsulated by e-government. Here the digitisation changes refer to the shift in communication from anachronistic to ICT-driven means in administrative processes.

These tools from the first era of digital governance heralded by no means trivial changes in government functions, but the digital technologies, in and of themselves, brought about no great and intrinsic transformation of the government toolkit (Perri 6, 2008). Utilising the new communication capabilities brought about by the Internet, the government tools from the first era of digital governance mostly resembled the digitalised versions of paper-based services and methods in detectors and effectors. Furthermore, digital transformation of government services primarily augmented capabilities for detection rather than effecting. The computerisation of paper-based services allows faster and more efficient ways of extracting information from society and businesses but using this information to feed in the government services requires running costs and capital investments in information management systems. These costs are not exclusive to IT infrastructures but are also about building HR capabilities to process and use the information in the actual policymaking and service provision areas. Moreover, and partially because of the former point, the integration of digital technologies in administrative processes inherently introduces new coordination challenges and strains on existing organisational and financial capabilities. Data silos, diverging protocols and visions between different sectors and departments, and cybersecurity issues are some of the operational challenges associated with digital transformation of administration. All in all, the impact of these digital technologies is bound to the empirical context of policy field and wider social organisation.

Now, we will use the same analytical framework to show the impact of new digital technologies on the government toolkit (Table 1.2).

The new era of digital governance is all about *data*. Unlike the communication-driven first digital era, the data-driven public services mark the government toolkit of new digital technologies. New digital technologies are more diverse and their multifarious influence on data ranges from

Table 1.2. Tools of government in the new era of digital governance.

Type of government tool	Detectors	Effectors
Nodality	Machine learning algorithms working on big data (e.g. social media, data held by private, non-public, or other governmental sources) to detect new patterns Edge and cloud computing systems for faster, more efficient, and scalable data processing	Automated agents for real-time interaction (e.g. chatbots) Personalised services and methods in citizen engagement (e.g. targeted advertisement)
Authority	Predictive analytics and credit scoring Self-sovereign identities (SSI) and decentralised identifiers (DID) Verifiable credentials	Regulatory technologies Lex cryptographia Blockchain-based authentication services
Treasure	Blockchain-based decentralised registry services (e.g. land-registry, asset registry systems) Fraud analytics in taxation, entitlement and social insurance programmes Automation of data collection from employer and state databases	Automated management of tax and social benefits Blockchained procurement, supply-chain, aid management systems Transparent and immutable blockchained systems to tracking inventory and assets
Organisation	Interoperability and open standards Data analytics officers, data protection officers	Open government data Digital platforms for citizen participation Decentralised autonomous organisations

collection to storing and processing. For instance, while big data and Internet of Things (IoT) signify the augmenting capabilities of data collection, AI technologies bring completely new sets of tools for data analysis and processing. Blockchain technologies, edge and cloud computing, make it possible to store and access a vast amount of data outside of government servers without adding significant strains on government resources. New digital technologies are also *multidimensional*, unlike the former sets of digital toolkits. The primary focus of ICT-driven government tools was striving for better, faster, and more efficient communication. Any trade-off condition to that end invokes financial, administrative, and managerial considerations about how best to implement the technology in a particular organisational context. The new digital technologies, however, unveil a new set of toolkits for governments to reflect upon issues, such as *automation, decentralisation, dehumanisation* of public services. These issues not only call for administrative, financial, managerial questions but also invoke ethical and moral challenges and trade-off conditions where more technology is not always in the best interests, either of administration or of society.

At this point, one clarification is needed in our use of the term 'new' digital technologies. In our verbiage, we do not mean that new digital technologies replaced or expunged 'older' digital technologies in government functions. On the contrary, the advancements in new digital technologies are built on the advancements achieved in the first era of digital governance. For example, the advancements in broadband connections and faster computing capabilities in public administration enabled the deployment of big data and AI-driven applications in government services. Nor do we mean that the innovations and advancements in ICT-driven government tools are halted or become inconsequential. What we mean by 'new' digital technologies is that these tools provide government with new ways of extracting information from the public and private domain and turning them into more insightful, proportionate, and faster public services. What makes the difference is that information is obtained through the exploitation of ubiquitous data in the public and private domain rather than the advancements in the transfer of information from one end to another.

Having said that, we think that there are three fundamental differences between the first and new digital-era governance regarding the role of digital technologies in government functions. First, a particular characteristic of the first digital-era governance was the enhanced coordination and centralised control achieved by the deployment of digital technologies. For instance, Dunleavy *et al.* (2006) identify nine reintegration components in the post-NPM era where digital technologies play an instrumental role in creating consolidated or shared services. In the new era of digital governance, we observe, however, a reversed trend or at least an obfuscated prospect in the reintegration of government services. For instance, there is a growing trend in the migration of government data to public clouds of third-party providers. A 2020 FedRAMP study published by Maximus and Genesys found that 49% of state/local governments in the USA have most of their systems and solutions in the cloud, while 56% of federal government offices have some cloud-based solutions and systems.[3] Similarly, a 2020 Gartner study[4] foresees double-digit growth in government use of public cloud services with spending forecast to grow on average 17.1% through 2021. The outsourcing trend is not only towards third-party service providers but also towards citizens and citizen groups. Setting open standards in the share, use and reuse of public datasets and making government data accessible to outside parties have become widespread practices in many Western countries. For instance, Data.gov of the US government, established in 2009, provides access to machine-readable datasets and other forms of open data generated by different levels of government. Similarly, Data.Europe.eu hosts more than 1.3 million datasets published by EU institutions and national governments, providing a point of access to public data for commercial and non-commercial purposes. Why are governments sharing their data freely with the public? One answer is, to improve the transparency and accountability of governments. Another answer is, to leverage the innovation potential of the public to explore new ways and means to exploit the data for commercial and societal ends. One of the central claims of the first era of digital governance is creating an 'intelligent centre', where data flows to a central node in government

[3] Available at: https://solutionsreview.com/cloud-platforms/half-of-u-s-state-and-federal-governments-heavily-use-the-cloud/.

[4] Available at: https://www.gartner.com/smarterwithgartner/understanding-cloud-adoption-in-government/.

organisation for better decision-making and efficient coordination of service processes. However, in the new digital-era governance, the predominant strategy appears to be utilising the 'hivemind' or 'collective intelligence' of people to develop better policies and practices. Although centralised data exchange platforms are still widely used, new decentralised technologies, such as distributed ledger technology and blockchain, create alternative decentralised data governance platforms. These technological advancements grow with theoretical advancements in community-based and self-governance systems for the management of public goods and resources. For example, the conceptualisation and theorisation of data as commons and the creation of data commons that collectively create, own, and use their data for the production and delivery of public services signal a reverse trend in reintegrated government services.

The second distinction is about the way services are provided to citizens and businesses. In the first digital-era governance, digital technologies allowed governments to develop more holistic organisations and policies that could address citizens' needs in (close to) real time. Creation of one-stop digital or physical shops, integration of service providers in a single digital place, and real-time access to centralised databases facilitated the production and delivery of services closer to citizens. The improved information capabilities of governments and service providers supplemented co-production and co-creation practices that were better attuned to citizens' needs. Here the efficiency and cost reduction in public services are attained by the speed and ease of communication across public institutions and citizens. In the new era of digital governance, better and more efficient public services are attained by better predictive models and automated services that are built upon past data. Unlike the first digital-era governance, faster and real-time access to information is not the central aim in the delivery of services; instead, access to more and better data[5] is the central aim to develop citizen-centric services. Predictive models are used to anticipate citizen and business needs in order to allocate the resources with better organisational and service efficiencies. Higher predictive power also allows proportional and real-time automated systems in citizen engagements, identification of risks, process optimisation, and regulatory compliance. In such a system, efficiencies and cost reductions are attained by the optimisation of business processes and replacement of human agents with artificial agents.

The third distinction concerns the way digital technologies affect the organisational processes and the overall formation of digital government. Since access to better and more data is the central aim of the new digital-era governance, reorganisation of public service provision is not focused on optimisation of information exchange but *optimisation of data exchange*. Albeit seemingly similar, the two concepts differ ontologically. Information contains an intelligible meaning in and of itself and provides the receiver with meaningful insight without any inference. Data, on the other hand, contains a collection of facts and representation of concepts in a formalised manner. To provide information, data requires processing. Another distinction between data and information is about their verification processes. To verify a piece of information, someone needs to crosscheck the source of information. Data verification, however, is more complicated. Depending on the means

[5] We use 'better data' to indicate the type of data that provides better insight into the reality and is less prone to processing error.

and methods used in processing, verification may require control of criteria used for data processing, control of the source of data, and control of the measures used for the collection of data. For certain machine-learning systems, sometimes it is simply not even possible to verify how certain information is obtained by a given source of data. Going back to the initial discussion point, in the former digital-era governance, the digitisation of services and administrative processes basically contained the translation of anachronistic methods in information exchange into the digital realm. Websites, e-mails, chatrooms, and message services made information collection and transfer cheaper, faster, and easier. However, except for the creation of IT service departments to ensure the maintenance of computation and communication systems, there was no drastic change in the administrative organisation. Conversely, new digital-era governance has brought a completely new set of roles and responsibilities in administrative units. Data analysts, data protection officers, machine-learning specialists are some new roles that have been added to administrative positions. Furthermore, these roles are not secluded – as in the case of IT – in separate technical units, but are instead often embedded in different policy units to ensure that data is properly exploited to gain insight. Another organisational change happens with restructuring of the managerial systems and rules in cross-sectoral and cross-border services in ensuring the quality standards for data exchange. Interoperability has become a key objective for public administrations. The original understanding of interoperability as the technical compatibility between different IT systems has evolved over time to include the semantic, organisational, and regulative standards to create integrated systems in data exchanges.[6] Optimisation of data exchange procedures across public sector organisations and with stakeholder organisations also support the proliferation of open standards. For example, general transit feed specification (GTFS) follows an open format and is used as the de facto standard for public transit in apps like Google Maps to find public transport schedules or in ride-sharing apps. Similarly, the US Environmental Protection Agency uses open standards for environmental data and a central data exchange to ease the data exchange among 65,000 active users. Moreover, the advancements in blockchain technology have paved the way towards alternative forms and means of verification of data held by individuals and public organisations. The traditional role of government as the purveyor and validator of public data may take on a new meaning with the advancements in self-sovereign identity systems and decentralised identifiers. For instance, the European Self-Sovereign Identity Framework (ESSIF), to be built on the European Blockchain Services Infrastructure (EBSI), creates a new means of data verification without the need for a centralised exchange through government offices but automated data verification through verifiable credentials. Verifiable credentials are on open standards and can be issued by anyone about anything and be verified by everyone.

1.4 Changing roles of digital technologies in public value creation

Public value can be described as the (added) value that an organisation provides for society. Ever since Max Weber, public value creation has been articulated as a central feature of public administration. Despite the earlier onslaught of NPM theorists against a distinctive role of the

[6] For instance, the new European interoperability framework (ISA^2) identifies legal, organisational, semantic, and technical interoperability as the four pillars of integrated public service governance.

public sector in public value creation, a reaction grew in time against the stricter focus of NPM on efficiency and parsimony principles in public management. Mark H. Moore, a principal scholar influential in the revival of public value creation, associated managerial success in the public sector 'with initiating and reshaping public sector enterprises in ways that increase their value to the public in both the short and the long run' (Moore, 1995: 10).

Although public value creation is nowadays acknowledged as fundamental in public management and governance, there is no unified view on how digital technologies and investment in digital technologies in the public sector create value for the public. Cresswell and his colleagues (2006) suggest that IT investment generates public value by improving the value of the government itself from the perspective of the citizens, and by delivering specific benefits directly to persons, groups, or the public at large. To this end, government creates financial, political, and social values. Financial value results from lowering the cost of increasing the efficiency of government or delivering financial benefits to citizens. Political value consists of increasing participation, fairness, transparency, legitimacy, or conferring political capital to elected officials or citizens. Social value includes increased social status, stronger relationships, or opportunities such as increased safety, trust in government, and economic advantage. Moreover, Panagiotopoulos and his colleagues (2019) condition the public value creation by technologies on the moderating effect of organisational capacities in service production. If organisations fail to deploy the technologies or the required capabilities, public organisations may need to extract resources externally through co-creation, platform organisations, and other collaborative multi-actor arrangements (Cabral *et al.*, 2019; Crosby *et al.*, 2017).

Furthermore, the literature shows that public value creation is not only dependent on how government creates value but also on the individual and collective views of citizens on the creation of public value (Bozeman, 2009). Timo Meynhard (2009) articulates this notion as 'public value starts and ends within the individual'. This view prioritises the perception, needs, and satisfaction of citizens and perceives public value creation as an evolutionary process linking political, social, and juridical obligations to the co-production of values in societies searching for coordination, legitimisation, and meaning (Meynhard, 2009). Bryson and his colleagues identify four dimensions of public values from citizens' view: value related to moral-ethical, value related to political-social, value related to utilitarian-instrumental, and value related to hedonistic-aesthetical (Bryson *et al.*, 2014).

Some scholars attribute this two-sided relationship to the dual nature of public value creation (Ju *et al.*, 2019; Nabatchi *et al.*, 2017; Wang *et al.*, 2021). For instance, citizen participation may satisfy the individual motivation of citizens in participation (e.g. feeling useful, socialisation processes), whereas the act itself can indirectly increase the welfare of society (e.g. justice, trust in government) (Wang *et al.*, 2021). Hence, public value creation does not need to follow linear processes; a particular act may lead to different types of value deliverance.

Having outlined public value creation processes in broad strokes, the way digital technologies create public values remains largely unexplored (Panagiotopoulos *et al.*, 2019). There have been several theoretical and empirical studies that explored the way ICT enabled government services such as websites (Karkin and Janssen, 2014; Samoilenko and Osei-Bryson, 2019), e-government functions (Cordella and Bonina, 2012; Seltsikas and O'Keefe, 2010), social media (Criado *et al.*, 2013; Grube, 2017), and mobile government (Walravens and Ballon, 2013; Wang, 2014) create public value. Furthermore, the application of ICT in open data sharing, service co-production, government platforms and cloud applications have been associated with public value creation (Cordella and Palletti, 2018, 2019; Panagiotopoulos *et al.*, 2019; Pereira *et al.*, 2017). Some recent academic research has, albeit nascently, also explored the way artificial intelligence and big data solutions affect public value creation (Chatfield and Reddick, 2018; Sun and Medaglia, 2019; Wang *et al.*, 2021; Wirtz and Müller, 2019).

In the given academic research, what can we say about the main differences between the first and new era of digital governance in public value creation? We see three fundamental differences in the way digital technologies affect public value creation: (1) shaping political and social trust; (2) shaping public service provision; and (3) shaping citizens' expectations about government.

1.4.1 Political and social trust

> The implementation of ICT in the public sector can be conceived as a tool to build public trust, to enhance confidence and to promote a more participatory citizen-government relationship, as well as a means for equitable ICT policies. (Cordella and Bonina, 2012)

Public value creation through ICT deployment has been associated with two specific goals, namely improving public governance, and enhancing the democratic process (Brewer *et al.*, 2006). While the former prioritises the value-enhancement of IT services in administrative and managerial processes, the latter emphasises the new means and ways of using ICT to foster citizen participation in decision-making processes. Through these advancements, citizens get enhanced access to the information systems of government, less administrative burden of paper-based services, and transactional and real-time exchanges with public officials. All these achievements are anticipated to engender trust in government. And indeed, studies show that the use of e-government tools and websites helps to build trust in government by improving interactions with citizens and creating perceptions of responsiveness (Mahmood *et al.*, 2020; Morgeson and Petrescu, 2011; Tolbert and Mossberger, 2006). Improved public services with lower transaction costs in participation and communication, openness and transparency of government services, are some key public value contributions associated with the first digital-era governance.

The first digital-era governance was nonetheless not without its strains on trust dimensions. Mass surveillance of citizens and privacy breaches of government agencies through backdoor programmes had significantly deteriorated trust in digital technologies and government's usage of ICT-driven surveillance techniques. Particularly, Edward Snowden's disclosures concerning the National Security Agency (NSA) and other complicit foreign agencies and a global surveillance system has created a public outcry about the establishment of democratic mechanisms to control government actions on mass surveillance. The disclosure has paved the way for several lawsuits against governments[7] and a global public debate calling for a new legal and technical framework to ensure the privacy rights of citizens against malevolent government actions.[8]

New digital technologies have in a way amplified the benefits and challenges brought by ICT on trust dimensions. On the one hand, enhanced insight from data analytics has enabled faster and more cost-effective administrative and managerial processes in public governance. Similarly, decentralised technologies and blockchain have allowed more transparent and secure means of processing personal data and citizen participation in democratic governance processes.[9]

On the other hand, new digital technologies brought up new types of trust challenges for governments. The new types of trust challenges are not only about how to keep governments accountable in the deployment of ICT tools. Rather, governments have found themselves in a precarious position to address the risks associated with the gigantic amount of data held by private service providers. In particular, the true challenge was revealed when the advancements in machine learning and other data analytics techniques showed that unscrupulous data held in various commercial and public databases could be used to gain better predictions about human behaviour and individual psychological responses. The algorithms employed by social media platforms are found to be prone to manipulation by malicious and coordinated attacks to distort the reality and thus undermine democratic processes.

In the notorious Cambridge Analytica scandal, Cambridge Analytica was able to exploit Facebook policy to access personal data from 87 million Facebook profiles to provide analytical assistance to the 2016 presidential campaigns of Ted Cruz and Donald Trump.[10] The political turmoil following the scandal led to various government actions all around the world, from the fining of Facebook by the United Kingdom's Information Commissioner's Office to probing investigations by the US Senate, Indian and Brazilian governments about Facebook's data policies. The political scandals surrounding Facebook were not only limited to the exploitation of the company's data policies by commercial actors. The special investigation report, the so-called 'Mueller report', by the Federal Bureau of Investigation (FBI) revealed that Russian government-linked Internet Research Agency (IRA) had used concentrated efforts to provoke and incite political and social discord among US citizens through various social media channels. The Mueller Report found

---

[7] For example, American Civil Liberties Union (ACLU) filed a lawsuit in 2013 against Director of National Intelligence James Clapper alleging that the NSA's phone records' program was unconstitutional (see ACLU v. Clapper No. 13-3994).

[8] As part of the political reactions, the Committee on Legal Affairs and Human Rights of Council of Europe called for a new legal and technical framework to ensure privacy rights of citizens and to protect whistleblowers against misuse of digital technologies by private and public organisations. Available at: https://assembly.coe.int/nw/xml/XRef/Xref-XML2HTML-en.asp?fileid=21583&lang=en.

[9] For instance, the city of Barcelona has tested decentralised ledger technology to integrate into the city's participatory democracy platform 'Decidim.Barcelona' to share a petition with participants, sign it with a mobile application and count the signatures in a cryptographically secure and completely anonymous way (Tan and Rodriguez Müller, 2020).

[10] https://www.nytimes.com/2018/04/04/us/politics/cambridge-analytica-scandal-fallout.html.

that the IRA spent $100,000 on more than 3,500 Facebook advertisements[11] from 2015 to 2017, which were geographically targeted and focused on socially divisive issues.[12] The disinformation campaign was grounded in the algorithmic design employed by Facebook that made it easier to target the most susceptible groups. It is difficult to gauge the actual impact of this incident on the political and societal trust of US citizens, but in the aftermaths, 'fake news' has become a moniker for distrust in any type of information released by traditional or social media sources. A 2018 Edelman report found that in a year trust in government had plummeted 30 percentage points among the informed public in a time of relative economic prosperity. The implications are not limited to the USA. The same study found that 'nearly 70% of respondents globally were concerned about 'fake news' being used as a weapon and 63% said they weren't sure how to tell good journalism from rumour or falsehoods'.[13] Concurrently, enhanced imaging techniques with deep learning have created 'deepfake' programs, to portray realistic images and motions of politicians and celebrities. US Senator Marco Rubio reacted to the advancements of deepfake programs during an interview in 2018 as:

> In the old days, if you wanted to threaten the United States, you needed 10 aircraft carriers, and nuclear weapons, and long-range missiles... Today, you just need access to our internet system, to our banking system, to our electrical grid and infrastructure, and increasingly, all you need is the ability to produce a very realistic fake video that could undermine our elections, that could throw our country into tremendous crisis internally and weaken us deeply.[14]

In this new world, the political trust challenges are not only associated with the invasive use of ICT by governments, but also with the predictive and manipulative power of data analytics and deep learning programs over individual users and behaviours. Governments are expected to take the legal and technical measures to curb the powers of private service providers that hold vast amount of data and enact the regulations and rules that ensure data protection of citizens. In 2018, in one of the first pervasive data regulation acts, the EU implemented the General Data Protection Regulation (GDPR). GDPR aims to protect European citizens against the processing of their personal information without their consent. Not only regulatory actions about data rights, but new administrative positions, such as data protection officers (DPOs) and privacy enhancing technologies are becoming more widespread in public sector organisations.

In a similar vein, the Digital Service Act (DSA) proposed by the European Commission (EC) in 2020 aims to bring content moderations on social media platforms and introduce new obligations for digital service providers to disclose how their algorithms work on content removal and on

---

[11] http://time.com/4930532/facebook-russian-accounts-2016-election/.

[12] https://www.washingtonpost.com/politics/facebook-says-it-sold-political-ads-to-russian-company-during-2016-election/2017/09/06/32f01fd2-931e-11e7-89fa-bb822a46da5b_story.html.

[13] https://www.theatlantic.com/international/archive/2018/01/trust-trump-america-world/550964/.

[14] https://www.dailysignal.com/2018/07/19/deep-fake-technology-is-a-threat-to-national-security-politics-and-the-media-rubio-says/.

the way that advertisers target users. With this regulation, the EC aims to bring a new liability to very large online platforms[15] about digital services. More recently, governments have started to explore self-sovereign identity and blockchain-based systems to engender trust in public data infrastructures and systems.

How can governments build trust for the use of new digital technologies? Better watchdog actions against misuse of personal data, opening up the government data and individual data used in public processes for outside surveillance, implementing the measures and systems where data is protected as a private property, and putting the systems in place that mitigate the exposure of data to private purposes are some new ways governments can build trust in public governance and democratic participation processes. We will take a closer look at these methods later when we talk about the citizens' expectations of governments.

1.4.2 Public service provision

The first era of digital governance has improved public service provision by transferring paper-based systems into digital domains, and by providing holistic solutions. In a similar vein, data-driven policies of the new era of digital governance have enabled individualised, tailor-made services that yield better process efficiency and higher citizen satisfaction. Both processes cut the administrative burden and allow for more agile systems in government. In a sense, the new digital-era governance has augmented the 'detectors' of the first digital-era governance and enhanced needs-based holism.

In a different vein, the new digital era has brought up better 'effectors' that were not the case in the first digital-era governance. Regulatory technologies, machine learning, and blockchain technologies allow more effective policy enforcement mechanisms. For instance, in blockchain-based systems, the rules of engagements are encoded in the system, activated by the group actions of miners, self-executed by smart contracts and a user action cannot be reversed by a third-party involvement.[16] Similarly, machine learning allows machines to implement better predictive mechanisms in application processing or fraud detection. These systems can cut the transaction costs of services by cutting the costs for monitoring, processing, and regulatory compliance, but deterministic systems, discriminatory or malpractices due to data biases, or absence of human judgements can inadvertently affect citizen satisfaction and public value creation.

We see three types of value-laden risks associated with the use of new digital technologies in public service provision.

First, risks associated with intentional or unintentional discriminatory policy practices by data-driven decision-making. For instance, China is creating a standardised and unified national data sharing system across public organisations, individuals, and corporations to reinforce a social

[15] The definition applies to online platforms with more than 45 million average monthly active users in the EU.
[16] Wright and De Filippi (2015) call this code for deterministic regulatory environments 'lex cryptographia'.

credit system to score and rank all Chinese citizens. It is not entirely clear how intrusive the system will be once it is fully functional,[17] but in one pilot, scores are determined by trivial acts, such as 'getting a traffic ticket' or by ideological elements such as 'committing a heroic act'. The promotion of people with high social credit scores as 'civilised families' and public statements stressing 'the trustworthy to roam everywhere while making it hard for the discredited to take a single step' emphasise the discriminatory nature of the intended system (Susskind, 2020: 212). Discriminatory policies can also occur by the biases embedded in the training data of algorithms. For instance, the algorithm employed by Dutch agencies to regulate childcare benefits has mistakenly labelled over 20,000 parents as fraudsters. An investigation[18] led by Dutch Data Protection Authority found out that a self-learning algorithm was used to classify the benefit claims. The algorithm disproportionally marked people with immigration backgrounds and dual citizenship as a high risk of being false. Even though the algorithm was supposed to serve as a first filter for officials, it turns out that officials just hastily approved the claims.

The second risk is that automated systems may create high overhaul and political costs in case of a problem. Above all, opaque, self-learning systems hinder the detection of problems and replacing a faulty system can sometimes be more costly than the benefits of an automated system. The automated systems do not allow quick managerial fixes in case of a problem. Once again, in the SyRI scandal surrounding the childcare benefits, the government had to resign following the incident by taking responsibility for the systematic failure and the scale of injustices caused by the algorithmic decision-making. In another example, an AI chatbot developed by Microsoft was released to learn from interacting with users on Twitter. The experiment quickly went wrong, when the chatbot started creating racist slurs and misogynistic responses.[19] The project team intervened soon after by deleting some of the inflammatory tweets, and within 16 hours the account was suspended for adjustments.[20] Soon after that, Microsoft decided to cancel the experiment indefinitely, and shelved the project. A replacement model released later was designed to avoid any potentially offensive subjects be it positive, negative, or neutral.[21] Similar issues are not only limited to machine-learning systems. For instance, following the hack of $50 million worth of funds invested in an automated investment application created on the Ethereum blockchain, the Ethereum community was divided on how to fix the problem, leading to a permanent split of Ethereum blockchain into two blockchains (Finck, 2018).

The third risk is associated with public accountability and ethical use of AI. First, opacity and unpredictability of machine-learning systems pose technical challenges as regards the extent of transparency, explainability, and accountability that can be achieved (Lim and Taeihagh, 2019; Mittelstadt *et al.*, 2016). Second, the complexity of extensive datasets used by machine-learning

---

[17] https://thediplomat.com/2021/03/chinas-social-credit-system-speculation-vs-reality/.

[18] https://www.autoriteitpersoonsgegevens.nl/sites/default/files/atoms/files/onderzoek_belastingdienst_kinderopvangtoeslag.pdf.

[19] https://www.bbc.com/news/technology-35890188.

[20] http://blogs.microsoft.com/blog/2016/03/25/learning-tays-introduction/#sm.00000gjdpwwcfcus11t6oo6dw79gw.

[21] https://qz.com/1340990/microsofts-politically-correct-chat-bot-is-even-worse-than-its-racist-one/.

algorithms makes it nearly impossible to identify and remove all variables that are correlated with sensitive categories of personal data (Carabantes, 2020; Goodman and Flaxman, 2016; Taeihagh, 2021). Third, keeping the humans in the loop does not necessarily ensure safeguards against harms resulting from defects in the system. Behavioural factors, lack of technical expertise, or managerial factors appear to downplay the role of humans in creating liability and accountability to AI systems (Kim, 2017; Kroll *et al.*, 2016).

1.4.3 Citizen expectations from government

Another distinction in public value creation is about the way digital technologies shape expectations from governments. Some of these dimensions have already been covered above, such as ensuring the transparency and safety measures in the deployment of digital technologies both within and outside the public sector. Citizens expect better, faster, cheaper services as value additions from digital technologies, but these benefits should not facilitate privacy breaches, discriminatory practices, or data leakages.

So, in what new ways can governments create public value in the new digital-era governance? We identify three emerging roles for governments to address citizens' expectations:

1. *Open government*. The phrase 'open government' goes back to post-WWII and the meaning of the term has evolved over time (Yu and Robinson, 2012). According to the OECD, open government is based on the principles of transparency, integrity, accountability, and stakeholder participation. Today, open government not only denotes the transparency of government-held information and documents but also refers to making publicly accessible data available to citizens through government databases. More specifically, open government data (OGD) refers to a set of policies that promotes transparency, accountability, and value creation by making government data available to all.[22] OGD might include datasets about business information, registers, patent and trademark information, public tender databases, geographic information, legal information, meteorological information, social data, and transport information. Public institutions share their datasets with the public not only to build accountability but also to encourage the use and reuse of datasets to promote business creation and innovative use of data in public governance.
2. *Government as a platform*. Government as a platform (GaaP) first appears in a 2011 article by Tim O'Reilly. Inspired by the success of the platform-business models employed by companies such as Amazon, Apple, Alibaba, and Airbnb among others, O'Reilly proposed a new role for government that can go beyond the dichotomy of government providing services to citizens and leaving everything to the private sector. In his definition, government as a platform provider builds essential infrastructure, creates core applications that demonstrate the power of the platform and inspire outside developers to push the platform further, and enforces the rules of the road that ensure that applications work well together. Similarly, Richard Pope from Harvard Kennedy School (2019) defines GaaP as 'the whole ecosystem of

[22] https://www.oecd.org/gov/digital-government/open-government-data.htm.

shared APIs and components, open-standards and canonical datasets, as well as the services built on top of them and governance processes that (hopefully) keep the wider system safe and accountable'.

There are different services provided by governments called platform models. For instance, GOV.UK Platform as a service (GOV.UK PaaS) provides a platform as a service for public sector organisations to host their digital services in a cloud. According to their website, as of 2021, 131 organisations run 1,652 applications on the platform and a joined-up live model for live service assessment is provided to participating users.[23] Similar platform systems are employed in other countries such as Estonia, India, and Singapore.[24]

GaaP is not only about making data accessible to people and organisations; it is also about designing a centralised place where governments can provide tools and rules for users in an ecosystem to create value from the data and platform functions. The platform functions could be about providing a single point of access to various government services or enabling activities and places for joined-up actions and activities by user groups for specific goals. For example, DataSF, the data platform of the city of San Francisco, has hosted a series of data jams, hackathons, and app competitions centred on specific challenges, from transportation to sustainable development (Parker *et al.*, 2016: 282). Platforms create value using resources they don't own, and they derive much of their value from the communities they serve (Parker *et al.*, 2016: 15). An important role of government in public value creation is about the design of platform governance. Platform governance entails decisions concerning type of behaviours to promote and sanction, the rules of engagement among users, and rules of inclusion and exclusion. Choices about openness, security, and privacy policies change the user experiences and the value GaaP creates for public.

3. *Data watchdogs*. Data watchdogs are public organisations and agencies that ensure that companies in a particular industry are obeying laws on protection of personal information. The increasing value of data in commercial and public use, and the astronomical rise of companies that build their business models on personal data exploitation have led to regulative and monitoring activities by governments all around the world. We have already touched upon some prominent scandals centred around the misuse of private data. Governments took different actions on how best to protect their citizens' interests and privacy. For instance, the office of European Data Protection Supervisor (EDPS) – an independent data protection authority established by the EU – has a key mission to monitor new technology that may affect the protection of personal information.[25] EDPS has developed an increasingly influential role in EU policies and national data watchdogs towards big tech companies following the GDPR's entry into force. A series of probes have been initiated by data watchdogs of the EU member states leading to the fining of big tech companies, such as Amazon, Google, Facebook, and Twitter.[26] Similar to DSA, some other countries passed new laws and regulation to control the

[23] https://gds.blog.gov.uk/2021/03/31/podcast-gov-uk-platform-as-a-service/.
[24] X-road in Estonia https://e-estonia.com/, Singapore government tech stack https://www.tech.gov.sg/products-and-services/singapore-government-tech-stack/, and Indian Mygovin https://www.mygov.in/.
[25] https://edps.europa.eu/about/about-us_en.
[26] Amazon case, Google, others.

data and cash flow of big tech companies. For instance, Australia introduced new regulations that ban the sharing and publishing of news content on social media channels. Following the law, Google negotiated with news companies over payment for content.[27] In a similar case, French antitrust authorities fined Google 500 million euros for failing to negotiate in good faith to reach a licensing deal with news publishers on search results.[28] Turkey, on the other hand, passed a social media law that requires sites with more than a million daily users to appoint a local representative to enforce court orders to remove content, and global platforms to store user data from Turkey within Turkey.[29] Despite initial resistance, many tech companies, including Facebook, Google, Twitter, eventually complied with the new law. Erdogan's government in Turkey has been notorious on media freedom, and the new law is a cautionary tale for personal data protection laws that act not only against the commercial interests of big tech companies but also may serve as a tool for political suppression.

To wrap up, governments are expected to fulfil new roles in the new digital-era governance concerning public value creation. Governments can act as an enabler in the access to data held by public institutions and agencies to foster innovative services and business opportunities for citizens and by citizens. Governments can act as designer of platform-based services for proprietary or open-source applications to improve user experience in government services. Governments can act as regulators of a data ecosystem by enforcing rules in data policies or designing the rules of engagements in government platforms. In a way, these new roles of governments complement the needs-based holism and reintegration of the first digital-era governance. Yet unlike the former era of digital governance, governments share the roles and responsibilities in data-driven value creation with stakeholders and citizens. The value proposition of governments is not necessarily about how effectively they process the data collected by public institutions for public needs, but they create streamlined and secure data-sharing ecosystems for innovation and public use.

1.5 Changing roles of digital technologies in human resources and management

Now, let's look at how new digital technologies are changing the roles and competences of civil servants. We will compare old and new digital technologies in shaping human resources and expectations about the managerial skills in public administration. We will start first with human resources and specific skills necessary for the integration of digital technologies in public administration. Next, we will continue comparing the influence of new and old digital technologies on managerial decisions and skills.

[27] https://www.theguardian.com/media/2021/feb/18/facebooks-australia-news-ban-what-is-the-social-media-giant-up-to-and-how-will-you-be-affected.

[28] https://www.nytimes.com/2021/07/13/business/media/google-france-news-content-fine.html.

[29] https://www.politico.eu/article/turkeys-social-media-law-a-cautionary-tale/.

1.5.1 Human resources and skills for digital technologies

Dunleavy and his colleagues distinguish four types of skills that organisations of all kinds must acquire in order to handle IT (Dunleavy *et al.*, 2006: 13). First, any medium to large agency needs IT specialists, who have the ability to develop, operate, and maintain IT systems. Second, many organisations require some 'advanced users', referring to a particular level of competency to use advanced and sector-specific software tools. Third, a basic level of competency for all those employees working with word processing programs, spreadsheet packages, and the Internet. Fourth, a new set of operational skills to better use Internet-based services and new media that are influential in the policymaking and competitiveness of organisations. Following this categorisation, we will trace the changing roles of the technology specialist and the general workforce, and the new operational skills expected by organisations.

We will first start by comparing the impact of old and new digital technologies on the role of specialised and technical staff. Most IT staff in the former era of digital governance provided the support services and daily maintenance services for ongoing systems. Most of the time, these services were purely technical in nature, such as building internal computer networks and providing helplines for computer-related problems in day-to-day business. Consequently, it was common practice to centralise the IT services in big government IT centres in periphery locations or physically separate the IT units from the rest of the organisational hierarchy (Dunleavy *et al.*, 2006). Only a minority of the IT staff has been integrated in the administrative hierarchy to support change management for new business processes built on the Internet and to liaise with middle and top management for Internet-based services (Dunleavy *et al.*, 2006).

In the new digital-era governance, data is everywhere, and data is integral to decision-making processes. Data scientists, data analysts, data entry supervisors and clerks, data centre operators, database administrators, data(base) architects, data quality specialists, data protection officers, are just some of the specialised positions that can be found in data-related services. Specialised and technical staff in the new digital era not only provide support services, but they often take on an integral role in the management systems and decision-making processes.

Not all data jobs are created by the advent of new digital technologies, but their job definitions have enhanced and transformed in relation to the digital technologies. For instance, interoperability of government information systems used to be an integral task of IT services. However, advancement in digital technologies and new government policies, such as open data, transformed the meaning of interoperability and the associated skills. A case in point is the European Interoperability Framework (EIF) that was revised in 2016 following its initial creation in 2010. The new interoperability framework focusses on openness and information management, data portability, interoperability governance, and integrated service delivery. In the new understanding of interoperability, skills required for the technical interoperability of computer systems have been expanded with skills to ensure legal, semantic, and organisational interoperability.[30] New specialised roles have also emerged with the changing role of data in government services. For instance, data protection officers (DPOs) are some of the new data-related jobs that have become an integral part of public sector organisations in Europe following GDPR. DPOs ensure that an organisation applies the laws protecting individuals' personal data.

What about general skills and competencies required for the deployment of digital technologies in day-to-day services? In the former digital-era governance, every worker and unit had to acquire some basic skills to run and work with office software programs (e.g. word processing, spreadsheets, presentation, e-mailing) and websites on the Internet. In the 'second wave' of digital-era governance,[31] these operational skills have expanded into communication skills through social media tools, and using social media tools has become part of the organisational visibility and public relations policies (Margetts and Dunleavy, 2013).

In the new digital-era governance, these general digital skills have been largely preserved but data literacy and digital proficiency with new digital tools have moved in the direction of the general skillset. In particular, the Covid-19 pandemic and the resulting government actions all around the world accelerated the integration of digital collaboration software programs and applications, such as video conferencing, online digital spaces, and cloud-based document sharing and editing services in day-to-day business activities. Furthermore, surveys show that operational skills around data work and digital technologies are becoming the most sought-after competencies by organisations. The 2020 'Future of jobs' report published by the World Economic Forum shows that almost all top jobs with increasing demands are related to either new digital technologies or data work.[32] Conversely, the top 10 jobs where demand is decreasing most heavily are disrupted by the advent of the digital technologies. For example, customer service workers are no longer sought-after positions due to the advancements in chatbots. Not only the technical skills, but the implications of these technologies on soft skills are also changing. Critical thinking on data interpretation and information sources, collaborating and working through digital platforms, and problem solving using digital tools are emerging soft skills in the new digital-era governance.

[30] For details, see ISA[2] https://ec.europa.eu/isa2/eif_en.

[31] The second wave of digital governance is a term used by Margetts and Dunleavy (2013) to expand their theoretical propositions for digital-era governance with the advent of social media tools. Not to be confused with new digital-era governance.

[32] The top jobs with increasing demand, in order, are as follows: data analysts and scientists, AI and machine learning specialists, big data specialists, digital marketing and strategy specialists, process automation specialists, business development professionals, digital transformation specialists, information security analysts, software/applications developers, and internet of things specialists. The top 10 jobs where demand is decreasing most heavily, in order, are as follows: data entry clerks, administrative and executive secretaries, payroll and bookkeeping clerks, accountants and auditors, assembly and factory workers, business services and administration managers, client information and customer service workers, general and operations managers, mechanics and machinery repairers, and stock-keeping clerks.

The Digital Competence Framework 2.0 published by the EU science hub provides a useful guideline and overview of new digital competence areas and competencies for citizens. For the brevity of the argument, we will share the framework below to highlight the key digital competence areas and competencies for citizens to improve their digital skills in the new digital-era governance (Table 1.3).

1.5.2 Managerial skills and competencies for digital technologies

In the former era of digital governance, only a minority of IT staff had a contributing role to the management of public administration. As mentioned before, these tasks focussed largely on providing investment analysis, specification and change management for new business processes, and liaising with middle or top management (Dunleavy *et al.*, 2006). Dunleavy and his colleagues assigned the role of IT staff in in-house services and in the administrative diagram to a mediocre place at best. They even hypothesised that 'there is unlikely to be a director of IT on the management board.'

Table 1.3. DigComp.2.0 – the digital competence framework for citizens (Vuorikari *et al.*, 2016).

Digital competence area	Competencies
1. Information and data literacy	1.1 Browsing, searching and filtering data, information and digital content
	1.2 Evaluating data, information and digital content
	1.3 Managing data, information and digital content
2. Communication and collaborating	2.1 Interacting through digital technologies
	2.2 Sharing through digital technologies
	2.3 Engaging in citizenship through digital technologies
	2.4 Collaborating through digital technologies
	2.5 Netiquette
	2.6 Managing digital identity
3. Digital content creation	3.1 Developing digital content
	3.2 Integrating and re-elaborating digital content
	3.3 Copyright and licences
	3.4 Programming
4. Safety	4.1 Protecting devices
	4.2 Protecting personal data and privacy
	4.3 Protecting health and well-being
	4.4 Protecting the environment
5. Problem solving	5.1 Solving technical problems
	5.2 Identifying needs and technological responses
	5.3 Creatively using digital technologies
	5.4 Identifying digital competence gaps

With new digital technologies, we see the increasing influence of technology and technology specialists on managerial decisions. Access to data and gaining insight from data have already become an integral objective in managerial planning. Furthermore, protection of data and ensuring the standards on the quality of data have become key managerial tasks. Data scientists, analysists, and protection officers now have more input in managerial decision-making processes, all the way from strategy building to day-to-day administrative tasks. For instance, many cities[33] and agencies[34] have introduced the role of Chief Data Officer (CDO), a person in charge of implementing data strategies and establishing data governance in public sector organisations.

With the enlarging and enhancing role of digital technologies and data in the management processes, we see there are some developing issues concerning the impact of new digital technologies on the management skills and competencies in public administration.

To begin with, public managers need to develop the skills and competencies to integrate new digital technologies inside the organisation. Integration of information systems in public administration requires capital, planning of procurement processes, and employment of specialised people for successful deployment and maintenance. As Dunleavy and his colleagues put it, IT investment is 'costly', 'complex', and 'difficult' and 'throwing money at the problem will not help to overcome fundamental IT limits in the same way that it can do with normal administrative constraints'. Indeed, annual market studies conducted by International Data Corporation (ICD) show that global IT investment has been growing at less than 5% per year since 2009. However, this trend seems to change from 2019 onwards. Figures show that while spending on tradition digital technologies is stagnating or falling, the share of new digital technologies, such as IoT, robotics, augmented and virtual reality, and AI, in global IT investment is rising. For instance, while global expenditure on new digital technologies increased by about 16% from 2019 to 2020, the expenditure on traditional ICT decreased by about 3%.[35] The shrinking investment in traditional ICT can partly be attributed to the impact of the COVID-19 pandemic but the growing new technology market suggests changing investment patterns in technology. The investment in new digital technologies is expected to grow further, reaching 30% of overall global IT investments in 2023 forecasts. Hence, the first imminent challenge for public managers is to prepare their organisations financially, strategically, and administratively for the investment in new digital technologies. Research shows that IT investments in public administration do not take place in quick fixes or repetitious, piecemeal improvements of existing information systems. Rather improvements in IT systems occur in junctures, where the overall system is overhauled by the replacement of hardware and software, as well as a change in the accompanying system processes (Dunleavy *et al*, 2006).

[33] See https://joinup.ec.europa.eu/sites/default/files/document/2020-10/JRC_Policy_brief_DigiGov.pdf.
[34] For instance, the federal CDO council in the USA was established in 2019 following The Evidence Act and Federal Data Strategy, to bring together all CDOs from different federal agencies.
[35] The findings from the study can be consulted at https://www.idc.com/promo/global-ict-spending/forecast.

The replacement and renewal of IT systems is unquestionably challenging and costly for all organisations alike. But can we say that the new digital technologies bring challenges that differ in the way that investment processes are managed in the replacement and maintenance of IT systems? The empirical findings about the digital transformation of governments point out specific managerial challenges associated with new digital technologies.

The first challenge is about hiring new types of digital talents in the public sector and having the competence to integrate digital tools, methods, strategies, and culture not only into investment strategies but also into daily habits (OECD-OPSI, 2021). This challenge is exacerbated by the lack of specialised talents in new digital technologies that are often hired by private sector organisations.

The second challenge is about the technical challenges associated with the information infrastructure of the organisation and the challenges associated with the transfer of critical data from the legacy systems into cloud solutions and supplier lock-in systems.

The third challenge is about business-IT alignment, especially the development of formalised procedures for the successful integration of data-driven solutions in business processes and the alignment of data-driven insight with street-level insight into public policy processes.

The fourth challenge is about the technological features of new digital technologies such as AI and blockchain. The challenge here is three-fold. First, these technologies invoke different value trade-offs concerning transparency, efficiency, effectiveness, security, and accountability in public services processes. Second, the effective implementation of these technologies necessitates strategical planning that is both internal and external to organisational processes. For instance, deployment of AI-based systems raises questions about the type of data to use in feeding into the system (e.g. training data, input data, feedback data), the way that data will be retrieved from public or private sources, cybersecurity and data protection protocols, and the precautions against data bias and behavioural limitations on the use of data insight in policy-making processes. Third, the integration of these technologies into integral organisational infrastructure is different. While AI-based solutions may demand more in-house solutions for effective integration of the technology, blockchain-based solutions necessitate effective coordination of stakeholder groups and outside actors in digital solutions.

Another managerial issue associated with new digital technologies is how to manage automation. Automation of administrative tasks through the deployment of technological innovation is not necessarily a new phenomenon. In the past, technological innovation has made old technologies and techniques obsolete on numerous occasions, leading to drastic changes in terms of employment. For instance, at the end of the 19th century, it is estimated that about 200,000 horses were put to work in New York city.[36] By 1912, the city had more cars than horses for the first time, and in 1917 the last horse-drawn tram was decommissioned by the city administration (Susskind, 2020: 2).

[36] https://nypost.com/2020/08/01/how-horses-shaped-nyc-and-now-thrive-in-it-in-surprising-ways/.

There are similar accounts in the history of digital technologies. In the 1960s and 1970s, the integration of computers in business processes allowed for the scrapping of many clerical jobs both in private and government organisations. Likewise, the advancements in processing technologies in 1980s and 1990s cut down on some middle-management positions providing risk assessment analysis (Dunleavy *et al*, 2006). However, for many, the type of automation prospect resulting from the increasing capabilities of new digital technologies is unprecedented, and its implication for the workforce will be tremendous. Studies gauging the risk of job elimination in the near future estimate that 30 to 90% of existing occupations and workforce[37] are at risk of automation.

There are conflicting views about the impact and prospect of artificial intelligence and automated systems on the future of work and the human workforce. For some, AI threatens to replace jobs based on repeatable, pattern-based tasks (Agrawal *et al.*, 2018; Eggers *et al.*, 2017). In this scenario, the manager's role is to bring in the value judgements in policy processes (Agrawal *et al.*, 2018). For others, the increasing capabilities of new AI systems also threaten to replace jobs that are reliant on human cognitive and creative skills (McAfee and Brynjolfsson, 2017). Daniel Susskind sees the risk of automation not necessarily in terms of the disappearance of the jobs but the replacement of the tasks and the changing nature of the work. Pointing to the 271 jobs that were considered at risk of automation during the 1950 US census, only elevator operators have disappeared due to automation (Susskind, 2020: 41).

Similarly, there are different expectations about the way the workforce will evolve with the changes in the task definition. Some expect that the advent of new digital technologies will create new tasks that may compensate for the jobs and positions at risk. Critics of this rosy picture argue that the sheer scale of the displaced people could disrupt the working processes of an organisation even before any replacement can take place, and they question the ability of people to develop the new skills and competencies to replace the new tasks. Furthermore, will the new jobs contribute to a sense of purposefulness or they will just serve as filler positions to accommodate the need for employment? Ajay Agrawal and his colleagues, and Daniel Susskind argue that there are significant reasons to surmise that new jobs will be less about the value of the work but more about the need for the fulfilment of working. In a sector such as the public sector, where the value and purpose of work are attached to the distinctiveness and the prestige of public services, the threat to the meaning of the work may have even more grave consequences.

So, what do tactful public managers need to successfully automate tasks in public management processes? First, the automated tasks can improve the efficiency of public services, but the scale of job replacements and displacement can disrupt overall administrative processes. Public managers need to contemplate to what extent automated systems and artificial agents complement or substitute the human workforce. Second, technological change may affect not only the amount of work but also the nature of that work. Public managers need to consider both the qualitative and quantitative impact of

[37] For instance, PwC estimates one third of jobs, and Oxford University and McKinsey Global Institute put about half of the jobs at automation risk (Wirtz *et al.*, 2019). Vivek Wadhwa from Carnegie Mellon University estimates about 80-90% of current jobs will be eliminated by 2030 (https://phys.org/news/2017-03-tech-world-debate-robots-jobs.html).

automated systems on the workforce. Third, not all organisations and not all public administration systems are at a similar risk of automation. For example, a 2018 OECD study shows that jobs in Slovakia are six times more at risk of automation[38] than jobs in Norway (Nedelkoska and Quintini, 2018). We infer from this study, depending on the organisational processes and management systems in place, that public administration may be affected by automation in different ways and degrees. The severity of automation may vary according to application; therefore, public managers need to assess the disruptive nature of automation case by case.

1.6 Changing roles of digital technologies in public governance

ICT revolution and the Internet have paved the way for an abundance of data about every dimension of human life. New digital technologies have created new ways and means of creating value from these data, influencing every aspect of public governance. Hence, the last and probably the most distinct difference between the old and new digital-era governance is observable in the way digital technologies affect the role of government and society at large, in shaping public governance processes.

In this concluding section about the differences between the two eras of digital governance, we will elucidate the changing roles of digital technologies in public governance processes. Dunleavy and his colleagues (2006: 247) trace the impact of the digital-era governance on government in 'simplifying the internal workings of state apparatus, cutting institutional complexity, and increasing the inter-visibility of government organisations to each other'. They anticipate that the fully implemented digital-era governance changes will have the effect of increasing citizens' and businesses' capacity to solve their own problems by simplifying information searches and transactional processes, and by reducing the perceived institutional complexity of the government. In this view, the impact of digital technologies on policy outcomes is indirect, either through changing organisational processes in government, or changing information exchanges and behaviour in society. Our argument is that the impact of new digital technologies and data-driven public services are not only direct but also transformative in the way that public governance takes place. In particular, our argument rests upon the fact that new digital-era governance is not only about the governance *of* digital technologies, but also about governance *by* digital technologies. In what follows, we first make the distinctions between these two governance dimensions and elaborate on their implications. Later, we proceed with our reflections upon the possible transformative pathways for governments in the new digital-era governance.

1.6.1 Governance of digital technologies

'Governance of' refers to the rules of governance concerning the use of digital technology in the information infrastructure. Different choices made by system designers and policymakers about information system architecture and technical standards in place may shape the way digital

[38] Automation risk shows the percentage of jobs in an economy that have a greater than 70% chance of being automated.

technologies are used in public governance processes. Such governance decisions often invoke certain trade-offs between the enhanced capabilities in public governance processes and the influence of the design choices on organisational culture, administrative processes in place, and policy outcomes. For the design of ICT systems, these trade-off conditions are mostly relevant for inside the administration. New ICT systems often prompt behavioural and operational changes in day-to-day work and in citizen interactions creating trade-off conditions between the improved outcomes, such as operational efficiencies and user experience, and the cost of the investment – financial or transactional. However, we assert that the trade-off conditions associated with new digital technologies are fundamental and determinant to policy outcomes. For instance, the introduction of machine-learning algorithms has implications concerning the efficiency, accountability, and morality of policy outcomes. Similarly, system design choices with blockchain-based systems are associated with trade-off conditions among privacy, usability, performance, flexibility, security, transparency, rule compliance, and community control (Kannengiesser *et al.*, 2019). All in all, system design choices in new digital-era governance are political in nature, and as such, cannot just be treated as administrative or technical decisions.

Furthermore, new digital technologies have different singular and aggregated impacts. As mentioned earlier, different digital technologies, such as AI and blockchain have different affinities and strengths in improving public governance processes. For example, blockchain can create trust in the way that data is handled and managed in public governance processes, whereas AI can lead to faster and deeper data insight to use in public governance processes. The way these technologies are employed in data governance systems can have an enhancing or delimiting impact on those value propositions. Yet again, these technologies can jointly be used in data governance systems that can lead to an aggregated impact on the way public governance is organised. For instance, with AI, blockchain, and IoT systems, it is possible to create increased functionalities and automated systems in smart city configurations. In a theoretical case published in the thematic report of European Blockchain Observatory Forum (2021), a new way of public governance is described where IoT sensors and other devices are gathering data on people and traffic patterns, population densities, infrastructure usage and the like, feeding that through the blockchain to the AI to analyse and make predictions for future use. Moreover, blockchain-based e-voting and citizen participation platforms involve local communities and neighbourhoods, providing a platform for gathering feedback and voting on proposed projects[39] in urban planning. Policy and system designers need a holistic approach to the possibilities of digital technologies whilst foreseeing the distinct challenges and implications of tasks carried out by individual digital technologies.

The aggregated impact through the convergence of these technologies cannot take place without a solid data environment to collect, share and analyse big data. Therefore, governments are supporting the development of data infrastructures to ensure the provision of reliable and high-quality data that can be shared with a wide range of users in a robust and accessible way (Taeihagh, 2021). To that end, governments need the standards and operating systems that

[39] Such participatory systems are already being tested or are in place in cities like Barcelona and Reykjavik.

limit data fragmentation and lack of interoperability between systems of governance. These standards cannot just be about technical interoperability. They should also contain operating and binding rules about the organisational and legal processes of data governance that allow seamless data transactions and data traceability in public governance processes. Here again, new digital technologies provide different alternatives on the way these systems can be developed and enforced. For instance, verifiable credentials and self-sovereign systems on blockchain infrastructure allow secure, transparent, and reliable monitoring of the use of personal data in public governance processes by data subjects. A design choice on how data governance systems is regulated between different public and private systems can drastically affect the way citizens, administration, and businesses interact and engage with each other.

1.6.2 Governance by digital technologies

'Governance by' includes all policy design choices about digital technologies outside the technical infrastructure that still affect the way the users interact in public governance processes. Unlike the technical emphasis of 'governance of' decisions, these type of governance decisions are concerned with the techno-social power dynamics, role ascriptions in the governance ecosystem, and accountability of the digital technologies employed in public governance processes.

To clarify our argument, we list a range of probes that we consider part of the 'governance by' decisions in the new era of digital governance:

- What type of accountability mechanisms are in place? If something goes awry with the use of digital technology in public governance, who is responsible?
- What is the level of automation in public governance? Are the decisions made by or through digital technologies automatically carried out? Or does a governing body make the final decision?
- What is the role of public managers in governance processes? Is it a human-in-the-loop or a human-over-the-loop system?[40]
- What is the role of users and stakeholder groups in governance processes? Do they control the data shared within the system? Do they contribute to the day-to-day governance of the operating system?
- Who controls the technology? Is it a proprietary or open-source system? Who can modify or make changes in the functionalities of the system?
- How inclusive are governance processes? How does the digital divide affect the inclusiveness of governance processes?
- What are the ethical and public value implications of the design choices with digital tools employed in public governance processes?

[40] 'Human-in-the-loop' refers to the decision-making processes whereby a human makes the final decision about an insight delivered by a data analytics system. 'Human-over-the-loop' refers to the decision-making processes where a human actor has an overseeing role. In these systems, insights from data analytics may be automatically carried out, but the responsibility for policy outcomes lies with the human actor overseeing the process.

- How is the power distribution in the data governance ecosystem? Is it a decentralised, polycentric, or centralised system? Who makes the decisions and who implements them?
- How is the system regulated? Are the governance rules encoded in the operating system (e.g. *lex cryptographia*)? Are there prescriptive protocols that regulate user behaviour and engagement through and with digital technologies?
- How are the decisions carried out in public governance? Through the technology or around the technology?

Clearly, the type of governance questions associated with new digital technologies go beyond the way information and communication technologies are used in public governance processes. Automation, dehumanisation, decentralisation are some new dimensions to consider in the new digital-era governance. Depending on the governance decisions, the role and importance of external actors in policymaking and policy implementation processes may vary. Chapter 2 will elaborate further on these dimensions, but for the moment, it is important to emphasise that 'governance by' new digital technologies and understanding the implications of design choices in the accountability and power dynamics in public governance processes is a key challenge for public administration.

1.6.3 Future of digital public governance

Based on the changes in the way public governance takes place, the possibilities for the future of public governance are wide and different. Different qualities of emergent technologies and changing regulative topographies open up different possibilities and prospects for the future of government and societal actors. Unlike the superseding vision of 'digital-era governance' to NPM,[41] we presume that new digital technologies can lead to different digital futures in public governance. The digital future in public governance does not necessarily have to be progressive or transformative as well. The disruptive implications of new digital technologies may lead to technology conservatism, limiting the uptake of new digital technologies or their abandonment. ICT and data-based technologies both provide a pathway to information, and they are often complementary. But if the disruptive cost of certain technologies, such as AI or blockchain, outweighs the benefits, it is imaginable that governments may take action either to limit their uptake in social processes, or regulate the technologies in a way that certain disruptive features are tamed. The regulative developments about self-learning and automated AI systems, or cryptocurrencies and token economics already hint at such a prospect. Without disregarding this possibility, we hypothesise three possible prospects for the future of digital public governance.

1. Digital New Public Management: this scenario is based on three assumptions. First, technology diffusion in society continues without a major regulative or other form of disruption that hampers the technology development or uptake in emergent digital technologies. Second, governments continue in their current trajectory of digitalisation that allows private sector controlled digital solutions to be integrated in the public governance processes. Third,

[41] The title of the seminal work of Dunleavy and his colleagues is 'NPM is dead – long live digital-era governance' (Dunleavy *et al.*, 2005).

efficiency and performance focused NPM ideologies remain as the key driver in public governance processes. In this futuristic scenario, we expect governments to experience a constantly encroaching public policy landscape, where the tools and decisions in public governance processes gradually become dependent on the rules and technologies developed by technology companies. The 21st century has shown us the meteoric rise of big technology firms that control vast amounts of data and the advancements in new digital technologies. The technological solutions and gadgets developed by 'big tech' have changed every aspect of society. Big tech companies that continue in an undisruptive path of technological progress and data accumulation on human behaviour and expectation will eventually provide better public service solutions reliant on performance and efficiency considerations. In this scenario, we expect governments and public agencies gradually to withdraw from public service areas and make way for private service providers. Public sector organisations specialise in market regulations and monitory actions, whereas public services are carried out through the digital platforms controlled by private companies or technologies provided by these companies. The rules and techniques developed by these companies shape and dictate every aspect of citizen and government interaction.

2. Digital Neo-Weberianism: the second scenario presumes a big government outcome in digital public governance. Digital technologies and solutions are not only integrated in the internal administrative processes but are also used to shape how the citizen and government interactions take place. The difference is that the system infrastructure and technological solutions are either developed by the state or outsourced to private technology providers, which allows the state to control and shape the governance of the system infrastructure. In this scenario, government and government-sanctioned organisations keep the role of maintenance and development of the technology infrastructure, and governments prioritise the technologies that allow them to control the data governance processes even if this is at the expense of efficiency and performance considerations. Digital public governance takes place in a controlled and closed data ecosystem, where the access of private technology providers to data infrastructure is limited or heavily regulated.
3. Digital Communitarianism: the third scenario presumes that open-source technologies along with increased data consciousness and digital capabilities in societies can lead to the formation of digital communities and commons that self-govern or co-govern public service domains. Community members exercise control over the data governance system through self-sovereign identity systems and determine the policies of data governance. Citizens develop solutions to developing governance issues through or around the digital platforms to exchange ideas, to share propositions, and to vote on them. The community members either preserve a fettered control in the way decisions are implemented in public service processes or they can be automatically implemented based on pre-agreed rules. In this scenario, community-led platforms can be used for the management of public decisions, public accounting, management of financial sources, and implementation of public services. Transparency in public governance processes, democratic and equitable participation of community members, and preservation of privacy rights are expected to be the leading public values in the formation of public policies and political debates.

These scenarios do not presume a utopian or dystopian future. We also do not associate a particular technology with a type of digital public governance. Rather, our assumption is that the choices of governance actors can lead to either positive or negative political outcomes. This depends on the value propositions of the actors in the political context and the strength of checks and balance systems in place. Yet in each scenario, the repercussion of policy failures or abuse of new digital capabilities can be detrimental for society. For instance, digital NPM can lead to a future where the actions of big tech providers can no longer be adequately checked by public sector agencies. Digital Neo-Weberianist policies may end up with a pervasive system of governance that allows the control of daily aspects of life by non-democratic and ideologically driven governments. Or digital communitarianism may end up in a highly dispersed system of governance, where governance actors cannot provide immediate actions to emerging crises or problems. The signs of these dystopian prospects are already here.

1.7 Summary of arguments

In this section, we outline the summary of arguments about what is 'new' in the new digital-era governance.

The advancements in new digital technologies undermine the existing theoretical assumptions about the role of digital technologies in public governance. The core distinction between the first era of digital governance and the new era of digital governance is that the former is driven by the ICT-centric digital transformation, the latter is focused on data-centric digital transformation. The distinctions in the two eras of digital governance are observable in the changing role of digital technologies in government functions, public value creation, human resource management, and governance.

In terms of government functions, the use of new digital technologies suggests a reversed trend or at least an obfuscated prospect in the reintegration of government services. Unlike the former digital-era governance, faster and real-time access to information is not the central aim in the delivery of services, instead access to more and better data is the central aim of developing citizen-centric services. Consequently, the reorganisation of public service provision and digitalisation of government is not focussed on optimisation of information exchange but optimisation of data exchange.

In terms of public value creation, the implications of new digital technologies differ in terms of political and social trust, value creation through public service provision, and the citizens' expectations of government. The trust challenges are not only associated with the invasive use of ICT by governments, but also with the predictive and manipulative power of data analytics and deep learning programs over individual users and behaviours. In public service provision, the new digital era governance brought up better 'effectors' that were not the case in the first digital-era governance. But new value-laden risks have emerged such as intentional or unintentional discriminatory policy practices by data-driven decision-making, high overhaul and political costs of failure in automated systems, and public accountability challenges associated with extensive datasets and machine-learning algorithms.

Governments can meet citizens' expectations in the new digital-era governance by acting as enabler in the access of data sources, as designer of platform-based services to improve user experience in government services, and as regulator of the data ecosystem. Unlike the former era of digital governance, government shares the roles and responsibilities in data-driven value creation with stakeholders and citizens. The value proposition of governments is not necessarily about how effectively they process the data collected by public institutions for public needs, but how to create streamlined and secure data-sharing ecosystems for innovation and public use.

In terms of skills and competences of public personnel and managers, new digital technologies have changed the organisational role of specialised and technical staff and new specialised roles on data and digital technologies have emerged. Data literacy and digital proficiency with new digital tools have moved in the direction of the general skillset. Furthermore, there is an increasing influence of technology and technology specialists on managerial decisions. To manage successful digital transformation in public administration, public managers need to acquire new skills and competences to manage investment decisions about the replacement and maintenance of IT systems. Public managers also need to develop new skills and competences to manage automation in decision-making and implementation processes.

The technical design choices with new digital technologies and their integration in governance processes have a direct influence on policy outcomes, differing from the indirect influences of ICT-based technologies. A set of new governance challenges have emerged concerning the techno-social power dynamics, role designations in the governance ecosystem, and accountability of the digital technologies employed in public governance processes. Furthermore, unlike the superseding vision of 'digital-era governance' to New Public Management, new digital technologies can lead to different digital futures in public governance. The prospect of digital public governance can be in the form of Digital New Public Management, Digital Neo-Weberianism, and Digital Communitarianism. The choices of governance actors can lead to either positive or negative political outcomes in each form of digital public governance.

1.8 Structure of the book

The book is structured around the questions of how and why we need to think differently about the role of new digital technologies in public governance. These questions will be reflected upon in individual chapters focussing on technological, managerial, legal, organisational, ethical, and societal dimensions influenced by the new digital technologies. Throughout the book, a mixture of theories and empirical findings will be used to substantiate our arguments about how new digital technologies are shaping public governance. The overarching objective of the book is to bring up new insights and practical recommendations about the new digital-era governance in the relevant academic and policy circles.

The main research input for the book comes from a two-year research project called DIGI4FED, funded by the Belgian Federal Science Policy Office to investigate the conditions for the use of AI, big data and blockchain technologies in the Belgian federal government. The policy focus of the project was about fraud detection in social security and taxation areas. Therefore, the empirical cases in individual chapters will be mostly focussing on these policy domains. The research has been conducted by a consortium of four Belgian universities bringing together scholars from political sciences, information sciences, administrative sciences, and law. In consecutive chapters, authors share their research insight from this project about how new digital technologies are shaping governance models, management systems and operational processes, legal obligations of public administration, citizen trust, policy innovation, and decision-making processes in public management.

This book comes in eight chapters including the Introduction (this Chapter). Chapter 2 provides a deeper understanding of the arguments presented in the Introduction. The chapter outlines the potential impact of big data, AI, and blockchain on the public sector by focussing on the theoretical and empirical insights gathered from the literature. The chapter explores the new features and types of services these technologies can bring in the public governance domain, elaborating further on the arguments set out in the Introduction. Specifically, the chapter addresses the questions of:

- How can big data, AI and blockchain technology affect the public sector?
- How can the convergence of these technologies affect public services?
- What are the advantages and associated challenges for the implementation of these technologies?
- How can governance of and by AI and blockchain affect digital public governance?

Chapter 3 focusses on the alignment problems occurring inside the public administration between the business and IT sides in big data analytics. Focussing on fraud analytics – a term indicating the use of big data analytics to detect fraud – the chapter formalises the fraud analytics process in public administration. The empirical findings stem from 21 interviews among the data specialists and business operators working in the taxation and social security domains within the Belgian federal government. The analysis identifies three crucial moments at which alignment between business and IT is needed: the identification of requirements of the business team before performing the analytics, the presentation of the output of the analytics to the business team, and the feedback of the business to the data scientists. The chapter presents a new methodological approach drawing from agile methods, participation methods and design thinking literature to foster business and IT alignment in fraud analytics.

Chapter 4 and 5 analyse the impact of the data protection and privacy laws at the national and EU level on the use of big data, AI and blockchain technologies in public governance. Whilst Chapter 4 specifically focusses on the legal framework for the use of blockchain, Chapter 5 elaborates on the legal conditions for the use of AI and automated systems in public governance. These chapters take a closer look at the existing Belgian situation, in terms of data collection and combination, notably via big data, and particularly focusses on data processing to fight social security infringements and tax fraud. In two subsequent chapters, the impact of the personal data protection and privacy rules on the use of blockchain and AI in Belgian governments are addressed. Additionally, legal issues concerning transparency and explainability, the right not to be subject to a fully automated decision unless with specific safeguards, and equal access to public services are investigated through the national and EU legislative frameworks. The chapters provide a legal reading and analysis about the issues at stake and highlight additional legal challenges concerning the development of algorithmic decision-making tools by the public administration.

Chapter 6 elaborates on the interrelationship of trust in government, digital governance, and transparency, specifically focussing on the potential disconnect between AI/blockchain systems used and citizens' perceptions thereof. Research shows that even well-designed big data and/or AI projects run the risk of losing political or public confidence when encountering salient difficulties. However, the existing body of literature does not account for the way that governments attempt to generate trustworthiness and how the trustworthiness of data analytics projects are perceived in practice. This chapter contributes to filling this research gap by the conceptualisation of trustworthiness dimensions and by qualitatively examining the imperfect translation of such strategies into externally perceived trustworthiness. An exploratory analysis based on interview data with governments and civil society actors provides new insights into: (1) how governments currently attempt to make their data analytics projects trustworthy; and (2) how external actors perceive the trustworthiness of data analytics in government. The findings suggest that governments place higher emphasis on trustworthiness and security than is frequently assumed, but these efforts do not always align with external expectations. The chapter proposes a more comprehensive focus and the incorporation of stakeholder consultation and external communication through both sector-wide policies and project-level initiatives to benefit from the perceived trustworthiness of data analytics projects.

Chapter 7 explores how public sector innovation through knowledge exchange, co-creation techniques, and participatory methods can act as an intermediary in the implementation of new digital technologies in public governance processes. More specifically, the chapter introduces the living lab approach as a participatory and deliberative methodology that fosters open innovation through knowledge exchange, co-creation techniques and participatory methods, while reflectively assessing the conduct of its implementation in the Belgian case for the governance of new digital technologies. The chapter presents a methodological and practical guideline about how to use living labs and scenario thinking as policy instruments in adopting new digital technologies. Drawing on the lessons gathered from the DIGI4FED project, the chapter assesses the strength of the proposed approach and provides theoretical and practical recommendations for further implementation.

We conclude in Chapter 8 by discussing how to create viable technology adoption strategies and policies in the new digital-era governance. Specifically, the chapter elucidates the policy processes and decisions that need to be taken to implement AI, blockchain, and big data solutions in public governance processes. Building on the research insight about the use of digital technologies in the fight against fraud, the authors expound on the drivers for the implementation of the AI, big data, and blockchain technologies in administrative decision-making processes. The chapter extrapolates new challenges and issues practitioners need to face in the new digital-era governance, and concludes with a new research agenda for the role of data-processing and new digital technologies in digital public governance.

References

Agrawal, A., J. Gans, and A. Goldfarb. 2018. *Prediction machines: the simple economics of artificial intelligence*. Harvard Business Review Press, Boston, MA, USA.

Bozeman, B. 2009. Public values theory: three big questions. *International Journal of Public Policy* 4 (5): 369. https://doi.org/10.1504/IJPP.2009.025077

Brewer, G.A., B.J. Neubauer, and K. Geiselhart. 2006. Designing and implementing e-government systems. *Administration and Society* 38 (4): 472-499. https://doi.org/10.1177/0095399706290638

Bryson, J.M., B.C. Crosby, and L. Bloomberg. 2014. Public value governance: moving beyond traditional public administration and the new public management. *Public Administration Review* 74 (4): 445-456. https://doi.org/10.1111/puar.12238

Cabral, S., J.T. Mahoney, A.M. McGahan, and M. Potoski. 2019. Value creation and value appropriation in public and nonprofit organizations. *Strategic Management Journal* 40 (4): 465-475. https://doi.org/10.1002/smj.3008.

Carabantes, M. 2020. Black-box artificial intelligence: an epistemological and critical analysis. *AI and Society* 35 (2): 309-317. https://doi.org/10.1007/s00146-019-00888-w

Chatfield, A.T., and C.G. Reddick. 2018. Customer agility and responsiveness through big data analytics for public value creation: a case study of Houston 311 on-demand services. *Government Information Quarterly* 35 (2): 336-347. https://doi.org/10.1016/j.giq.2017.11.002

Cordella, A., and C.M. Bonina. 2012. A public value perspective for ICT enabled public sector reforms: a theoretical reflection. *Government Information Quarterly* 29 (4): 512-520. https://doi.org/10.1016/j.giq.2012.03.004

Cordella, A., and A. Paletti. 2018. ICTs and value creation in public sector: manufacturing logic vs service logic. *Information Polity* 23 (2): 125-141. https://doi.org/10.3233/IP-170061

Cordella, A., and A. Paletti. 2019. Government as a platform, orchestration, and public value creation: the Italian case. *Government Information Quarterly* 36 (4): 101409. https://doi.org/10.1016/j.giq.2019.101409

Cresswell, A.M., G.B. Burke, and T.A. Pardo. 2006. Advancing return on investment analysis for government IT a public value framework. University at Albany, Albany, NY, USA. Available at: https://www.ctg.albany.edu/media/pubs/pdfs/advancing_roi.pdf

Criado, J.I., R. Sandoval-Almazan, and J.R. Gil-Garcia. 2013. Government innovation through social media. *Government Information Quarterly* 30 (4): 319-326. https://doi.org/10.1016/j.giq.2013.10.003

Crosby, B.C., P. 't Hart, and J. Torfing. 2017. Public value creation through collaborative innovation. *Public Management Review* 19 (5): 655-669. https://doi.org/10.1080/14719037.2016.1192165

Dunleavy, P., H. Margetts, S. Bastow, and J. Tinkler. 2005. New Public Management is dead--long live digital-era governance. *Journal of Public Administration Research and Theory* 16 (3): 467-494. https://doi.org/10.1093/jopart/mui057

Dunleavy, P., H. Margetts, S. Bastow, and J. Tinkler. 2006. *Digital era governance: IT corporations, the State, and E-government*. Oxford University Press, Oxford, UK.

Eggers, W.D., T. Fishman, and P. Kishnani. 2017. AI-augmented human services: using cognitive technologies to transform program delivery. Available at: https://www2.deloitte.com/content/dam/insights/us/arti%0Dcles/4152_AI-human-services/4152_AI-human-services.%0Dpdf.

European Blockchain Observatory Forum. 2021. Convergence of blockchain, AI and IoT. Available at: https://www.eublockchainforum.eu/sites/default/files/reports/Convergence%20Workshop%20report%20-%20Sept%202021.pdf.

Finck, M. 2018. *Blockchain regulation and governance in Europe*. Cambridge University Press, Cambridge, UK. https://doi.org/10.1017/9781108609708

Goodman, B., and S. Flaxman. 2017. European Union regulations on algorithmic decision-making and a right to explanation. *AI Magazine* 38 (3): 50-57. https://doi.org/10.1609/aimag.v38i3.2741.

Grube, D.C. 2017. Rules, prudence and public value: public servants and social media in comparative perspective. *Government and Opposition* 52 (1): 75-99. https://doi.org/10.1017/gov.2015.21.

Hood, C. 1983. *The tools of government*. Macmillan, Basingstoke, UK.

Hood, C., and H. Margetts. 2006. *The tools of government*. Palgrave Macmillan UK, Basingstoke, UK.

Ju, J., L. Liu, and Y. Feng. 2019. Public and private value in citizen participation in e-governance: evidence from a government-sponsored green commuting platform. *Government Information Quarterly* 36 (4): 101400. https://doi.org/10.1016/j.giq.2019.101400

Kannengiesser, N., S. Lins, T. Dehling, and A. Sunyaev. 2019. Mind the gap: trade-offs between distributed ledger technology characteristics (working paper). *ArXiv*: 1906.00861.

Karkin, N., and M. Janssen. 2014. Evaluating websites from a public value perspective: a review of Turkish local government websites. *International Journal of Information Management* 34 (3): 351-363. https://doi.org/10.1016/j.ijinfomgt.2013.11.004

Kim, S. 2018. Crashed Software: Assessing Product Liability for Software Defects in Automated Vehicles. *Duke Law and Technology Review* 16: 300-317.

Kroll, J.A., J. Huey, S. Barocas, E.W. Felten, J.R. Reidenberg, D.G. Robinson, and H. Yu. 2016. Accountable algorithms. *University of Pennsylvania Law Review* 165.

Lim, H.S.M., and A. Taeihagh. 2019. Algorithmic decision-making in AVs: understanding ethical and technical concerns for smart cities. *Sustainability* 11 (20): 5791. https://doi.org/10.3390/su11205791

Mahmood, M., V. Weerakkody, and W. Chen. 2020. The role of information and communications technology in the transformation of government and citizen trust. *International Review of Administrative Sciences* 86 (4): 708-728. https://doi.org/10.1177/0020852318816798

Margetts, H., and P. Dunleavy. 2013. The second wave of digital-era governance: a quasi-paradigm for government on the web. *Philosophical Transactions of the Royal Society A: Mathematical, Physical and Engineering Sciences* 371 (1987): 20120382. https://doi.org/10.1098/rsta.2012.0382

McAfee, A., and E. Brynjolfsson. 2017. *Machine, platform, crowd: harnessing our digital future*. W.W. Norton Company, New York, NY, USA.

Meynhardt, T. 2009. Public value inside: what is public value creation? *International Journal of Public Administration* 32 (3-4): 192-219. https://doi.org/10.1080/01900690902732632

Mittelstadt, B.D., P. Allo, M. Taddeo, S. Wachter, and L. Floridi. 2016. The ethics of algorithms: mapping the debate. *Big Data and Society* 3 (2): 205395171667967. https://doi.org/10.1177/2053951716679679

Moore, M. 1995. *Creating public value – strategic management in government*. Harvard University Press, Cambridge, MA, USA.

Morgeson, F.V., and C. Petrescu. 2011. Do they all perform alike? An examination of perceived performance, citizen satisfaction and trust with US federal agencies. *International Review of Administrative Sciences* 77 (3): 451-479. https://doi.org/10.1177/0020852311407278

Nabatchi, T., A. Sancino, and M. Sicilia. 2017. Varieties of participation in public services: the who, when, and what of coproduction. *Public Administration Review* 77 (5): 766-776. https://doi.org/10.1111/puar.12765

Nedelkoska, L., and G. Quintini. 2018. Automation, skills use and training. OECD Social, Employment and Migration Working Papers, No. 202, OECD Publishing, Paris, France. https://doi.org/10.1787/2e2f4eea-en

O'Reilly, T. 2011. Government as a platform. *Innovations: Technology, Governance, Globalization* 6 (1): 13-40. https://doi.org/10.1162/INOV_a_00056

Organisation for Economic Cooperation and Development (OECD). 2014. Recommendation of the Council on Digital Government Strategies. Available at: https://www.oecd.org/gov/digital-government/Recommendation-digital-government-strategies.pdf

Organisation for Economic Cooperation and Development Observatory of Public Sector Innovation (OECD-OPSI). 2021. Digital transformation. 2021. https://oecd-opsi.org/guide/digital-transformation/.

Panagiotopoulos, P., B. Klievink, and A. Cordella. 2019. Public value creation in digital government. *Government Information Quarterly* 36 (4): 101421. https://doi.org/10.1016/j.giq.2019.101421

Parker, G.G., M.W. Van Alstyne, and S.P. Choudary. 2016. *Platform revolution how networked markets are transforming the economy and how to make them work for you*. W.W. Norton and Company, New York, NY, USA.

Pereira, G.V., M.A. Macadar, E.M. Luciano, and M.G. Testa. 2017. Delivering public value through open government data initiatives in a smart city context. *Information Systems Frontiers* 19 (2): 213-229. https://doi.org/10.1007/s10796-016-9673-7

Perri 6. 2008. The tools of government in the digital age – by Christopher C. Hood and Helen Z. Margetts. *Public Administration* 86 (4): 1137-1138. https://doi.org/10.1111/j.1467-9299.2008.00756_4.x

Pope, R. 2019. Playbook: government as a platform. Ash Center for Democratic Governance and Innovation, Harvard Kennedy School, Cambridge, MA, USA. Available at: https://ash.harvard.edu/files/ash/files/293091_hvd_ash_gvmnt_as_platform_v2.pdf

Samoilenko, S., and K-M. Osei-Bryson. 2019. Representation matters: an exploration of the socio-economic impacts of ICT-enabled public value in the context of Sub-Saharan economies. *International Journal of Information Management* 49 (December): 69-85. https://doi.org/10.1016/j.ijinfomgt.2019.03.006

Seltsikas, P., and R.M. O'Keefe. 2010. Expectations and outcomes in electronic identity management: the role of trust and public value. *European Journal of Information Systems* 19 (1): 93-103. https://doi.org/10.1057/ejis.2009.51

Sun, T.Q., and R. Medaglia. 2017. Artificial intelligence and public healthcare service innovation: a service ecosystem perspective. In: *The 40th Information Systems Research Conference in Scandinavia*. Halden, Norway.

Susskind, D. 2020. *A world without work: technology, automation and how we should respond*. Allen Lane, London, UK.

Taeihagh, A. 2021. Governance of artificial intelligence. *Policy and Society* 40 (2): 137-157. https://doi.org/10.1080/14494035.2021.1928377

Tan, E., and A.P. Rodriguez Müller. 2020. The use of blockchain technology in digital coproduction: the case of Barcelona. In: *Proceedings of Ongoing Research, Practitioners, Workshops, Posters, and Projects of the International Conference EGOV-CeDEM-EPart 2020*, 125-134. CEUR. Available at: http://ceur-ws.org/Vol-2797/paper12.pdf

Tolbert, C.J., and K. Mossberger. 2006. The effects of E-government on trust and confidence in government. *Public Administration Review* 66 (3): 354-369. https://doi.org/10.1111/j.1540-6210.2006.00594.x

United Nations (UN). 2020. E-government survey 2020 digital government in the decade of action for sustainable development. With addendum on COVID-19 response. UN, New York, NY, USA.

Vuorikari, R., Y. Punie, S. Carretero Gomez, and G. Van Den Brande. 2016. DigComp 2.0: the digital competence framework for citizens. Update phase 1: The conceptual reference model. Publications Office of the European Union, Luxembourg, Luxembourg. Available at: https://publications.jrc.ec.europa.eu/repository/handle/JRC101254

Walravens, N., and P. Ballon. 2013. Platform business models for smart cities: from control and value to governance and public value. *IEEE Communications Magazine* 51 (6): 72-79. https://doi.org/10.1109/MCOM.2013.6525598

Wang, C. 2014. Antecedents and consequences of perceived value in mobile government continuance use: an empirical research in China. *Computers in Human Behavior* 34 (May): 140-147. https://doi.org/10.1016/j.chb.2014.01.034

Wang, C., T.S.H. Teo, and M. Janssen. 2021. Public and private value creation using artificial intelligence: an empirical study of ai voice robot users in Chinese public sector. *International Journal of Information Management* 61 (December): 102401. https://doi.org/10.1016/j.ijinfomgt.2021.102401

Wirtz, B.W., and W.M. Müller. 2019. An integrated artificial intelligence framework for public management. *Public Management Review* 21 (7): 1076-1100. https://doi.org/10.1080/14719037.2018.1549268.

Wirtz, B.W., J.C. Weyerer, and C. Geyer. 2019. Artificial intelligence and the public sector – applications and challenges. *International Journal of Public Administration* 42 (7): 596-615. https://doi.org/10.1080/01900692.2018.1498103

World Economic Forum. 2020. The future of jobs report 2020. Available at: https://www3.weforum.org/docs/WEF_Future_of_Jobs_2020.pdf

Wright, A., and P. De Filippi. 2015. Decentralized blockchain technology and the rise of lex cryptographia. *SSRN Electronic Journal*. https://doi.org/10.2139/ssrn.2580664

Yu, H., and D.G. Robinson. 2012. The new ambiguity of open government. *SSRN Electronic Journal*. https://doi.org/10.2139/ssrn.2012489

Chapter 2

The role of big data, AI and blockchain technology in digital public governance

E. Tan

KU Leuven, Public Governance Institute, Parkstraat 45 – bus 3609, 3000 Leuven, Belgium; evrim.tan@kuleuven.be

Abstract

Artificial intelligence and blockchain technologies allow organisations to leverage big data and to gain insights in order to make sound public policy decisions. Public sector organisations need to learn about these technologies and develop the necessary skills and competencies to help their organisations stay competitive. One distinguishing characteristic of these emerging digital technologies is that they are not only tools for public sector organisations to enhance existing capabilities in public policy processes, but their technological features demand a complete revisit of the existing administrative systems and processes. This chapter focuses on the implications of these technologies for public sector organisations. Specifically, the chapter elucidates on how these technologies are currently employed in the public sector, what new features and capabilities can be unlocked through their conjoint use, and what opportunities and challenges are associated with these technologies. Furthermore, the chapter presents the specific governance requirements associated with these technologies in the transformation towards digital public governance.

Keywords: big data, AI, blockchain, governance of, governance by

2.1 Big data

The perception and description of big data (BD) keeps evolving (Allam and Dunny, 2019). The most common usage of the concept relies on the three 'v's, namely volume of information, a variety of the different data sources and types (structured and unstructured), and the velocity of data (i.e. the speed of creation, storage, and dissemination of data) in real time (Einav and Levin, 2014; Pencheva *et al.*, 2020; Zikopoulos *et al.*, 2012). Some authors even add other 'v's to the definition, which correspond to the value of data (Kimble and Milolidakis, 2015) and veracity of data (Demchenko *et al.*, 2014). Kitchin and McArdle (2016) add to these characteristics the dimensions of exhaustivity, resolution, indexicality, relationality, extensionality, and scalability.

Not only its features, but the scope of BD often varies according to the disciplinary backgrounds. In managerial sciences, BD corresponds to unstructured content generated from a plurality of sources, such as Internet clicks, mobile transactions, user-generated content, and social media,

DOI 10.3920/978-90-8686-930-5_2,

as well as purposefully generated content through sensor networks and business transactions, such as sales queries and purchase transactions (George *et al.*, 2014). In public policy, BD is associated with new formats, quality, and availability of administrative data (Pirog, 2014). In political sciences, BD refers to the technological innovations to gather either new types of data, such as social media data, or vast quantities of traditional data at less expense (Clark and Golder, 2014). In information and technology sciences, BD is defined through the big and openly linked data (Janssen and Van den Hoven, 2015), and by the massive quantities of information produced by and about people, things, and their interactions (Boyd and Crawford, 2012). In computer sciences, BD refers to the second-by-second picture of interactions over extended periods of time, providing information about both the structure and content of relationships (Lazer *et al.*, 2009). Finally, in public administration, Mergel *et al.* (2016) identify BD as (1) data created by private citizens through their interactions with each other online (such as social media data), (2) data automatically generated by sensors and automatically transmitted online, and (3) data that is automatically collected by public entities in the course of their operations.

BD is expected to bring value creation at almost every stage of policy design processes. First, BD-based solutions can improve agenda setting and policy formulation processes. BD allows a quick reaction to and incorporation of collective information from a variety of sources (Höchtl *et al.*, 2016; Mergel, 2017). It provides greater accuracy, efficiency and speed of the administrative processes (Pencheva *et al.*, 2020), It allows predictive analysis of which policies will work and under what conditions (Clarke and Margetts, 2014; Cook, 2014). It allows for the design of policies better matched to the preferences of stakeholders (Stritch *et al.*, 2017; Taeihagh, 2017). It improves the legitimacy of policy formulation processes by enabling citizens and governments to engage in more meaningful dialogue and to collaborate in policy design (Desouza and Bhagwatwar, 2012; Schintler and Kulkarni, 2014).

Second, BD can improve policy implementation processes. BD ensures compliance with regulations, fraud prevention and detection, and criminal investigation (Janssen and Kuk, 2016). It allows quick responses to emergencies, prevention of damages, and fewer casualties (Hondula *et al.*, 2018). It improves operative efficiency by exposing and eliminating redundancies (Gamage, 2016; Shindelar, 2014), and improves effectiveness in the delivery of public services by better resource allocation modelling and real-time operation optimisation (Daniell *et al.*, 2015). It increases public sector efficiency by delivering saving and boosting productivity (Agostino and Arnaboldi, 2017; Johnes and Ruggiero, 2017). It helps policymakers to better understand and segment users for personalised policies (Pencheva *et al.*, 2020). It facilitates better supervision of the implementation processes through the detection of irregularities (Maciejewski, 2017).

Third, policy research and evaluation can be augmented via BD-based solutions. BD improves policy analysis and helps identify areas for reform (Decker, 2014). Through BD, policymakers can understand the long-term effects of interventions on citizens in policy areas and develop new interventions for hard-to-reach populations (Blume *et al.*, 2014; Cook, 2014). BD provides superior insight and decision-making capabilities (Pencheva *et al.*, 2020). BD enables holistic evaluation of policy outcomes with the capacity to handle time-series data from multiple, diverse sources and the ability to simultaneously observe individual and aggregate variables (Jarmin and O'Hara, 2016; Pencheva *et al.*, 2020). BD allows a rapid and real-time evaluation process (Höchtl *et al.*, 2016). Through BD-based solutions, it is possible to experiment with new business models and organisational performance techniques (Arinder, 2016), and supplement traditional techniques in policy evaluation such as surveys (Bachner and Hill, 2014; Pandey *et al.*, 2017; Zhu *et al.*, 2017).

Fourth, new modes of data-driven public governance can be formulated by improving mutual government-citizen understanding (Clarke and Margetts, 2014). The use of data-smart governance models (Goldsmith and Crawford, 2014) and open data governance (Millard, 2018) models depend in particular on the effective use of BD-based solutions. The adoption of these new modes of digital public governance relies on the development of new policies and programmes describing what type of data will be produced and how it should be analysed (Pirog, 2014). Furthermore, data and time-conscious evaluative frameworks can emphasise evidence-based decision-making and longitudinal cost-benefit analytics as part of policy-making processes, making public policy processes more predictable and transparent, and thereby changing the administrative culture in public sector organisations (Arinder, 2016: 394).

2.2 Artificial intelligence

There are a variety of definitions for artificial intelligence (AI) in the literature. Russell and Norvig (2016) define the term AI to describe systems that mimic cognitive functions generally associated with human attributes such as learning, speech and problem-solving. For Haenlein and Kaplan (2019), AI is a system's ability to interpret external data correctly, to learn from such data, and to use those learnings to achieve specific goals and tasks through flexible adaptation. In an even broader conceptualisation, Tegmark (2017) defines AI as the ability of a non-organic, mechanical entity to accomplish complex goals. According to Dwivedi *et al.* (2019), the common thread amongst various definitions of AI is the increasing capability of machines to perform specific roles and tasks currently performed by humans within the workplace and society in general.

The use of AI technologies in public policy and administration processes is not a new phenomenon but, over the last decade, there have been dramatic advances in core AI technologies like machine learning, natural language processing, deep learning (or neural networks), virtual agents, and computer vision. AI literature distinguishes between 'narrow AI' and 'general AI' (Bullock, 2019). Narrow or weak AI refers to systems capable of carrying out tasks that require single human capabilities, e.g. visual perception, understanding context, probabilistic reasoning and dealing with complexity (Russell and Norvig, 2016). In public administration, narrow AI is often used in boosting the capabilities of data analytics and data mining techniques to deliver deeper and better insights beyond what human analysts can do. General or strong AI refers to (hypothetical) systems with human or superhuman intelligence, which simulate the complex human ability to think and execute intelligent tasks such as ethical judgements, symbolic reasoning, managing social situations, and ideation (Brynjolfsson and McAfee, 2014). Despite some noteworthy advancements in unlocking general AI technology (e.g. ALPHAGO Zero developed by DeepMind), most real-world applications of AI fall under the category of narrow AI.

The early promise of AI was largely viewed in terms of providing decision support for public managers (e.g. Hadden, 1986; Hurley and Wallace, 1986; Jahoda, 1986; Masuch and Lapotin, 1989). The latest advances in AI allow computers to learn from past experiences and understand the world through a hierarchy of concepts (Goodfellow *et al.*, 2016) that can lead to automation of tasks (Bailey *et al.*, 2016; Barth and Arnold, 1999; Pencheva *et al.*, 2020). The use of BD has enabled algorithms to deliver excellent performance for 'narrow AI', rather than the more human level 'general AI' where the complexities of human thinking and feelings have yet to be translated effectively (Pencheva *et al.*, 2020; Russell and Norvig, 2016). A key contributing factor to the increasing maturity of AI technologies and the viability of AI application to public policy and administration was the availability of data that can be used in the machine-learning process. At the same time, without the underlying analytical technology, the data-driven public governance can be viewed simply as a shift in the scale of the available data rather than as a transformational change (Pencheva *et al.*, 2020). Therefore, advanced analytics and other AI technologies are essential for the usage of BD in the public sector (Mergel *et al.*, 2016; Pencheva *et al.*, 2020).

There are a multitude of AI applications currently in use in the public sector.[1] First, we see the use of AI in knowledge management. Generation and systemisation of knowledge, development of expert systems supporting the codification of the knowledge, and use of neural networks to analyse, distribute, and share knowledge with others are some examples of AI applications in knowledge management (Wirtz *et al.*, 2019).

Second, AI applications are used in process automation systems. Automation of standard tasks through rule-based assessment, workflow processing, schema-based suggestions, data mining, and case-based reasoning improve the pace and quality of administrative services (Chun, 2007; Wirtz *et al.*, 2019). Automated AI systems can also support complex human action processes or repetitive tasks by leveraging the ability of software robots or AI-driven workers to mimic human interaction with user interfaces of software systems (Wirtz *et al.*, 2109; Jefferies, 2016).

Third, AI-based systems can also create higher-level cognitive functions in building autonomous systems and replacing human agents. For example, the use of electric-powered autonomous vehicles for public transport (Jefferies, 2016) or robot-assisted surgeries (Collier *et al.*, 2017) are some areas in which AI-based agents can replace humans.

Fourth, AI solutions based on speech analytics and natural language processing are used in creating virtual agents (e.g. chatbots, avatars), digital assistants, and forming complex recommendation systems to replace or support human agents in internal and external human relations management (Cortés-Cediel *et al.*, 2017; Mehr, 2017; Wirtz *et al.*, 2019; Zheng *et al.*, 2018).

[1] A list of AI-use cases in the European public sector can be obtained at: https://joinup.ec.europa.eu/collection/elise-european-location-interoperability-solutions-e-government/news/143-ai-cases-public-sector-are-available-open-data.

Fifth, by leveraging BD, AI solutions such as machine learning, deep learning, predictive analytics are used to create complex predictive and prescriptive models in fraud detection (Hemken and Gray, 2016), preventive actions in crime and terror threats (Power, 2016), and forecasting models in natural resource and water management (Kouziokas *et al.*, 2017).

2.3 Blockchain technology

Blockchain and distributed ledger technology (DLT) belong to a class of technologies known as blockchain technologies (BCT), which create a transparent, autonomous and decentralised data governance system that gives users confidence that archived information has not been tampered with (Beck *et al.*, 2018). In a blockchain, the information is archived in a distributed ledger that is shared across a network of users where participants, called nodes, keep a copy of the ledger and record all transactions. By agreeing on a form of consensus mechanism, nodes validate the transactions, provide an immutable (or nearly immutable) record of transactions, and ensure traceability. Each validated transaction is registered in a block, which is time-stamped and includes a digital trace (called hash) of a prior block, together forming a blockchain. Every block in the system continuously synchronises with other blocks. In this way, blockchain ensures nearly unhackable decentralised systems of encrypted data, without the need for a centralised authority to ensure the continuity of the system.

BCT is not a monolith technology, and it covers several underlying technologies (e.g. smart contracts, token technology, decentralised applications, etc.) that can jointly support a decentralised and automated information infrastructure. In the context of public governance, BCT has the potential for facilitating direct interactions between public institutions, citizens, and social and economic agents. At the most basic level, BCT can be used as an information infrastructure for the more efficient management of public information and exchanging information between public administrations. At a more advanced level, BCT can leverage the features of decentralised information management and automated execution of algorithmic decisions through smart contracts in support of BD and advanced analytic techniques to create autonomous organisations that can replace the human agency and traditional forms of public sector organisations in public governance.

There are three categorical areas where BCT is currently in use in the public sector. First, BCT-based solutions are used in identity verification and in supporting self-sovereign digital identity systems for individuals and businesses. For instance, European Blockchain Services Infrastructure (EBSI) adopts the use of self-sovereign electronic identification, authentication and trust services (ESSIF) as one of the use cases (European Commission, 2021). Another example is a blockchain-based identity system that is used for refugees in Finnish camps (Hempel, 2018). Furthermore, BCT-supported identity verification systems are also used to develop more secure e-voting systems. For example, uPort decentralised digital identity in the Swiss city of Zug is used for e-voting and renting e-bikes (Berryhill *et al.*, 2018). In another case, the Digital Democracy and Data Commons pilot in the city of Barcelona developed a DLT-based voting system to support petition mechanisms in Barcelona's e-governance platform called Decidim.Barcelona (López, 2019).

Second, BCT is used in asset registries. Land titling, patents, health data, diplomas are some areas where BCT is currently used to improve the reliability, traceability, and security of registered data in public and private databases, and to reduce the operational costs and verification times (Allessie *et al.*, 2019). Some examples are the notary and diploma use cases for EBSI, Exonum land title registry in Georgia (Berryhill *et al.*, 2018), Equity Platform, to register and validate individual energy transactions between batteries and the grid (Janssen, 2020), and e-health records in Estonia to register health records at the national level (Einaste, 2018).

Third, BCT is used in automating and tracking high-risk transactions and improving the traceability and transparency of supply-chain management systems. Currently, there are various use cases in logistics, food tracking and energy sectors for blockchain-based supply-chain management systems (Yafimava, 2019).

2.4 Convergence of technologies in public governance

So far, we have covered the individual implications of BD, AI, and BCT in public governance. However, their true disruptive impact is contingent on their convergence with each other and with other digital technologies such as the Internet of Things (IoT), sensing technologies, 5G systems, cloud computing, robotics, and 3D printing. Here, the convergence refers to the joint applications and seamless interactions of these digital technologies in the systems of public governance. The idea is that with convergence, separate digital technologies will complement each other, and their disruptive impact will be augmented in creating a world where individuals, organisations, and machines can freely interact with one another with little friction and at a fraction of the current costs (Voshmgir, 2020: 109). In this new world, public services can be organised more efficiently and effectively without any interference from intermediaries, and new forms of autonomous and self-governing systems of public governance can be envisaged.

Following the work of Hassani *et al.* (2019), Table 2.1. sets forth certain augmented features and outcomes associated with the convergence of AI, BD and BCT.

In a more specific look at the public sector, the convergence of these technologies is expected to unlock new features and capabilities in digital governance and in overall organisational and institutional rearrangements in public governance. Nevertheless, any disruptive influence brings with it challenges and obstacles to be overcome during the implementation of these technologies. While these obstacles are expected to vary according to the specific institutional and contextual conditions, we can draw on certain opportunities and challenges for the public sector resulting from the convergence of these technologies.

2.4.1 Opportunities for public governance

More data availability, better big data management & analytics: public blockchains are expected to improve AI algorithms and market prediction solutions since data will be available via a public ledger. This will allow scalable and more accurate solutions and better AI models (McConaghy, 2016)

Table 2.1. Convergence of artificial intelligence (AI), big data (BD) and blockchain technology (BCT).

Technology stack	Augmented features	Outcome
Big data + blockchain	BCT makes BD even bigger and contributes towards making BD more secure and valuable for data analytics	A bigger, more structured, and secure data-sharing ecosystem
	BCT's decentralised data storage and computing can create a more efficient and seamless management of BD	Improved blockchain management and functioning
	BCT enables verified, structured, and secured BD for better analysis and prediction of consumer behaviour	
	More efficient governance mechanisms and strategic decision-making in BCT-based systems due to real-time BD sharing and analytics capabilities	
	Blockchained BD mitigates a single point of failure in data governance and introduces an extra level of security against cybercrimes	
Big data + AI	Through intelligent modelling techniques, raw data can be transformed into business value data	More intelligent, customised modelling, prediction, and decision-making
	Through deep-learning systems, raw input data is used to extract higher-level features that can mimic human cognitive abilities	Better data training and more intelligent AI functions
	Smart data administration and data structuralising	
Blockchain + AI	Automated decentralised autonomous organisations (DAOs) and intelligent smart contract management systems can be created on blockchain	Secured AI marketplace, crowd-sourcing AI models, techniques and algorithms, and expanded structured data resource for training
	Data-sharing in trustless, decentralised blockchain networks can augment the abilities of machine-learning techniques	More efficient mining, improved blockchain maintenance, management, and intelligent smart contract initiating
	AI can promote the efficiency of mining in blockchain and significantly reduce energy consumption whilst increasing the sustainability of BCT	
AI + blockchain + big data	A decentralised infrastructure for intelligent data analytics can be created	Transparent, secure, automated data governance ecosystems that can generate useful insights that satisfy the diverse interests of the different parties
	Well-trained intelligent automation can largely boost the processing capacity and accuracy of blockchain-based systems	
	Blockchained big data can provide transparency and potentially a degree of traceability of data in a machine learning model and could contribute to the explainability of AI decisions and outcomes	

within multiple contexts, enhancing the possibilities of data analytics (Casino *et al.*, 2019). The secure and verifiable blockchain structure may be used to ease big data management (Karafiloski and Mishev, 2017) and for efficient risk management and strategic decision-making owing to real-time BD sharing and analytics capabilities (Hassani *et al.*, 2019).

Higher transparency and reliability of AI solutions in data analytics: BCT can be used to train and test AI systems and to ensure protection against misuses by creating a safe and socially desirable AI solution (Dwivedi *et al.*, 2019). Dwivedi *et al.* (2019) propose an AI certification transparency and scorecard blockchain (ACTS-B) to integrate the information about the training dataset used for an AI system, thereby tracking whether the training dataset meet certain criteria such as diversity, equity, etc. For the ACTS-B to engender trust in the AI-based decision-making, it should be a universal publicly viewable blockchain. This would create a transparent mechanism for rating and understanding AI solutions before putting them into use. Blockchain can also provide a tamperproof record of the changes made, along with the authorisation details, making all changes traceable.

Better auditability: the adoption of deep learning in conjunction with faster machines and larger storage spaces have paved the way for modern auditing, which is already being enhanced by blockchain (Appelbaum *et al.*, 2017; Issa *et al.*, 2016). However, machine-learning algorithms are characterised by their opaque features. Their opacity most commonly stems from the large number of possible features included in a classifier which prevent us from understanding and explaining decisions made by AI (Burrell, 2016). Justifying why a specific choice was made by AI creates many complications due to the enforcement of GDPR and its derived requirement of explainability.[2] Additionally, concerning automated decision-making, a data subject has the right to be provided with meaningful information about the logic involved.[3] Blockchains can overcome some of these regulative challenges by providing auditable trails to prove why a particular decision was made by an AI system and resolve the discrepancies raised by the non-linear use of numerous factors and use of randomisation (Casino *et al.*, 2019). Furthermore, having a clear audit trail will also eventually increase every machine-to-machine interaction through IoT and transaction, providing a secure way to share data and coordinate decisions (Corea, 2019).

Customer-centric services: AI and human collaborative automation through BCT can improve working efficiency and prioritisation of human-agent workflow, as well as support human-agents with intelligent advice and information (Hassani *et al.*, 2019). BCT provides greater control of data shared for public service processes and decentralised data processing. The convergence of AI, BD and BCT-based applications can enable real-time data processing and the formation of customer-specific public services.

Automated decentralised autonomous organisations: AI technology can improve the effectiveness of BCT-based systems by providing higher efficiency in mining processes and governance of blockchain, better scalability solutions, increased security of applications, and coding of more

[2] Art. 22, GDPR.
[3] Art 13, *ibid.*

complicated applications for autonomous agents and DAOs (Corea, 2019). AI supported DAOs can autonomously or semi-autonomously operate without a centralised control or third-party intervention (Wang *et al.*, 2019). These new forms of organisations can redefine the mechanisms of control and coordination in public governance. For example, units specialised in streamlining, regulatory and network governance can be replaced by DAOs that are (semi)controlled by other public sector organisations. Furthermore, new forms of management structures with non-governmental organisations can be established as DAOs replacing traditional forms of public-private partnerships and collaborative governance mechanisms.

2.4.2 Challenges for public governance

Difficulty of applying risk management approach and challenges with change management: applying risk management to digital security and other digital risks is challenging for most organisations, in particular where the rights of third parties are involved (e.g. the privacy rights of individuals and the intellectual property rights of organisations and individuals) (OECD, 2019). Especially, insufficient budget and a lack of qualified personnel have been identified as major obstacles to the effective use of risk management to address trust issues (OECD, 2017). Not only the lack of organisational capacities but also the technological features of AI and BCT complicate the application of effective risk management and change management mechanisms. For example, in decentralised blockchains, the absence of a centre to enforce governance policies may hinder the use of effective sanctions against malevolent behaviours or the reparations of harm due to malevolent behaviours. Similarly, the black box features of self-learning AI applications carry the risk of biased decisions in public policies, undercutting effective risk management in data exploitation. For the moment, our knowledge base lacks the determinants of effective risk and change management in the use of advanced AI and BCT solutions.

Increased operating costs: data analytics using blockchain structure incur too many overheads (Casino *et al.*, 2019). One of the identifying features of blockchains is that one can neither delete nor purge specific records/files in blockchains. The ever-increasing number of records can in the long run overburden the cost of storage, operation, and archiving (Liu, 2016). Nevertheless, blockchain-based architectures for BD storage already exist (Kumar and Abdul Rahman, 2017), and intermediate or efficient auxiliary structures may be implemented via cloud solutions, thereby increasing the overall efficiency. Furthermore, the environmental impact of high energy-consuming digital solutions (e.g. proof-of-work (PoW)-based blockchains) may delimit the scalability of solutions or lead to additional operational costs for compliance with environmental standards (De Vries, 2018). Yet again, AI-based solutions and/or alternative BCT consensus models (e.g. proof-of-stake (PoS), delegated proof-of-stake (DPoS)) can improve the energy efficiency of system infrastructures limiting the negative externalities of high energy consumption. For a better estimate of the impact of these technologies on the operational costs in public administration, experimentations with new technologies and empirical data from use cases are necessary.

Need for new interoperability standards and public governance mechanisms: while AI algorithms are more adaptable in existing information management systems, the use of BCT solutions requires new interoperability standards between different information management systems, and the development of legitimate governance and administrative mechanisms concerning the reading, editing and use of data. In particular, the technological and sectoral challenges of blockchain interoperability still need to be resolved (Janssen *et al.*, 2020). Furthermore, AI and BCT may bring increasing levels of effectiveness and efficiency, but political and administrative legitimacy of these technologies remain key concerns (Salamon, 2002), and the impact of these technologies on equity, accountability, and democracy is far from clear at the moment (Barth and Arnold, 1999; Bullock, 2019). Even if a completely unbiased and trustworthy AI solution can be developed aided by BCT, a balance needs to be found between the amount, type and variety of data processed and the goals involved in public governance processes. Not only the technical interoperability standards, but also legal interoperability challenges (e.g. different transparency standards, different standards in data access and sharing) create barriers in cross-border data flows (OECD, 2019) and the scaling up of AI and BCT solutions for cross-border services. Advancements in the related legislation and interoperability policies, developments with cross-border data infrastructures (e.g. European Blockchain Services Infrastructure), and advancements in technical solutions and governance models will shape the future usage of these technologies in public governance.

Cultural transformation in administrations: data verification and implementing effective BCT and AI solutions often require a cultural transformation in administrative processes. This cultural transformation includes changing working habits in system operations and in managerial processes (Bean, 2020). The effective usage of digital solutions proposed by AI, BCT and BD depends on the openness of data and exchange of information inside and outside administrations (European Commission, 2020). Transformation toward openness can be more challenging in siloed organisations. Another challenge is that such transformations may inherently be resisted in some administrations due to incumbent civil servants distrusting the new techniques or political distrust of citizens, which may contribute to low alignment. Furthermore, it is not entirely clear what the implications of these technologies will be on public values. For instance, the operations of BCT-based systems often require the re-allocation of certain prerogatives in public governance concerning accountability, leadership in change management, and decision-making with organisations outside the administrative hierarchy. AI-based systems, on the other hand, are output-oriented, easy to implement and user-friendly to interpret the results, yet machine-discretion may undercut the accountability mechanisms and significantly affect the trust vested in public sector organisations. There is a need for further research to better understand the implications of these technologies on public values.

Automation vs human discretion: a challenge faced by public managers is deciding on the tasks to be automated in AI and BCT-supported systems. The difference between two technologies is that, while AI automates more decision-making processes, BCT automates execution processes (e.g. smart contracts). Bullock (2019) anticipates that tasks that are high in complexity (more deviations from the norm) and high in uncertainty (less analysable) are likely to remain as a discretionary task to be completed by humans. Nevertheless, Bullock acknowledges the limitations of human weaknesses in situations of uncertainty and admits AI's relative advantage for tasks that are high in uncertainty but low in complexity. Although Bullock anticipates that humans will retain a relative advantage in identifying highly complex and abstract patterns across task sets, not all AI experts share the same sentiment that domains requiring creative and abstract thinking will remain in the hands of humans. For example, McAfee and Brynjolfsson (2017) emphasise the creative abilities of AI applications by giving examples of designing novel industry models, composing music, and advancing scientific hypotheses. Their expectation is that machines will generate initial proposals that people can extend and improve with their better attunement to human conditions. Certainly, the decisions to introduce automation are not only managerial or technological decisions. Regulative, social, economic, behavioural, and political factors concerning the involvement of human agents in administrative processes will determine to what extent technologically available automated solutions are implemented in public governance. The sharing of responsibilities among machines and humans in the future of public governance, and the determinants on the role division decisions are other unexplored research areas with which we can expand our current knowledge base.

2.5 Governance of and by digital technologies

A key challenge for the implementation of digital technologies in public governance processes is to configure the 'governance of' and 'governance by' digital technologies (De Filippi and Lovelock, 2016; Ølnes *et al.*, 2017; Tan *et al.*, 2022). 'Governance of' refers to the design choices associated with the digital technologies in accessing and using the underpinning data infrastructure. The 'governance of' dimension focuses on how the data infrastructure affects the usage of the technology and links the data acquisition and data processing mechanisms in overall public policy processes. The rules and standards in acquiring and processing the available data, and the way that the feature of the digital technology affects data processing and exploitation fall under the 'governance of' considerations.

'Governance by' refers to the use of a digital technology in the policymaking and policy implementation. Unlike the technical emphasis of the former approach, 'governance by' prioritises the techno-social power dynamics and control mechanisms in the use of a digital technology in public policy processes by looking into the role ascriptions of automated systems and human agents in the overall digital public governance.

In the following subsections, I will elucidate what 'governance of' and 'governance by' AI and BCT entails in public governance, and shed light on the possible choices to be made by the decision-makers and system designers to effectively implement these digital technologies in the public governance processes.

2.5.1 Governance of artificial intelligence

Data acquisition

The AI system acquires data by either conducting a primary data ascertainment through sensor systems and human data input, or by accessing available databases (e.g. machine logs, clouds, or Internet databases) in a secondary data ascertainment (Wirtz and Müller, 2019). A data acquisition system samples the data input and transforms it into machine-readable data, while the software processes the acquired data for storage or presentation. The data feeding the system and the median technologies and storage mechanisms integrated in the system are key governance considerations in data acquisition. Challenges concerning the database size, data integration, data quality, and data standards (i.e. how and what data is collected, and what format it is stored in) can undercut the effectiveness of data processing and AI predictions.

For a successful implementation of AI strategies and programs, organisations must have access to a base set of data and must maintain a constant source of relevant data to ensure that AI can be useful in the selected policy domains. The input data can be in a multitude of formats such as text, audio, images, and videos. The wide range of sources to collect and store this data adds to the governance challenges. For successful predictions, all the relevant data must be integrated in a manner that the AI can understand and transform into useful results. A technological challenge for AI-based systems is to analyse unstructured data. For example, in healthcare, medical imaging represents a large share of relevant data, which even the most advanced AI-based systems (e.g. Watson) cannot read directly (Sun and Medaglia, 2018). This means that, depending on the data source, the AI system may need to be complemented with human experience.

Data quality is another core challenge in data acquisition. AI performs best when it has a good amount of quality data available to it. Therefore, AI solutions built on BD augment the performance of AI-based predictions. But BD pools different data from different origins that needs contextualising for analyses and reports. Greg Hanson from Informatica argues that for well-curated data, enterprises should build a catalogue of data assets and use engineering tools with AI-built in the backend (Ross, 2019).

In the public sector, another challenge for data quality is the rules and standards employed in publicly available data. For example, GDPR requires the purpose limitation principle on data acquisition, which may limit the pattern recognition functions of machine-learning systems. Therefore, not only the quality of data but the variety of available data can affect the performance of AI applications.

The performance of AI is also related to the quality of the training data. Here, bias embedded in the training data is one of the biggest challenges that AI faces (Rosso, 2018). Often, data sources are laced with racial, gender, communal, or ethnic biases (Manyika *et al.*, 2019). Biases embedded in the training data could easily lead to discriminatory and unjust consequences in policymaking and implementation processes. For example, the 'Correctional Offender Management Profiling for

Alternative Sanctions' (COMPAS), a system to predict whether a defendant would re-offend, was found to be as successful (65.2% accuracy) as a group of random humans (Dressel and Farid, 2018) and to produce more false positives and fewer false negatives for black defendants (Müller, 2020).

Furthermore, data privacy in data acquisition, and how to achieve an appropriate balance between privacy and data acquisition, is another pressing issue in AI adoption (Begg, 2009). Hence, the ethical challenges accompanying data acquisition processes need to be addressed for the wider adoption of AI-based systems in public services.

Data processing

The acquired data is stored in data servers for data processing. For security or privacy issues, some organisations may need to store data on in-house data servers. Those organisations need to maintain the cost of in-house data servers and a technical support team. Cloud-based alternatives, centralised government data silos, or digital crossroad data centres present more cost-benefit friendly solutions and effective upscaling of AI-based systems. However, the interoperability considerations and administrative burden in data acquisition can undermine the appeal of these alternative data storage options for organisations.

The acquired data is processed by algorithms and machine-learning techniques. Choosing the algorithm (supervised or unsupervised learning), selecting optimisation standards,[4] and checking and rechecking the performance are all required to configure an AI-based solution for a policy objective. The quality of the human resources and available software solutions in data analysis are some key considerations for the public sector organisations. The massive computing power necessary to process BD to build an AI system, and to utilise data-intensive machine-learning systems such as deep learning, can also technically and financially challenge organisations. The effectiveness of AI-based solutions requires high-end processing power, and the cost of large infrastructure requirements are considered as impediments to the adoption of AI technology (Roberts, 2017). In addition, there is a high demand for a limited number of AI experts, which is associated with an increasing cost of education and salaries (Bughin *et al.*, 2017). Cloud computing environment and outsourcing can mitigate some maintenance costs. Nevertheless, organisations need to plan in advance the cost and technical requirements to maintain higher computational speed requirements along with higher availability of data to scale up their AI-based systems.

---

[4] For instance, in a game developed by two university students in China, two wolves and six sheep would be placed at random within a game space by the computer. The wolves would have to catch all the sheep in 20 seconds while avoiding some boulders within the space. If a wolf caught a sheep, 10 points were awarded. If he hit a boulder, one point was deducted. To encourage them to catch the sheep as fast as possible, 0.1 points were deducted from the wolves for every second that passed. The goal was to see if the AI wolves could, through training and retraining, figure out a way to maximise their scores. After 200,000 iterations, the researchers found that the wolves simply rammed themselves against the boulders to commit suicide most of the time. The researchers discovered that the wolves were unable to catch the sheep most of the time, leading to heavy time penalties. The AI wolf then decided that, if at the start of the game it was close enough to the boulders, an immediate suicide would earn it fewer point deductions than if it had spent time trying to catch the sheep. In this perverse self-terminating manner, the AI wolf was 'optimising' its performance by minimising the time penalties incurred (Ng, 2021).

The output procured by the machine needs to be presented in an easily interpretable way by the user. A caveat on data visualisation is the phrasing of wording and visuals in the results of analysis. The confidence intervals of the results may vary depending on the employed algorithm or statistical techniques, but studies on human cognitive bias show that the framing of information and behavioural factors tend to affect the interpretation of results by the users (Kahneman, 2011). There are ongoing technical efforts to detect and remove bias from AI systems, but these efforts are in the early stages (Brownsword *et al.*, 2017; Yeung and Lodge, 2019). A mathematical notion of fairness and technological fixes have their limits in overcoming systematic and cognitive biases in each social context (Selbst *et al.*, 2019; Whittaker *et al.*, 2018).

2.5.2 Governance of blockchain technology

Data acquisition

Data acquisition in a BCT-based system varies according to the type of blockchain used in the system infrastructure. Blockchains can have permissionless/permissioned and public/private forms (Miller, 2020). The difference between public and private blockchains is about who owns the data infrastructure. The difference between permissionless and permissioned systems is about the restrictions imposed on network participants in terms of read, write, and audit/commit functions. In the former, anyone can participate in the network and validate the transactions taking place in the platform, and in the latter, only selected entities are authorised to validate the transactions. These classifications are not exclusive to each other, and it is possible to have a public and private blockchain with varying degrees of permission models. For instance, the two most known blockchain platforms, Bitcoin and Ethereum, are public and permissionless blockchains. Hyperledger Fabric, R3Corda, and Quorum are some examples of technology developers that provide built-in tools for permissioned systems applicable to public and private blockchains. The merits of different types of blockchains vary technically concerning decentralisation, security, scalability, speed, throughput, privacy/confidentiality, trust, and finality dimensions. On the one hand, public and permissionless blockchains engender better trust and security for data infrastructure while experiencing a lack of scalability and performance issues. On the other hand, private and permissioned blockchains are largely developed and preferred by private enterprises (e.g. IBM, Microsoft) as they allow a certain degree of control of the data privacy and the governance of the system.

The choice between permissioned and permissionless blockchains calls for several trade-offs regarding policy priorities (e.g. privacy, throughput, security, trust, etc.). Furthermore, the design choices of blockchain architects may inherently prioritise one system function over others (e.g. throughput over privacy). These trade-offs are context-dependent and should not necessarily have the same significance for public sector organisations. For example, the anonymity of transactions could be a vital factor for private enterprises, but in democracies, the public sector organisations – excluding the security and intelligence forces – are already expected to be transparent in public governance. Similarly, higher throughput and scalability are vital concerns for cryptocurrencies, but in the public sector, the volume of transactions and the number of users are often bound to the jurisdiction area and administrative scope of the services. This network predictability may allow predetermination of the scalability and throughput needs in data acquisition.

Moreover, Kannengiesser *et al.* (2019) identified 23 endogenous trade-offs among 7 DLT properties (usability, performance, flexibility, security, transparency, law and regulation, and community) that vary according to the design choice with blockchain infrastructure. For example, if the purpose in data acquisition is to enhance the trust among users, the system architecture should prioritise transparency, which can come at the expense of other properties (e.g. usability, performance, flexibility). Therefore, any choice in the system architecture is a political act that involves optimising the trade-offs between various DLT properties, and permissioned and permissionless systems.

Data processing

In BCT-based systems, data-processing functions can be enhanced by decentralised applications (DApps) and smart contracts. DApps are open source coded digital applications or programs that exist and run on a blockchain or P2P network of computers. Through DApps, users can access blockchain networks and engage with other users for different purposes (e.g. storage of data space). Nonetheless, each DApp comes with certain advantages and disadvantages (e.g. finality, speed, scalability, energy consumption, etc.) based on their consensus and incentive mechanisms, which makes choosing a DApp inherently a governance decision. It is also possible through modular DApps to anchor values in a non-blockchain system or a private blockchain to a public blockchain. For instance, the land registry use case in Sweden employs a two-layered blockchain approach, whereby the summary of hashes from a private blockchain is stored in Bitcoin blockchain (Kairos Future, 2017). Therefore, it is possible to circumvent certain trade-offs between DLT properties (e.g. performance vs security) through the use of DApps.

Smart contracts are mechanisms that contain the digital assets of two or more involved parties, where assets are distributed automatically according to predefined response actions when trigger conditions are met (Governatori *et al.*, 2018). DApps access the blockchain network through smart contracts, which enforce the term of the agreement between two parties. Through the combination of smart contracts and DApps, it is possible to create DAOs where the operational rules are encoded on blockchain in the form of smart contracts, and DAOs can autonomously or semi-autonomously operate without a centralised control or third-party intervention (Wang *et al.*, 2019).

In data processing, another important consideration is the interoperability of the BCT-based system. The interoperability within the BCT context refers to the ability of blockchain networks to share, see, and access information across existing data management systems in the public and private domain, without the need for an intermediary to do the exchange. In that sense, an important decision concerning interoperability is assessing to what extent BCT-based systems comply with the existing mechanisms, technologies, regulations, and standards in the digital governance landscape. There are three types of challenges that can be identified concerning the interoperability of blockchain. First, there is a lack of standardisation from technical protocols to smart contracts that connect auxiliary technologies such as AI and automated agencies in blockchain platforms (Janssens *et al.*, 2020). Moreover, for the moment, stored information on blockchains is not necessarily in line with market standards and practices, such as ISDA (International Swaps and Derivatives Association), FPL (FIX Protocol Ltd), or ISO (International Organization for Standardization) (Janssens *et al.*, 2020). Second, blockchains are, by design, not interoperable with each other (Lafourcade and Lombard-Platet, 2020). Several solutions have been proposed to facilitate cross-chain transactions, such as facilitating messaging via a management chain, using cross-chain cryptocurrency for atomic asset transfers, or direct communication via hardware connections using TLS or smart contracts, each with their unique set of problems to solve (Johnson *et al.*, 2019). Third, interoperability challenges can arise from the specific needs in the public service domain. For example, Zhang *et al.* (2017) identify particular interoperability challenges in the healthcare sector regarding the evolvability, flexibility, extent, and continuity of data exchange across various data providers.

Interoperability challenges can affect governance choices in two possible ways. Firstly, they can preclude the scalability of BCT-based solutions and thereby limit the use of blockchain only to specific functions (e.g. data verification through verifiable credentials). The lack of scalability can further undermine the applicability of decentralised and permissionless systems. Secondly, the technology choices of stakeholder organisations in a blockchain ecosystem can elevate one blockchain infrastructure, and its associated governance choices, over others and establish it *de facto* as the system architecture of choice in the public sector. National and international policy initiatives can especially be a trendsetter in that regard. For example, Estonia aims to solve the interoperability challenges by creating one standardised system in public services based on keyless signature infrastructure (KSI). KSI does not rely on the joint actions of the users to verify the data, and unlike PoW or PoS consensus mechanisms, uses the proof of time of the signature to verify the integrity of the data.[5] Another example is the European Blockchain Services Infrastructure (EBSI),[6] which aims to provide cross-border public services in the EU by utilising blockchain. As part of the EBSI, member states are expected to establish the necessary infrastructure for the transition and to assign nodes (i.e. trusted and capable entities in every member state) to support the proof-of-authority (PoA) consensus mechanism. In case there is a nation-wide data-sharing platform (e.g. digital solutions in the Estonian government) or if a European or global level blockchain platform (e.g. EBSI) is adopted, the interoperability requirements may delimit available technical design choices.

One final governance consideration is to what extent BCT-based system infrastructure can incorporate BD and IoT systems existing in the digital public governance ecosystem. The digital transformation in the public sector has introduced several technologies and solutions, and BCT, as a disruptive technology in public services, has a higher chance of success if it complements the

[5] See the use case at https://digiexpo.e-estonia.com/Solutions/guardtime-ksi-blockchain-stack/.
[6] See https://ec.europa.eu/digital-single-market/en/blockchain-technologies.

existing technologies rather than replaces them. Policy makers and system designers need to assess beforehand the needs of the existing systems, to determine how and what type of BCT solutions can improve the system capabilities in digital public governance. Furthermore, blockchain can confirm the genuineness of a transaction, but whether the rules on the transaction are genuine and fairly established is beyond the technical scope of blockchain. Therefore, 'who is authorised to code the protocols?' and 'how do the rules/procedures encoded in the system affect public policy processes?' are important questions for the adoption of BCT in public governance. The answers to these questions are context-dependent and may call for a combination of political, administrative, and technological solutions. Here, the challenge for system designers and policy makers is creating a public governance design that does not undermine the effectiveness and efficiencies of transactions, while safeguarding the privacy and autonomy of the users and maintaining institutional principles and regulations.

2.5.3 Governance by artificial intelligence

Policymaking

According to Eggers *et al.* (2017), AI can change the role of humans in the policymaking processes in four ways. First, AI can relieve public workers by taking over repetitive tasks, allowing public servants to focus on more valuable tasks. Second, AI can help to break up a job into smaller tasks and can take over as many of them as possible, improving the efficiency in policymaking processes. Third, AI can replace a human agent by automating policymaking processes. Fourth, AI can augment the performance of public servants by complementing their skills and improving the effectiveness of policymaking processes.

However, according to Scherer (2016), AI presents two risks in policymaking, namely the loss of predictability and the loss of control. Loss of predictability is caused by the huge processing capabilities' gap between AI and human agents, whereby AI can process huge amount of data at high speeds beyond the capabilities of human agents, making the results no longer comprehensible and verifiable for humans. Thierer *et al.* (2017), therefore, define AI-led information processing systems as a 'black box' for human end-users, turning their role into one of data feeder and recipients of results without the ability to control the validity of methods and criteria in policymaking. Loss of control refers to the dislocation of human control in influencing the system operations. The self-learning mechanisms of AI allow it to reprogram itself for process optimisation, and the personnel responsible for maintenance and monitoring the system can partially or completely lose the ability to realign the system with policy objectives.

Furthermore, AI-led policymaking is stranded by the legal, moral and ethical framework conditions. Human decisions in public policy are political in nature and soft skills such as ethical trade-offs, social rules, empathy, humanity, and conscientiousness have a conclusive influence on the outcome of decision-making processes and its subsequent evaluation (Wirtz and Müller, 2019). For the moment, AI-technology lacks these human cognitive qualities and has limitations when it comes to taking over human roles in public decision-making processes (Andersson *et al.*, 2012). Conversely, in a study on 758,027 bail decisions, Kleinberg and his colleagues (2018) show that their machine-learning algorithm was able to better predict the flight risk of detained defendants than actual judges. The researchers calculated that if the risk threshold is set to reduce the number of people detained on bail as much as possible without increasing crime, the crime rates could be reduced by up to 24% whilst the number of people detained could be reduced by up to 42%. The results show that AI-led decisions not only performed better than human judgements but also led to fairer decisions.

The use of AI in policymaking also raises concerns about its behavioural impact on human-led decision-making processes. For instance, a study on an automated profiling system for unemployment claims in Poland found that fewer than 1 in 100 decisions made by the algorithm were questioned by the responsible clerk (Misuraca and Van Noordt, 2020). Behavioural factors such as lack of time to ponder the details, fear of repercussions from the supervisors, and a belief in the objectivity of the process appear as driving influences behind the behaviour of clerks, making human-led decision-making processes practically automated systems (Kuziemski and Misuraca, 2020).

In terms of using AI for predictive policies, Kahneman and his colleagues (2021) provide new insights about the use of AI in policymaking processes. Their research on judicial cases shows that AI outperforms intuitive human judgement, but this improvement may not justify the marginal cost necessary to replace a human-led system with automated decision-making systems. From their research, two conclusions can be drawn. One, some decisions and some policy areas are more adaptable for algorithmic decision-making (e.g. where the improvement of prediction is vital). The second conclusion is that managers believe in their decisions; something comes from the expectations of the position, and expectations of people of themselves. In decisions where it is not the prediction but the inclusion judgement that is essential, algorithmic decisions will very likely be secondary to human judgement.

By taking into consideration these potential challenges, Janssen and Kuk (2016) envisage that governance by AI can at best be used only for mundane tasks. Similarly, Eggers *et al.* (2017) stress that AI is most suited to handling repetitive, highly structured, and regulated work processes. They recommend that organisations issue a work structure and process analysis on a regular basis to draw up the respective areas of application for AI in policymaking.

Policy implementation

AI can replace human agents in the delivery of public services, and engage with end-users, not only in policymaking but also in policy implementation. AI-led bots and AI-powered digital interfaces can especially replace public servants in repetitive and predictable tasks. Many authors even highlight the risk of the increasing replacement of human agents in healthcare, education, logistics, and organisational processes, which raises the threat of mass unemployment in various public sector areas (Batko and Szopa, 2017; Susskind and Susskind, 2016; Ford, 2013).

Nonetheless, the wider usage of AI in policy implementation can also create new jobs and roles in public services. There is already a high demand for skills in algorithms and use of AI in systems on the job market (CEDEFOP, 2020). Not only the technical staff but also public managers are required to expand their working capabilities with AI usage and develop a better understanding of how AI can supplement the workforce to achieve better results faster (Eggers *et al.*, 2017). The challenge, however, is that the technical and managerial staff or the people working in the jobs at risk do not necessarily have the required skills and formal training to ease the pressure on human resource management in public administration.

The impact on the workforce is one governance dimension for the usage of AI in policy implementation. AI safety and end-user behaviours are other concerns in governance by AI. AI safety refers to assuring the secure performance and impact of AI (Boyd and Wilson, 2017). These safety concerns relate not only to issues of information security, but also to complex and safety-critical situations resulting from circumstances where the AI may learn negative behaviour from its environment and misunderstand its surrounding (Conn, 2017). Taking the necessary precautions and safety measures is important for the scope of AI applications. Bostrom and Yudkowsky (2014) underline the need for AI technology to be resilient to adverse manipulation by humans. For AI applications based on reinforcement learning and automated executions, safety measures need to be in place to avoid catastrophic consequences. It is also important to avoid negative side effects in the working environment during the learning process of AI applications (Amodei *et al.*, 2016).

Lastly, studies show that the uptake of new services by AI-bots has not been particularly high, leading authorities to believe that some form of proactive marketing would be necessary to change citizens' behaviour (Kuziemski and Misuraca, 2020). Kuziemski and Misuraca (2020) assert that in order to pursue such projects further, local authorities need to develop user experience and awareness. However, user experience and awareness cannot be enhanced solely by marketing practices, and trust in public administration is important to facilitate the transition. In particular, recent studies suggest that public service processes have a significant impact on citizens' trust in public administration, and the absence of corruption is a particularly strong institutional determinant in trusting public administration (Migchelbrink and Van De Walle, 2019). Therefore, holistic approaches are needed in service design to leverage AI technology in service provider and citizen interactions.

2.5.4 Governance by blockchain technology

Policymaking

A BCT-based system relies on interactions among different types of users, such as miners, verifiers (or node operators), core developers, token holders, content producers, and network users. In policymaking, governance by BCT or blockchain governance needs to ensure that the wishes of decision-makers are executed at the operational level of the network. Policymaking depends on how collaborative actions among these actors are managed, how incentive mechanisms support system maintenance, and how consensus mechanisms affect blockchain governance.

The collaborative action mechanism varies between on-chain and off-chain governance processes (Reijers *et al.*, 2018). In on-chain governance, stakeholders participate in discussions and decisions through the protocol itself, and when a decision is reached through a voting procedure, the protocol automatically adopts the decision. In on-chain governance, the way that the engagement and decision-making take place is encoded directly in the underlying infrastructure. The strength of on-chain governance is its enforcement mechanism, because when a decision is agreed upon, it is implemented by following the rules embedded in the code. However, the downside of this mechanism is that it automates a winner-takes-all rule. The winning majority of a voting process determines all system functioning, and as such, it risks alienating political dissidents of the process. Off-chain governance refers to the endogenous and exogenous rules and processes of deliberations around the protocol that contribute to the operations and development of the blockchain-based systems (Reijers *et al.*, 2018). Off-chain governance decisions are introduced to blockchain networks often through the improvement proposals or forks (e.g. soft fork, hard fork). Although off-chain governance presents a more democratic alternative and allows an incremental adaptation, it presents an inherent security risk in permissionless systems. The miners of the system are rent-seekers and they do not necessarily possess the technical expertise to evaluate a proposal. Therefore, an initially benign-looking software modification initiated by a malicious actor can create security risks in the future (Finck, 2018). It is also possible to develop hybrid systems, where some decisions can be taken through on-chain processes, whereas others can be taken through off-chain processes.

Additionally, depending on the blockchain protocol, some actors may hold veto rights over certain decisions, or some decisions can be open only to the vote of certain actors. Here, the institutional framework, the area of application, and country-specific conditions may dictate that different actors take a more prominent role in the on-chain and off-chain governance processes. Since a core added value of blockchain is to establish confidence in processes where there is a lack of it, the policymaking mechanisms may need to prioritise the involvement of the relevant actors and increase the level of transparency in the design and operation of BCT-based solutions. For that, a social layer can be introduced in blockchain governance, where core developers, stakeholders, and societal actors deliberate on the political and functional implications of the system design choices. The deliberations that take place at the social layer can foster on-chain processes to create a politically legitimate process and can serve as a dynamic constitution of the BCT-based system. The difficulty of establishing an inclusive decision-making mechanism is to create legitimacy in blockchain governance without deadlocking the collaborative action mechanism. A democratic constitution of blockchain governance might be necessary to establish the rules and roles of the verifiers, developers, and users of the system.

The arrangement of incentive mechanisms is also important in the usage of BCT in policymaking processes. Both in Bitcoin and Ethereum blockchains, this mechanism relies on monetary rewards in the form of blockchain cryptocurrency, which aligns the individual rent-seeking behaviour with the overall benefit of the platform. For public sector organisations, using monetary incentives for user engagement can create ethical challenges. However, not all incentive mechanisms need to leverage the monetary value of cryptocurrencies. For example, allocations of tokens and reputation scores, which grant the users enhanced access to platform functions and weighted-voting rights in decision-making processes, can be some non-monetised rewards.

Another related policy decision is about the disincentivising effect of transaction fees. Transaction fees are the cost of actions in a particular blockchain, such as transactions or executions of smart contracts (e.g. 'gas' in Ethereum). The disincentivising effect of transaction fees and the incentivising effect of rewards to miners create trade-offs and entail rational choice calculations by the users and miners before engaging with blockchain. If the costs outweigh the anticipated benefits of engagement, some blockchain applications can lose their attractiveness or miners might find less value in supporting the blockchain altogether, adversely affecting the security of the blockchain (De Filippi and Wright, 2018: 41).

Moreover, a consensus mechanism is necessary to coordinate the decentralised actions of users in deciding which information can be added in the blockchain. Different consensus mechanisms exist, but most blockchains either rely on Nakamoto consensus (or PoW) that ties mining capability to computing power or Byzantine consensus that uses staking to assign miners, such as PoS and DPoS. There are also PoA systems, where a lower number of nodes or masternodes take the role of transaction validators. Each of these consensus mechanisms has advantages and disadvantages over others, and their affinity with decentralised or centralised governance structures varies.

PoW is mostly used in permissionless public blockchains, such as Bitcoin and Ethereum, whereby the users validate the transactions by solving complex mathematical puzzles through the computing power of the hardware. The advantage of the PoW system is its stability to deter cyberattacks (e.g. denial-of-service attacks) while maintaining a distributed system governance. However, its vulnerability to 51% attack, high energy cost, increasing centralisation of mining operations, and low transaction throughput are some of the significant challenges for PoW-based systems. Moreover, the increased strain on the environment imposed by the high energy demands of PoW-based systems reduces their likelihood of being chosen for public sector projects.

In PoS, the mining power is attributed to nodes in the proportion of tokens (or coins) held by nodes instead of their computing powers. Node operators lock away a stake for the right to participate in block creation, and nodes with bigger stakes have a higher chance of being selected to verify transactions. A transaction fee is paid to the node operator in return for the transaction verification. In the case of fraudulent transactions or any misbehaviour, the node loses the right to participate in staking. The difference in DPoS is that the token holders choose a small number of nodes as delegates by staking tokens with different candidates. If the delegate is chosen for block creation, a fraction of the reward is allocated to those who voted for the delegate. In some systems, decision-making power is associated with the delegation of tokens where delegates may have the capability to monitor and amend network parameters such as fees, block size, block rewards, and the length of the transaction cycle (Karjalainen, 2020). The advantage of PoS is that it is less susceptible to 51% attacks and more scalable with higher transaction throughput, but it is assumed to be less secure than completely decentralised PoW systems (Edchain, 2018).

In PoA systems, unlike the PoS and PoW systems, the identity of the validator is known, and the assumption is that the reputation of the validator plays the role of stake. The advantage of PoA systems is that they can achieve much higher throughput as a result of the lower number of validators and they are suited for both private networks and public networks. However, the downside is that PoA is not imbued with the sense of security derived from decentralised consensus mechanisms and the nodes therefore need to be kept uncompromised. Hence, algorithmic trust typical of Bitcoin or Ethereum is not present in a PoA-based blockchain. Rather, an *ex-ante* trust is required in node selection. For instance, as a PoA-based blockchain, the nodes in EBSI are selected from key public or private institutions that had already preserved a central role in digital governance (e.g. BELNET and SMALS in Belgium).

The differences among consensus mechanisms also affect the role of users in blockchain governance and in overall public governance. For instance, PoW or PoS as a mechanism empowers the role of the users and miners as *demos*, and makes them indispensable for the continuity, safety, and relevance of the system. In both mechanisms, the users are co-producers of the blockchain governance. In the PoA, the users are the beneficiaries of the system whereas some organisations hold the roles of a public agency to produce public services. In public governance literature, this distinction is associated with the beneficiary or client role of citizens in traditional public administrations and new public management regimes, and the co-producer role of citizens in new public governance systems (Osborne and Strokosch, 2010).

Lastly, blockchain governance needs to be integrated into legitimate and transparent political and administrative decision-making processes. Blockchain technology requires the government and public sector organisations to give up certain prerogatives in policymaking. To what extent the government needs to give up its prerogatives through BCT is associated with the level of trust vested in the public sector organisations and society. Permissionless-public blockchains are expected to be desirable in public service areas where there is a low level of trust among users (Meijer and Ubacht, 2018). However, the literature on trust suggests that when there is a low level of trust in society or other stakeholders, decision-makers are more inclined to keep the control through direct or indirect means (Bijlsma-Frankema and Costa, 2005). Understanding the mechanisms of citizen trust in government, stakeholder organisations and digital technologies are important to create effective governance mechanisms by blockchain. Establishing a political consensus mechanism, improving user experience, safeguarding the survival of the system, and ensuring interoperability with existing infrastructures, and the regulatory and institutional frameworks are some core considerations for system designers and decisionmakers in order to build trust in blockchain governance.

Policy implementation

Fair distribution of power among stakeholder organisations and citizens and establishing legitimate accountability and control mechanisms are needed to effectively use blockchain technology in policy implementation processes.

Power distribution in policy implementation relates to the allocation of decision-making rights among the actors participating in the blockchain network. Decision-making in blockchain governance can comprise systemic changes (e.g. hard and soft forks), rules of operations, conflict resolutions, and system functions (e.g. DApps). It is possible to have centralised, semi-centralised (or hybrid), polycentric, and decentralised forms of policy implementation. Centralised structures are those where the decision-making rights are held by a specific group of people or organisations, and the policy implementation processes can be organised through off-chain or on-chain processes. A semi-centralised or hybrid governance structure exists when some governance decisions (e.g. conflict resolution) are taken only by a centralised board of directors, and some other governance decisions (e.g. concerning the network of users or platform functions) are taken with an additional on-chain voting procedure by the platform users. Here, the impact of the users' vote on the actual implementation of decisions can vary. For example, an absolute or qualified majority in a caucus of users can be required, or different weighting options can be used for the result of voting processes to supplement decisions taken by the board of directors.

In polycentric governance, different clusters of actors (e.g. miners, developers, nodes) hold different roles in blockchain governance, and decision-making requires taking into account what the others are doing (Stephan *et al.*, 2019). In a decentralised structure, decision-making processes are not dominated by a single actor or a group of actors. Rather, the decisions are taken by the majority of users operating in blockchain networks through on-chain voting processes or coordinated mining action. Here, the choice architecture of voting processes and the openness of propositions to the public may influence the policy implementation processes. Furthermore, in decentralised systems, improvement proposals and DApps can be introduced directly by network users without the interference of an intermediary actor (e.g. developers).

It is important to underline that certain governance structures may have different degrees of affinities with system design choices. For example, while a centralised governance structure might be more suitable for permissioned systems, a polycentric or decentralised structure may be more suitable for permissionless systems. Moreover, while PoA may be more suitable for semi-centralised systems, PoS and PoW may be more suitable for polycentric or decentralised systems. It is also possible that the power relations of actors may change an initially different-looking governance structure into another. For example, if on-chain governance processes in PoW or PoS-based systems are dominated by a few large operators who control most of the mining resources and/or token holdings, an initially decentralised governance structure may act as a semi-centralised or polycentric governance structure. Alternatively, if off-chain governance processes are dominated by a few specialised influential players, a centralised governance structure may act like a semi-centralised or polycentric structure.

The rules of transactions in blockchain governance are set at the design phase, and once a decision is made, a change in the system requires the consent of authorised users. Therefore, even minor editing requirements in the data structure can cause significant transaction costs. Consequently, establishing necessary accountability mechanisms in blockchain governance is important because blockchain as a technology re-establishes the importance of trust in the initial design phase instead of the implementation phase. Different contexts may require different user roles for accountability mechanisms with private-permissioned, public-permissioned, or hybrid design solutions. An important rule of thumb with accountability mechanisms can be whether a design solution creates legitimacy in blockchain governance.

In the blockchain context, there are soft and hard regulatory approaches to accountability mechanisms. One approach is described as *lex cryptographia* (De Filippi and Wright, 2018: 32). Lex cryptographia means that the rules of exchange are inbuilt codes, and, as such, code becomes the law in blockchain governance. Through the use of smart contracts, rewards and sanctions are executed automatically, creating a deterministic system of governance. The challenge of transposing law into code is that codes are written *ex-ante*, in strict and formalised language, and therefore, code-based rules need to be predictable and leave no room for interpretation (De Filippi and Wright, 2018: 56). This inherently limits the applicability of code-based rules in areas where the contingencies and conditionalities cannot be determined *a priori*.

A more gradual change management relies on soft and hard forks. A soft fork refers to a change to the software protocol where previously valid blocks/transactions are made invalid. A soft fork does not change the structure but rather modifies the functions of blockchain. The implementation of soft fork relies on the coordinated action of the majority of users to implement the suggested changes. The changes enter into force only if the majority of the network's mining power adopts them. Otherwise, the soft fork fails, and the old chain remains unchanged. In the case of political dissonance among different groups of users in a blockchain network, soft fork presents a flexible instrument with which to modify the system. Soft forks are mostly used for the introduction of improvement proposals and DApps in a blockchain network. A hard fork occurs when a rule change is adopted in blockchain protocol and the nodes of the newest version of a blockchain no longer accept the older version of the blockchain. In the case of a hard fork, all nodes are meant to work by the new rules to upgrade their software. Otherwise, a permanent split from the previous version of the blockchain occurs and two different blockchains are created.

Lastly, new digital technologies and advanced data analytics techniques supplemented by exogenous data sources can create complex automated decision-making and control mechanisms using BCT. As mentioned earlier, DAOs are some forms of new organisations that can replace the human agency in policy implementation. A decision towards the removal of the human element is closely linked with the embedded public values in public sector organisations and the sectoral area of application. The salience of efficiency and cost-effectiveness principles may insinuate the replacement of certain functions and organisations in public services with DAOs or automated agents. Yet again, replacing a large number of public sector organisations and administrative positions with automated agents may be politically unfeasible. One downside of DAOs is that the notion of accountability can be blurred in public administrations. Here, the cultural and behavioural reservations may suggest that certain roles and decisions, which can be automated, should be left in the hands of actual people. Furthermore, the adaptation of automated solutions needs to comply with the existing capacities and practices at the societal level regarding digital governance. A core consideration here is assessing to what extent people can manage their digital identities and assets, and whether they need custodian organisations in digital public governance.

2.6 Conclusions

The technological evolution of BD, AI and BCT hold new implications for the future of the public sector. From smart city solutions to innovative fraud detection and prevention, there is a plethora of areas in which these technologies are currently used or tested to improve the quality of service production and delivery. Furthermore, these technologies will uproot many processes, organisations, and practices in public administration and management. There are, however, many issues and challenges to be overcome before they become widely accepted by the end users. The full potential of these technologies in transforming public governance processes is contingent upon developing innovative solutions to a myriad of legal, technological, organisational, cultural, ethical, behavioural, and institutional challenges. This chapter provided an overview of the possibilities that can be provided by these technologies and elaborated on the associated challenges along the way.

Applications and use cases in the public sector reveal several key insights about the role of these technologies in the future of public governance. First, at least for the moment, none of these technologies are mature enough to fully disrupt the existing governance processes. Governments can harness big data through the use of AI and data analytics techniques to improve productivity, performance, innovation in service delivery and policymaking processes. However, advanced AI solutions are opaque and are tainted by biased datasets to train algorithms. Blockchained BD systems emerge as new innovative solutions to improve the quality and traceability of data analytics, but privacy, interoperability, and scalability issues still need to be addressed.

Second, existing legal, socio-political, and institutional framework conditions need to be adjusted to the technological conditions of AI and BCT-driven services. Token technology, smart contracts, blockchain interoperability, machine-learning algorithms, human-machine interfaces, and automated systems for decision-making present fundamental contradictions to the existing regulations and norms about data protection, accountability, human dignity, public values, and power structures between public and private sector organisations. The future of these technologies will not only be determined by the technological possibilities but also the political preferences of citizens and public bodies. For example, the draft proposal published by the EU commission for an AI act[7] recommends a single future-proof definition of AI and prohibits certain AI practices in public governance as contravening EU values. For instance, AI systems used by public authorities, or on their behalf, for the evaluation or classification of the trustworthiness of natural persons based on their social behaviour or known or predicted personal or personality characteristic, or real-time, remote biometric identification systems in publicly accessible spaces for the purpose

[7] The draft act was published in April 2021. It is accessible at: https://digital-strategy.ec.europa.eu/en/library/proposal-regulation-laying-down-harmonised-rules-artificial-intelligence.

of law enforcement are classified as prohibited practices. Furthermore, specific restrictions and safeguards are proposed in specific 'high-risk' areas where invasive AI technologies may pose a threat to the fundamental rights of Union citizens. For some specific AI systems, only minimum transparency obligations are proposed, in particular when chatbots or 'deep fakes' are used. For other non-high-risk AI systems, only very limited transparency obligations are imposed, for example in terms of the provision of information to flag the use of an AI system when interacting with humans. We do not know to what extent these regulative precautions and restrictions will limit the diffusion of AI solutions in 'high-risk areas' in the public sector, but we can anticipate that the disruptive potential of these technologies will be mitigated or at least postponed during the adjustment of the legal and normative framework conditions.

Third, the technological features of AI and BCT call for the re-evaluation of existing governance models and practices in the design and implementation of public services. In this chapter, the governance requirements of AI and BCT are studied through the lenses of 'governance of' and 'governance by' approaches in the digital governance literature. 'Governance of' refers to the system design requirements for the application of a digital technology. 'Governance by' refers to the use of a digital technology to replace the existing mechanisms in the public governance processes. These governance dimensions are not detached from each other, and design choices in one affect available governance options in other dimensions. In particular, clear role definitions for data users and consumers, transparent power allocation across technology providers, end users, service designers, and other supporting organisations for system infrastructure, safeguarding the accountability and control mechanisms, development of innovative change and risk management approaches, and improving user awareness and experience are required to ensure the successful transition of public governance processes.

References

Agostino, D., and M. Arnaboldi. 2017. Social media data used in the measurement of public services effectiveness: empirical evidence from twitter in higher education institutions. *Public Policy and Administration* 32 (4): 296-322. https://doi.org/10.1177/0952076716682369

Allam, Z., and Z.A. Dhunny. 2019. On big data, artificial intelligence and smart cities. *Cities* 89 (June): 80-91. https://doi.org/10.1016/j.cities.2019.01.032

Allessie, D., M. Sobolewski, L. Vaccari, and F. Pignatelli. 2019. Blockchain for digital government. *Joint Research Center*. Vol. 29677. https://doi.org/10.2760/93808

Amodei, D., C. Olah, J. Steinhardt, P. Christiano, J. Schulman, and D. Mané. 2016. Concrete problems in AI safety, June. http://arxiv.org/abs/1606.06565

Andersson, A., Å. Grönlund, and J. Åström. 2012. 'You can't make this a science!' – analyzing decision support systems in political contexts. *Government Information Quarterly* 29 (4): 543-552. https://doi.org/10.1016/j.giq.2012.05.007

Appelbaum, D., A. Kogan, and M.A. Vasarhelyi. 2017. Big data and analytics in the modern audit engagement: research needs. *AUDITING: A Journal of Practice and Theory* 36 (4): 1-27. https://doi.org/10.2308/ajpt-51684

Arinder, M.K. 2016. Bridging the divide between evidence and policy in public sector decision making: a practitioner's perspective. *Public Administration Review* 76 (3): 394-398. https://doi.org/10.1111/puar.12572

Bachner, J., and K. Wagner Hill. 2014. Advances in public opinion and policy attitudes research. *Policy Studies Journal* 42 (April): S51-70. https://doi.org/10.1111/psj.12052

Bailey, S.J., A-V. Anttiroiko, and P. Valkama. 2016. Application of Baumol's cost disease to public sector services: conceptual, theoretical and empirical falsities. *Public Management Review* 18 (1): 91-109. https://doi.org/10.1080/14719037.2014.958092

Barth, T.J., and E. Arnold. 1999. Artificial intelligence and administrative discretion. *The American Review of Public Administration* 29 (4): 332-351. https://doi.org/10.1177/02750749922064463

Batko, R., and A. Szopa. 2017. *Strategic imperatives and core competencies in the era of robotics and artificial intelligence.* Advances in Computational Intelligence and Robotics. IGI Global, Hershey, PA, USA. https://doi.org/10.4018/978-1-5225-1656-9

Bean, R. 2020. Why culture is the greatest barrier to data success. *MIT Sloan Management Review*, September 2020.

Beck, R., C. Müller-Bloch, and J.L. King. 2018. Governance in the blockchain economy: a framework and research agenda. *Journal of the Association for Information Systems* 19 (10).

Begg, R. 2009. Artificial intelligence techniques in medicine and health care. In *Intelligent Information Technologies*, 1750-1757. IGI Global, Hershey, PA, USA. https://doi.org/10.4018/978-1-59904-941-0.ch098

Berryhill, J., T. Bourgery, and A. Hanson. 2018. Blockchains unchained: blockchain technology and its use in the public sector. OECD Working Papers on Public Governance, No. 28, OECD Publishing, Paris, France. https://doi.org/10.1787/3c32c429-en

Bijlsma-Frankema, K., and A.C. Costa. 2005. Understanding the trust-control nexus. *International Sociology* 20 (3): 259-282. https://doi.org/10.1177/0268580905055477

Blume, G., T. Scott, and M. Pirog. 2014. Empirical innovations in policy analysis. *Policy Studies Journal* 42 (April): S33-50. https://doi.org/10.1111/psj.12050

Bostrom, N., and E. Yudkowsky. 2014. The ethics of artificial intelligence. In *The Cambridge handbook of artificial intelligence*, edited by K. Frankish and W.M. Ramsey, 316-334. Cambridge University Press, Cambridge, UK. https://doi.org/10.1017/CBO9781139046855.020

Boyd, D., and K. Crawford. 2012. Critical questions for big data: provocations for a cultural, technological, and scholarly phenomenon. *Information Communication and Society* 15 (5): 662-679. https://doi.org/10.1080/1369118X.2012.678878

Boyd, M., and N. Wilson. 2017. Rapid developments in artificial intelligence: how might the New Zealand government respond? *Policy Quarterly* 13 (4). https://doi.org/10.26686/pq.v13i4.4619

Brownsword, R., E. Scotford, and K. Yeung, eds. 2016. *The Oxford handbook of law, regulation and technology.* Vol. 1. Oxford University Press, Oxford, UK. https://doi.org/10.1093/oxfordhb/9780199680832.001.0001

Brynjolfsson, E., and A. McAfee. 2014. *The second machine age: work, progress, and prosperity in a time of brilliant technologies.* W.W. Norton and Company, New York, NY, USA.

Bughin, J., E. Hazan, S. Ramaswamy, M. Chui, T. Allas, P. Dahlström, N. Henke, and M. Trench. 2017. How artificial intelligence can deliver real value to companies. McKinsey and Co., New York, NY, USA.

Bullock, J.B. 2019. Artificial intelligence, discretion, and bureaucracy. *The American Review of Public Administration* 49 (7): 751-761. https://doi.org/10.1177/0275074019856123

Burrell, J. 2016. How the machine 'thinks': understanding opacity in machine learning algorithms. *Big Data and Society* 3 (1): 205395171562251. https://doi.org/10.1177/2053951715622512

Casino, F., T.K. Dasaklis, and C. Patsakis. 2019. A systematic literature review of blockchain-based applications: current status, classification and open issues. *Telematics and Informatics* 36 (March): 55-81. https://doi.org/10.1016/j.tele.2018.11.006

Cedefop. 2020. Artificial intelligence in a post-pandemic world of work and skills. Cedefop. 2020. Available at: https://www.cedefop.europa.eu/en/news-and-press/news/artificial-intelligence-post-pandemic-world-work-and-skills.

Chun, A.H.W. 2007. Using AI for e-government automatic assessment of immigration application forms. In *Proceedings of the Twenty-Second AAAI Conference on Artificial Intelligence*. Vancouver, Canada.

Clark, W.R., and M. Golder. 2015. Big data, causal inference, and formal theory: contradictory trends in political science? *PS: Political Science and Politics* 48 (1): 65-70. https://doi.org/10.1017/S1049096514001759

Clarke, A., and H. Margetts. 2014. Governments and citizens getting to know each other? open, closed, and big data in public management reform. *Policy and Internet* 6 (4): 393-417. https://doi.org/10.1002/1944-2866.POI377

Collier, M., R. Fu, L. Yin, and P. Christiansen. 2017. Artificial intelligence: healthcare's new nervous system. Accenture, Dublin, Ireland. Available at: https://www.accenture.com/au-en/insights/health/artificial-intelligence-healthcare.

Conn, A. 2017. Artificial intelligence: the challenge to keep it safe. Future of Life Institute. Available at: https://futureoflife.org/2017/09/21/safety-principle/.

Cook, T.D. 2014. 'Big data' in research on social policy. *Journal of Policy Analysis and Management* 33 (2): 544-547. https://doi.org/10.1002/pam.21751

Corea, F. 2019. *An introduction to data*. Vol. 50. Studies in big data. Springer International Publishing, Cham, Switzerland. https://doi.org/10.1007/978-3-030-04468-8

Cortés-Cediel, M.E., I. Cantador, and O. Gil. 2017. Recommender systems for e-governance in smart cities. In *Proceedings of the International Workshop on Recommender Systems for Citizens*, 1-6. ACM, New York, NY, USA. https://doi.org/10.1145/3127325.3128331

Daniell, K.A., A. Morton, and D.R. Insua. 2016. Policy analysis and policy analytics. *Annals of Operations Research* 236 (1): 1-13. https://doi.org/10.1007/s10479-015-1902-9

De Filippi, P., and A. Wright. 2018. *Blockchain and the law*. Harvard University Press, Cambridge, MA, USA. https://doi.org/10.2307/j.ctv2867sp

De Vries, A. 2018. Bitcoin's growing energy problem. *Joule* 2 (5): 801-805. https://doi.org/10.1016/j.joule.2018.04.016

Decker, P.T. 2014. Presidential address: false choices, policy framing, and the promise of 'big data.' *Journal of Policy Analysis and Management* 33 (2): 252-262. https://doi.org/10.1002/pam.21755

Demchenko, Y., C. de Laat, and P. Membrey. 2014. Defining architecture components of the big data ecosystem. In *2014 International Conference on Collaboration Technologies and Systems (CTS)*, 104-112. IEEE. https://doi.org/10.1109/CTS.2014.6867550

Desouza, K.C., and A. Bhagwatwar. 2012. Leveraging technologies in public agencies: the case of the U.S. Census Bureau and the 2010 Census. *Public Administration Review* 72 (4): 605-614. https://doi.org/10.1111/j.1540-6210.2012.02592.x

Dressel, J., and H. Farid. 2018. The accuracy, fairness, and limits of predicting recidivism. *Science Advances* 4 (1): eaao5580. https://doi.org/10.1126/sciadv.aao5580

Dwivedi, Y.K., L. Hughes, E. Ismagilova, G. Aarts, C. Coombs, T. Crick, Y. Duan, *et al.* 2021. Artificial intelligence (AI): multidisciplinary perspectives on emerging challenges, opportunities, and agenda for research, practice and policy. *International Journal of Information Management* 57 (April): 101994. https://doi.org/10.1016/j.ijinfomgt.2019.08.002

EdChain. 2018. POW vs. PoS. a comparison of two blockchain consensus algorithms. EdChain, 12 June 2018. Available at: https://medium.com/@EdChain/pow-vs-pos-a-comparison-of-two-blockchain-consensus-algorithms-f3effdae55f5.

Eggers, W.D., T. Fishman, and P. Kishnani. 2017. AI-augmented human services: using cognitive technologies to transform program delivery. Available at: https://www2.deloitte.com/content/dam/insights/us/arti%0Dcles/4152_AI-human-services/4152_AI-human-services.%0Dpdf.

Einaste, T. 2018. Blockchain and healthcare: the Estonian experience. E-Estonia. 2018. Available at: https://e-estonia.com/blockchain-healthcare-estonian-experience/#:~:text=Estonia%2C home to one of,healthcare on a national scale.&text='We are using blockchain as,the integrity of health records.

Einav, L., and J. Levin. 2014. The data revolution and economic analysis. *Innovation Policy and the Economy* 14 (January): 1-24. https://doi.org/10.1086/674019

European Commission. 2021. EBSI. Accessed May 28, 2021. Available at: https://ec.europa.eu/cefdigital/wiki/display/CEFDIGITAL/EBSI.

European Commission. 2020. Shaping Europe's digital future: commission presents strategies for data and artificial intelligence. Available at: https://ec.europa.eu/commission/presscorner/detail/en/ip_20_273

Finck, M. 2018. *Blockchain regulation and governance in Europe*. Cambridge University Press, Cambridge, UK. https://doi.org/10.1017/9781108609708

Ford, M. 2013. Could artificial intelligence create an unemployment crisis? *Communications of the ACM* 56 (7): 37-39. https://doi.org/10.1145/2483852.2483865

Gamage, P. 2016. New development: leveraging 'big data' analytics in the public sector. *Public Money and Management* 36 (5): 385-390. https://doi.org/10.1080/09540962.2016.1194087

George, G., M.R. Haas, and A. Pentland. 2014. Big data and management. *Academy of Management Journal* 57 (2): 321-326. https://doi.org/10.5465/amj.2014.4002

Goldsmith, S., and S. Crawford. 2014. *The responsive city: engaging communities through data-smart governance*. Jossey-Bass, Hoboken, NJ, USA.

Goodfellow, I., Y. Bengio, and A. Courville. 2016. *Deep learning*. MIT press, Cambridge, MA, USA.

Governatori, G., F. Idelberger, Z. Milosevic, R. Riveret, G. Sartor, and X. Xu. 2018. On legal contracts, imperative and declarative smart contracts, and blockchain systems. *Artificial Intelligence and Law* 26 (4): 377-409. https://doi.org/10.1007/s10506-018-9223-3

Hadden, S.G. 1986. Intelligent advisory systems for managing and disseminating information. *Public Administration Review* 46 (November): 572. https://doi.org/10.2307/975579

Haenlein, M., and A. Kaplan. 2019. A brief history of artificial intelligence: on the past, present, and future of artificial intelligence. *California Management Review* 61 (4): 5-14. https://doi.org/10.1177/0008125619864925

Hassani, H., X. Huang, and E.S. Silva. 2019. *Fusing big data, blockchain and cryptocurrency*. Springer International Publishing, Cham, Switzerland. https://doi.org/10.1007/978-3-030-31391-3

Hemken, T., and C. Gray. 2016. Smart move: technologies make their mark on public service. Available at: https://www.slideshare.net/accenture/smart-move-emergent-technologies-make-their-mark-on-public-service

Hempel, J. 2018. How refugees are helping create blockchain's brand new world. *Wired*, 2018.

Höchtl, J., P. Parycek, and R. Schöllhammer. 2016. Big data in the policy cycle: policy decision making in the digital era. *Journal of Organizational Computing and Electronic Commerce* 26 (1-2): 147-169. https://doi.org/10.1080/10919392.2015.1125187

Hondula, D.M., E.R. Kuras, J. Longo, and E.W. Johnston. 2018. Toward precision governance: infusing data into public management of environmental hazards. *Public Management Review* 20 (5): 746-765. https://doi.org/10.1080/14719037.2017.1320043

Hurley, M.W., and W.A. Wallace. 1986. Expert systems as decision aids for public managers: an assessment of the technology and prototyping as a design strategy. *Public Administration Review* 46 (November): 563. https://doi.org/10.2307/975578

Issa, H., T. Sun, and M.A. Vasarhelyi. 2016. Research ideas for artificial intelligence in auditing: the formalization of audit and workforce supplementation. *Journal of Emerging Technologies in Accounting* 13 (2): 1-20. https://doi.org/10.2308/jeta-10511

Jahoda, M. 1986. Artificial intelligence: an outsider's perspective. *Science and Public Policy*, December. https://doi.org/10.1093/spp/13.6.333

Janssen, D. 2020. Power grid operators launch blockchain for home and car batteries. *Euractiv*, May 1, 2020. Available at: https://www.euractiv.com/section/energy/news/power-grid-operators-launch-blockchain-for-home-and-car-batteries/.

Janssen, M., V. Weerakkody, E. Ismagilova, U. Sivarajah, and Z. Irani. 2020. A framework for analysing blockchain technology adoption: integrating institutional, market and technical factors. *International Journal of Information Management* 50: 302-309. https://doi.org/10.1016/j.ijinfomgt.2019.08.012

Janssen, M., and G. Kuk. 2016. The challenges and limits of big data algorithms in technocratic governance. *Government Information Quarterly* 33 (3): 371-737. https://doi.org/10.1016/j.giq.2016.08.011

Janssen, M., and J. van den Hoven. 2015. Big and open linked data (BOLD) in government: a challenge to transparency and privacy? *Government Information Quarterly* 32 (4): 363-368. https://doi.org/10.1016/j.giq.2015.11.007

Jarmin, R.S., and A.B. O'Hara. 2016. Big data and the transformation of public policy analysis. *Journal of Policy Analysis and Management* 35 (3): 715-721. https://doi.org/10.1002/pam.21925

Jefferies, D. 2016. The automated city: do we still need humans to run public services? *Guardian*, September 20, 2016.

Johnes, G., and J. Ruggiero. 2017. Revenue efficiency in higher education institutions under imperfect competition. *Public Policy and Administration* 32 (4): 282-295. https://doi.org/10.1177/0952076716652935

Johnson, S., P. Robinson, and J. Brainard. 2019. Sidechains and interoperability. Available at: http://arxiv.org/abs/1903.04077.

Kahneman, D. 2011. *Thinking, fast and slow*. Farrar, Straus and Giroux, New York, NY, USA.

Kahneman, D., O. Sibony, and C.R. Sunstein. 2021. Noise: a flaw in human judgement. Little Brown Spark, New York, NY, USA.

Kairos Future. 2017. The land registry in the blockchain-testbed. Available at: https://static1.squarespace.com/static/5e26f18cd5824c7138a9118b/t/5e3c35451c2cbb6170caa19e/1581004119677/Blockchain_Landregistry_Report_2017.pdf.

Kannengießer, N., S. Lins, T. Dehling, and A. Sunyaev. 2019. Mind the gap: trade-offs between distributed ledger technology characteristics (working paper). *ArXiv*: 1906.00861.

Karafiloski, E., and A. Mishev. 2017. Blockchain solutions for big data challenges: a literature review. In *IEEE EUROCON 2017 – 17th International Conference on Smart Technologies*, 763-768. IEEE. https://doi.org/10.1109/EUROCON.2017.8011213

Karjalainen, R. 2020. Governance in decentralised networks. *SSRN Electronic Journal*. https://doi.org/10.2139/ssrn.3551099

Kimble, C., and G. Milolidakis. 2015. Big data and business intelligence: debunking the myths. *Global Business and Organizational Excellence* 35 (1): 23-34. https://doi.org/10.1002/joe.21642

Kitchin, R., and G. McArdle. 2016. What makes big data, big data? Exploring the ontological characteristics of 26 datasets. *Big Data and Society* 3 (1): 205395171663113. https://doi.org/10.1177/2053951716631130

Kleinberg, J., H. Lakkaraju, J. Leskovec, J. Ludwig, and S. Mullainathan. 2017. Human decisions and machine predictions*. *The Quarterly Journal of Economics*, August. https://doi.org/10.1093/qje/qjx032

Kouziokas, G.N., A. Chatzigeorgiou, and K. Perakis. 2017. Artificial intelligence and regression analysis in predicting ground water levels in public administration. *European Water* 57: 361-366.

Kumar, D.S., and M.A. Rahman. 2017. Simplified HDFS architecture with blockchain distribution of metadata. *International Journal of Applied Engineering Research* 12 (21): 11374-11382.

Kuziemski, M., and G. Misuraca. 2020. AI governance in the public sector: three tales from the frontiers of automated decision-making in democratic settings. *Telecommunications Policy* 44 (6): 101976. https://doi.org/10.1016/j.telpol.2020.101976

Lafourcade, P., and M. Lombard-Platet. 2020. About blockchain interoperability. *Information Processing Letters* 161 (September): 105976. https://doi.org/10.1016/j.ipl.2020.105976

Lazer, D., A. Pentland, L. Adamic, S. Aral, A-L. Barabasi, D. Brewer, and N. Christakis. 2009. Social science: computational social science. *Science* 323 (5915): 721-723. https://doi.org/10.1126/science.1167742

Liu, P.T.S. 2016. Medical record system using blockchain, big data and tokenization. In: K.Y. Lam, C.H. Chi, and S. Qing. (eds) *Information and Communications Security*. ICICS 2016. Lecture Notes in Computer Science, Vol. 9977. Springer, Cham, Switzerland. https://doi.org/10.1007/978-3-319-50011-9_20

López, A.C. 2019. Building alternatives: the digital democracy and data commons pilot. DECODE Project. Available at: https://decodeproject.eu/blog/building-alternatives-digital-democracy-and-data-commons-pilot.html.

Maciejewski, M. 2017. To do more, better, faster and more cheaply: using big data in public administration. *International Review of Administrative Sciences* 83 (1): 120-135. https://doi.org/10.1177/0020852316640058

Manyika, J., J. Silberg, and B. Presten. 2019. What do we do about the biases in AI? *Harvard Business Review*, 2019.

Masuch, M., and P. LaPotin. 1989. Beyond garbage cans: an AI model of organizational choice. *Administrative Science Quarterly* 34 (1): 38. https://doi.org/10.2307/2392985

McAfee, A., and E. Brynjolfsson. 2017. *Machine, platform, crowd: harnessing our digital future*. W.W. Norton and Company, New York, NY, USA.

McConaghy, T. 2016. How blockchains could transform artificial intelligence. Dataconomy. Available at: http://dataconomy.com/2016/12/blockchains-for-artificial-intelligence/.

Mehr, H. 2017. Artificial intelligence for citizen services and government. Ash Center for Democratic Governance and Innovation, Harvard Kennedy School, Cambridge, MA, USA. Available at: https://ash.harvard.edu/files/ash/files/artificial_intelligence_for_citizen_services.pdf.

Meijer, D., and J. Ubacht. 2018. The governance of blockchain systems from an institutional perspective, a matter of trust or control? *ACM International Conference Proceeding Series*. https://doi.org/10.1145/3209281.3209321

Mergel, I. 2017. Building holistic evidence for social media impact. *Public Administration Review* 77 (4): 489-495. https://doi.org/10.1111/puar.12780

Mergel, I., R.K. Rethemeyer, and K. Isett. 2016. Big data in public affairs. *Public Administration Review* 76 (6): 928-937. https://doi.org/10.1111/puar.12625

Migchelbrink, K., and S. Van de Walle. 2019. Deliverable D2.3 final recommendations to transform the public sector processes and services. H2020 CITADEL project. Available at: https://ec.europa.eu/research/participants/documents/downloadPublic?documentIds=080166e5c54ca53c&appId=PPGMS.

Millard, J. 2018. Open governance systems: doing more with more. *Government Information Quarterly* 35 (4): S77-87. https://doi.org/10.1016/j.giq.2015.08.003

Miller, A. 2020. Permissioned and permissionless blockchains. In *Blockchain for distributed systems security*, edited by S. Shetty, C. Kamhoua, and L. Njilla, 193-204. Wiley, Hoboken, NJ, USA.

Misuraca, G., and C. Van Noordt. 2020. AI watch artificial intelligence in public services: overview of the use and impact of AI in public services in the EU. JRC publications. https://publications.jrc.ec.europa.eu/repository/handle/JRC120399

Müller, V.C. 2020. Ethics of artificial intelligence and robotics. In *Stanford Encyclopedia of Philosophy*. Stanford University, Stanford, CA, USA.

Ng, L. 2021 (July). The AI wolf that preferred suicide over eating sheep. Medium. Available at: https://onezero.medium.com/the-ai-wolf-that-preferred-suicide-over-eating-sheep-49edced3c710

Organisation for Economic Cooperation and Development (OECD). 2017. *OECD digital economy outlook 2017*. OECD, Paris, France. https://doi.org/10.1787/9789264276284-en

Organisation for Economic Cooperation and Development (OECD). 2019. *Enhancing access to and sharing of data*. OECD, Paris, France. https://doi.org/10.1787/276aaca8-en

Osborne, S.P., and K. Strokosch. 2013. It takes two to tango? Understanding the co-production of public services by integrating the services management and public administration perspectives. *British Journal of Management* 24 (S3): 1-28. https://doi.org/10.1111/1467-8551.12010

Pandey, S., S.K. Pandey, and L. Miller. 2017. Measuring innovativeness of public organizations: using natural language processing techniques in computer-aided textual analysis. *International Public Management Journal* 20 (1): 78-107. https://doi.org/10.1080/10967494.2016.1143424

Pencheva, I., M. Esteve, and S.J. Mikhaylov. 2020. Big data and AI – A transformational shift for government: so, what next for research? *Public Policy and Administration* 35 (1): 24-44. https://doi.org/10.1177/0952076718780537

Pirog, M.A. 2014. Data will drive innovation in public policy and management research in the next decade. *Journal of Policy Analysis and Management* 33 (2): 537-543. https://doi.org/10.1002/pam.21752

Power, D.J. 2016. 'Big brother' can watch us. *Journal of Decision Systems* 25 (1): 578-588. https://doi.org/10.1080/12460125.2016.1187420

Reijers, W., I. Wuisman, M. Mannan, P. De Filippi, C. Wray, V. Rae-Looi, A.C. Vélez, and L. Orgad. 2018. Now the code runs itself: on-chain and off-chain governance of blockchain technologies. *Topoi* 40: 821-831. https://doi.org/10.1007/s11245-018-9626-5

Roberts, A. 2017. Five big challenges to ai adoption and success. ClickZ. Accessed at: https://www.clickz.com/five-big-challenges-to-ai-adoption-and-success/112795/.

Ross, A. 2019. The hardest part of AI and analytics is not AI, it's data management. Information Age. Available at: https://www.information-age.com/the-hardest-part-of-ai-analytics-is-not-ai-it-data-management-123486226/.

Rosso, C. 2018. The human bias in the AI machine. *Psychology Today*. 2018.

Russell, S., and P. Norvig. 2016. *Artificial intelligence: a modern approach*. 3th ed. Pearson, London, UK.

Salamon, L.M. 2002. *The tools of government: a guide to the new governance*. Oxford University Press, Oxford, UK.

Scherer, M.U. 2015. Regulating artificial intelligence systems: risks, challenges, competencies, and strategies. *Harvard Journal of Law & Technology* 29 (2). https://doi.org/10.2139/ssrn.2609777

Schintler, L.A., and R. Kulkarni. 2014. Big data for policy analysis: the good, the bad, and the ugly. *Review of Policy Research* 31 (4): 343-348. https://doi.org/10.1111/ropr.12079

Selbst, A.D., D. Boyd, S.A. Friedler, S. Venkatasubramanian, and J. Vertesi. 2019. Fairness and abstraction in sociotechnical systems. In *Proceedings of the Conference on Fairness, Accountability, and Transparency*, 59-68. ACM, New York, NY, USA. https://doi.org/10.1145/3287560.3287598

Shindelar, S. 2014. Big data and the government agency. *The Public Manager*. 2014. Available at: https://www.td.org/magazines/the-public-manager/big-data-and-the-government-agency.

Stephan, M., G. Marshall, and M. McGinnis. 2019. An introduction to polycentricity and governance. In *Governing Complexity*, edited by A. Thiel, W.A. Blomquist, and D.E. Garrick, 21-44. Cambridge University Press, Cambridge, UK. https://doi.org/10.1017/9781108325721.002

Stritch, J.M., M.J. Pedersen, and G. Taggart. 2017. The opportunities and limitations of using mechanical turk (MTURK) in public administration and management scholarship. *International Public Management Journal* 20 (3): 489-511. https://doi.org/10.1080/10967494.2016.1276493

Sun, T.Q., and R. Medaglia. 2017. Artificial intelligence and public healthcare service innovation: a service ecosystem perspective. In *The 40th Information Systems Research Conference in Scandinavia*. Halden, Norway.

Susskind, R., and D. Susskind. 2016. Technology will replace many doctors, lawyers, and other professionals. *Harvard Business Review*, October 2016.

Taeihagh, A. 2017. Crowdsourcing: a new tool for policy-making? *Policy Sciences* 50 (4): 629-647. https://doi.org/10.1007/s11077-017-9303-3

Tan, E., S. Mahula, and J. Crompvoets. 2022. Blockchain governance in the public sector: a conceptual framework for public management. *Government Information Quarterly* 39 (1): 101625. https://doi.org/https://doi.org/10.1016/j.giq.2021.101625

Tegmark, M. 2017. *Life 3.0: Being Human in the Age of Artificial Intelligence*. Alfred A. Knopf, New York, NY, USA.

Thierer, A., A. O'Sullivan Castillo, and R. Russell. 2017. Artificial intelligence and public policy. Mercatus Center, Arlington, VA, USA. Available at: https://www.mercatus.org/system/files/thierer-artificial-intelligence-policy-mr-mercatus-v1.pdf.

Voshmgir, S. 2020. *Token economy: how the web3 reinvents the internet*. 2nd ed. Token Kitchen, Berlin, Germany.

Wang, S., W. Ding, J. Li, Y. Yuan, L. Ouyang, and F-Y. Wang. 2019. Decentralized autonomous organizations: concept, model, and applications. *IEEE Transactions on Computational Social Systems* 6 (5): 870-878. https://doi.org/10.1109/TCSS.2019.2938190

Whittaker, M., K. Crawford, R. Dobbe, G. Fried, E. Kaziunas, V. Mathur, S.M. West, R. Richardson, J. Schultz, and O. Schwartz. 2018. AI Now Report 2018. AI Now Institute, New York University, New York, NY, USA. Available at: https://ainowinstitute.org/AI_Now_2018_Report.pdf.

Wirtz, B.W., and W.M. Müller. 2019. An integrated artificial intelligence framework for public management. *Public Management Review* 21 (7): 1076-1100. https://doi.org/10.1080/14719037.2018.1549268

Wirtz, B.W., J.C. Weyerer, and C. Geyer. 2019. Artificial intelligence and the public sector – applications and challenges. *International Journal of Public Administration* 42 (7): 596-615. https://doi.org/10.1080/01900692.2018.1498103

Yafimava, D. 2019. Blockchain in the supply chain: 10 real-life use cases and examples. Openledger. Available at: https://openledger.info/insights/blockchain-in-the-supply-chain-use-cases-examples/.

Yeung, K., and M. Lodge, eds. 2019. *Algorithmic regulation*. Oxford University Press, Oxford, UK. https://doi.org/10.1093/oso/9780198838494.001.0001

Zhang, P., J. White, D.C. Schmidt, and G. Lenz. 2017. Applying software patterns to address interoperability in blockchain-based healthcare apps. *ArXiv*: 1706.03700.

Zheng, Y., H.Yu, L. Cui, C. Miao, C. Leung, and Q. Yang. 2018. SmartHS: an AI platform for improving government service provision. In *AAAI Conference on Artificial Intelligence Thirty-Second AAAI Conference on Artificial Intelligence*.

Zhu, J., H. Huang, and D. Zhang. 2019. 'Big tigers, big data': learning social reactions to China's anticorruption campaign through online feedback. *Public Administration Review* 79 (4): 500-513. https://doi.org/10.1111/puar.12866

Zikopoulos, P.C., C. Eaton, D. DeRoos, T. Deutsch, and G. Lapis. 2012. *Understanding big data: analytics for enterprise class Hadoop and streaming data*. McGraw Hill, New York, NY, USA.

Chapter 3

Business-IT alignment in fraud analytics: fostering collaboration between domain experts and data scientists

A. Simonofski*, B. Vanderose and B. Frenay

Namur Digital Institute, University of Namur, Rue de Bruxelles 61, 5000 Namur, Belgium; anthony.simonofski@unamur.be

Abstract

Fraud analytics refers to the use of Big Data Analytics to detect fraud. In government, two groups of stakeholders are usually involved in this process: domain experts knowledgeable in fraud detection and data scientists knowledgeable in analytics. However, data scientists in government are rarely knowledgeable in the business domain (fraud detection) and domain experts do not always have an information technology (IT) profile. Thus, ensuring collaboration between the business and IT sides of fraud analytics is a key challenge for governments. Alignment is increasingly important to detect fraud efficiently, because the complexity of fraud, as well as the techniques used to detect them, keeps increasing and makes collaboration necessary. The goal of this chapter is to formalise the fraud analytics process and to illustrate this key alignment challenge. For this, we examine two case studies from the Belgian Federal government: the detection of tax fraud and of social security infringements. Data from these two cases has been collected from 21 interviews. As a result, we infer two fraud analytics processes and identify three crucial moments where alignment between business and IT is needed: the identification of requirements of the business team before performing the analytics, the presentation of the output of the analytics to the business team, and feedback from the business to the data scientists. In order to foster this alignment in fraud analytics, we suggest a methodology drawing from agile methods, participation methods and design thinking literatures.

Keywords: big data, artificial intelligence, agile, participation, design thinking

3.1 Introduction

The concept of big data can be defined in several ways and is still evolving. Daniell *et al.* (2016) and Klievink *et al.* (2017) talk about massive, complex collections of data that are constantly changing. Big data is characterised by three basic features: large volumes, high velocity (meaning rapidly generated data), and large variety (meaning large heterogeneity) (Daniell *et al.*, 2016; Kim *et al.*, 2014, 2017). Desouza and Jacob (2017) add a factor 'complexity' to these 'three V's',

DOI 10.3920/978-90-8686-930-5_3, © Wageningen Academic Publishers 2022

while others claim that big data is complex because of these three V's. The benefits of big data for government reveal themselves when applying analytics to support policymaking in diverse domains (Kim *et al.*, 2014).

Fraud analytics is one of these domains and refers to the use of advanced analytics to detect fraud, with a focus on big data analytics due to the large amount of data necessary to ensure detection. Numerous techniques, ranging from data mining to social network analysis, are developed and applied to detect various types of fraud. While fraud analytics offers the promise of more efficiency in fighting fraud, it also raises legal, technical and organisational challenges for governments. A key organisational challenge relates to the necessary alignment between the business and information technology (IT) side of fraud analytics. In their seminal paper, Henderson and Venkatraman (1999) define business-IT alignment as a means by which organisations can optimally use IT to achieve their business objectives on strategic and operational levels. This challenge is reinforced with the introduction of emerging digital technologies, such as big data or artificial intelligence (AI) that impact, even more, the way traditional organisations work. This challenge is particularly important for governments. Indeed, governments constitute a diverse ecosystem with multiple internal stakeholders, each of which has their own objectives. Among these stakeholders, there are different business teams and IT teams that do not always communicate with each other, leading to organisational silos. In the case of fraud analytics, the business teams are composed of jurists, tax specialists and inspectors, whereas the IT teams are composed of data scientists and data miners (Baesens *et al.*, 2015). Organisational silos may hamper the alignment of development projects in the organisation.

The goal of this chapter is thus to formalise the fraud analytics process, to illustrate the business-IT alignment challenge and to provide leads for solutions to foster this alignment. For this, we examined two case studies from the Belgian Federal government: the detection of tax fraud and of social security infringements. We chose these two case studies as they both apply fraud analytics in their daily work, performed by diverse business and data analysis teams. Furthermore, organisational silos are particularly present in large governmental organisations, such as the Federal Public Services in Belgium (Chantillon *et al.*, 2020), which makes them potential good cases to understand the problem of alignment. Data from these two cases have been collected from 21 interviews and relevant documents. As a way forward, this chapter then suggests a methodology to foster business-IT alignment based on three leads for a solution: agile methods, participation methods and design thinking.

The chapter is structured as follows. In Section 3.2, we summarise the literature related to the use of big data analytics to detect fraud and highlight the research gaps to be tackled in this chapter. In Section 3.3, we present the methodology we followed to formalise the fraud analytics process and illustrate the business-IT alignment challenge. In Section 3.4, we present this process in the Belgian use cases and identify the moments where the alignment challenge takes place. In Section 3.5, we suggest a methodology based on three concrete leads for a solution to foster the alignment between business and IT profiles in fraud analytics. Finally, in Section 3.6, we summarise the contributions of this chapter as well as the limitations and further research leads.

3.2 Background: unpacking the fraud analytics process

In this section, we unpack the fraud analytics process, defined as big data analytics applied in a fraud detection context. Then, we detail how the advanced techniques used in fraud analytics challenge the alignment between data scientists and fraud investigators and calls for the research performed in this chapter.

In order to detail the big data analytics process, we rely on the three-step formalisation suggested by two comprehensive models in the literature. On the one hand, Klievink *et al.* (2017) suggest a usage process for the analysis of big data and its relevance for policymaking that we follow in this section. First, there is the *pre-processing* stage where data is identified, collected from several sources, combined and cleansed (Section 2.1). Second, there is the *data analytics* stage (Section 2.2) where several techniques are applied to analyse the data. Third, there is the *post-processing* stage, where the output of the analysis is presented to the relevant stakeholders, interpreted and, in case of governments, has an impact on policymaking (Section 2.3). On the other hand, Baesens *et al.* (2015) suggest a process of big data analytics that also follows this overall structure, as shown below in Figure 3.1.

The following sections detail the available literature for each of these three key stages.

3.2.1 Pre-processing: data collection and combination

The first step of the pre-processing stage relates to the identification and understanding of the business problem to be tackled in the analytics project. In a digital government context, this problem can be related to several policy domains. This approach relates to the evidence-based approach, introduced by the UK government in the 1990s (Tsoukias *et al.*, 2013). This approach is about governing based on 'facts' and thus using 'evidence' to support decisions (De Marchi *et al.*, 2016). However, evidence-based policymaking fails to address some challenges like sufficiently solving the issue of mistrust between citizens and policy-makers or involving opinion groups and individual citizens enough (Tsoukias *et al.*, 2013). This failure can be the result of the fact that multiple conclusions can be drawn from the same 'fact', or because the 'evidence' that policy-makers use is subjectively put in a hierarchy by them, which makes the term 'evidence' non-neutral (De Marchi *et al.*, 2016). Policy analytics (Gil-Garcia *et al.*, 2018; Hamza and Mellouli, 2018) is a term describing a phenomenon where data and analytical techniques are used to support policy decisions. In this chapter, a specific policy domain (fraud detection) is investigated and will be detailed in Section 2.3.

Figure 3.1. Formalisation of the big analytics process (Baesens *et al.*, 2015).

The second step of pre-processing relates to the identification and selection of data. A lot of data can be used to support policymaking: ranging from data in simple, well-structured databases to unstructured data like pictures, audio or social-media data and other data from web 2.0 involving textual data, structural data (metadata) and temporal data (Chung and Zeng, 2018). From this data, things like policy-relevant information, knowledge and ideas from citizens can be deducted (Loukis, 2018). Data is a key ingredient for any analytical exercise and experience in different fields shows that more data is better for the analysis (Baesens *et al.*, 2015). The pre-processing of the data (from collecting to transforming) is thus essential to avoid the Garbage In – Garbage Out[1] principle. Data can come from numerous sources (Baesens *et al.*, 2015): transactional data, contractual data, sociodemographic data, surveys, behavioural data, expert-based data, textual data, etc.

Five differentiating characteristics of big data analytics are the following (Klievink *et al.*, 2017): making combinations of multiple large external and internal datasets and using these; combining and using structured and unstructured data while analysing; using incoming data streams in (near) real-time; applying advanced analytics and algorithms; and using existing datasets in an innovative way. This also implies that collections of data are in fact 'big data' if they cannot be handled by conventional data processing. Some frequently mentioned challenges about big data in government are data integration across departmental silos, developing data standards, analysing unstructured data, establishing sufficient control towers, implementing regulations regarding compliance and security, dealing with access rights, building the infrastructure, archiving and preservation, accepting change and addressing privacy issues (Bertot *et al.*, 2014; Joseph and Johnson, 2013; Kim *et al.*, 2014).

One possible source of data is the government itself. Open government data (OGD) can help big data projects succeed by ensuring publicly accessible datasets through managed processes. OGD is not necessarily big data and big data contains more than just OGD. Governments can publish so-called OGD catalogues or portals, in that way providing a single point of access for governmental data (Erickson *et al.*, 2013; Kalampokis *et al.*, 2013). OGD is data that anyone can freely (not necessarily for free) use, reuse and redistribute (Kalampokis *et al.*, 2013; Ubaldi, 2013). These OGD initiatives increase government transparency and accountability, but also involve many challenges (Kalampokis *et al.*, 2013). Kalampokis *et al.* (2013) also claim that, ideally, OGD is linked, meaning that the data should be machine-readable, with a explicitly defined meaning, and linked to other external datasets.

The last step of the pre-processing stage relates to the cleansing and transformation of the data so that the analysis and use can be properly performed (Baesens *et al.*, 2015). In this stage, the data can be gathered (e.g. in a data mart or data warehouse) so that exploratory analysis can be performed. This is usually followed by a cleaning of the data to remove potential inconsistencies

[1] This expression refers to the idea that, in computer science, inputs of poor quality (e.g. missing data) will produce outputs of poor quality.

such as duplicate data, missing values or outliers. For instance, missing value can occur because of undisclosed data or merging errors. Outliers are extreme observations that differ greatly from the dataset. They should be detected and, if necessary, treated.

As the pre-processing stage consists of many activities (data collection, cleaning or transformation) (Archenaa and Anita, 2015), it can be supported by diverse management tools or data processing applications. For instance, Hadoop is an open source java-based framework that can help to manage, integrate and store these big datasets (Archenaa and Anita, 2015; Joseph and Johnson, 2013; Kim *et al.*, 2014; Ubaldi, 2013). These management tools can also help analyse the big data, as described below in the 'data analytics' section. Janssen and Kuk (2016) argue that there is a need for transparency about how the algorithm operates.

3.2.2 Data analytics

In this stage, the pre-processed data will be employed to build an analytical model. Big data is increasingly being used by governments in a wide range of areas to identify and analyse problems (Bertot *et al.*, 2014). In order to do so, analytics make intensive use of the data described in the previous section. Analytics is a very broad umbrella term that encompasses several techniques (Davenport and Jarvenpaa, 2008). These include statistics, data mining, business intelligence, operational research and decision analysis, machine learning and computer science, as well as disciplines like sociology, psychology and economics (Daniell *et al.*, 2016; De Marchi *et al.*, 2016; Tsoukias *et al.*, 2013). The aim of analytics is to create useful information and knowledge with which to make good decisions. There are different types of analytics, like descriptive, predictive, explanatory and decisive analytics (Daniell *et al.*, 2016; Davenport and Jarvenpaa, 2008).

Several analytics techniques exist to exploit the data and support policymaking. These techniques are labelled under several terms. *Data mining* techniques are focused on finding existing patterns in datasets and making sense of the data to hand. Data mining is more of a manual process with human intervention and a focus on decision-making. When applied to business decisions, data mining is part of the global process called *business intelligence*. Negash (2004) defines business intelligence as 'systems that combine data gathering, data storage, and knowledge management with analytical tools to present complex internal and competitive information to planners and decision makers'.

Machine learning, a sub-field of AI, is a field in which the analysis aims to be automated by a machine. Machine-learning systems are autonomous and can operate without human intervention and can learn and identify patterns with which to make decisions and to reach different conclusions based on the analysis of different situations (De Sousa *et al.*, 2019). Bishop (2006) defines machine learning as 'the automatic discovery of regularities in data through the use of computer algorithms to take actions such as classification'. Tom Mitchell (1997) defines a machine learning problem as a computer program that learns from experience E (e.g. watching our classification of mails), with respect to some task T (e.g. classifying mails) and some performance measure P (e.g. number of mails correctly classified), if its performance in T, as measured by P, improves with experience E. Machine learning can be supervised or unsupervised (Han *et al.*, 2012). In *supervised* machine learning, a function maps an input to an output based on example input-output pairs (e.g. predicting the price of a house, based on the surface, based on existing price/surface mappings). The function can predict continuous values (regression) or discrete values (classification). Therefore, training examples are needed to 'teach' the algorithm to perform a task. In *unsupervised* learning, the aim is to establish the existence of classes or clusters in the data given a set of measurements, observations, etc. (e.g. Google News groups news into broad categories and topics automatically).

Compared to data mining, machine learning goes one step further and aims to predict future outcomes based on the pre-existing datasets. The two techniques remain complementary, as machine learning will use data mining techniques as resources. The understanding of these diverse analytics terms is essential in order to understand the foundations of fraud analytics, detailed in the following section.

3.2.3 Post-processing: use of big data analytics for fraud detection

According to the Oxford Dictionary, fraud can be defined as a 'wrongful or criminal deception intended to result in financial or personal gain'. De Roux *et al.* (2018) define tax fraud as the intentional act of lying on a tax return form with the intent to lower one's tax liability, and social security infringement as the intentional act of receiving social security payments one is not entitled to.

Traditionally, tax authorities tackled tax fraud using two approaches (Castellón González and Velásquez, 2013). The *auditor experience* approach randomly selects tax declarations and audits them based on experience and domain knowledge. The *rule-based system* approach applies 'if-then' rules to detect fraud cases. These rules are burdensome to develop, as experts have to review and generalise fraud characteristics after they identify them. The two main issues with these techniques are that they exclusively rely on past experiences and dismiss new fraud mechanisms and are based on the subjective judgement of experts.

Policy analytics have been applied in previous research to detect tax fraud (Van Vlasselaer *et al.*, 2017; Yu *et al.*, 2003) and has been labelled as 'fraud analytics' (Baesens *et al.*, 2015). The detection of fraud can be considered as the use of analytics in the specific step of 'implementation' in the policy lifecycle (Pencheva *et al.*, 2018). The use of analytics techniques can also be inserted in a more global approach consisting of (Baesens *et al.*, 2015):

- Fraud detection: applying detection models to new, unseen observations and assigning a fraud risk to every observation.
- Fraud investigation: a human expert is often required to investigate suspicious, flagged cases given the involved subtlety and complexity.
- Fraud confirmation: determining true fraud label, possibly involving field research.
- Fraud prevention: preventing fraud from being committed in the future. This might even result in detecting fraud even before the fraudster knows s/he will commit fraud.

As data mining allows for patterns and anomalies to be found in large amounts of data, it is helpful for fraud detection. However, traditional data mining techniques are not adequate to detect advanced types of fraud due to their inherent characteristics (Van Vlasselaer *et al.*, 2017):

- Uncommon: the datasets are skewed as a limited number of fraud cases are identified.
- Well-considered: frauds are planned, well-thought-through and probably present in more than one attribute in the data.
- Time-evolving: fraudsters are adaptive and learn from past mistakes.
- Carefully organised: fraudsters have allies and exchange knowledge with each other to commit fraud without being detected.
- Imperceptibly concealed: data can be overlapping as fraudsters may have the same characteristics as legitimate companies.

Therefore, advanced techniques based on machine learning are needed for the detection of more complex frauds as reported in De Roux *et al.* (2018) and Van Vlasselaer *et al.* (2017). Recent works in tax fraud detection are based on supervised machine-learning techniques with labelled, audit-assisted data. A typical example in a fraud detection setting using regression is predicting the amount of fraud. In classification, the target is categorical which means that it can only take on a limited set of predefined values. In binary classification, only two classes are considered (e.g. fraud versus no fraud) whereas in multiclass classification, the target can belong to more than two classes (e.g. severe fraud, medium fraud, no fraud). Predictive analytics (supervised learning) can however also be useful to help explain the anomalies found by descriptive analytics (unsupervised learning). However, the 'follow and auditing' approach can be slow and costly.

Indeed, several authors argue for the application of unsupervised techniques (De Roux *et al.*, 2018). For instance, unsupervised techniques could find behaviour that deviates from normal behaviour and discover outliers/anomalies (Baesens *et al.*, 2015). They are unsupervised, as they do not need observations to be labelled as fraudulent or non-fraudulent. Anomalies do not necessarily represent fraudulent observations. Hence, the usage of unsupervised learning for fraud detection requires extensive follow-up and validation of the identified, suspicious observations.

A third type of technique that can be used is social network analysis where fraudulent activities can be identified within a network of linked entities (Baesens *et al.*, 2015). Indeed, fraud is social in nature. The probability of someone committing fraud depends on the people he or she is connected to. This is the so-called guilt-by-association (Koutra *et al.*, 2011). Optimal fraud detection will make use, in a complementary manner, of these different techniques. A good example of big data analysis to detect tax fraud is the British Connect System[2] (Maciejewski, 2017).

[2] https://www.telegraph.co.uk/finance/personalfinance/tax/11697816/What-does-the-taxman-know-about-you-your-finances-and-your-lifestyle.html.

3.2.4 Research gaps

Following this background overview of the literature, we can identify two main research gaps this chapter intends to tackle.

First, there is a lack of a formalised process for fraud analytics by governments that integrate big data and AI solutions (Research Gap 1). Indeed, the big data analytics process has been well explored in the literature but its application in the fraud detection domain remains to be formalised. The existing knowledge base focuses more on the diverse techniques used in fraud analytics but less on describing the global processes that formalise the integration of digital technologies in business processes. This gap will be tackled in Section 4.1 by presenting the main available insights from two practical cases.

Second, this section showed that the big data analytics process is not only technical but also focuses on interactions between data scientists and domain experts. To make the fraud detection process as efficient as possible, a close collaboration should be ensured between IT experts in charge of the data analytics and the business domain experts, knowledgeable in fraud detection, investigation or legal constraints. Indeed, as (Viaene, 2013) emphasises, IT experts (data scientists, computer scientists, data miners, etc.) are not always knowledgeable in the *business domain* as they experiment with the data in the *model domain*. Domain experts should assess and validate the output of the data experiments and algorithms as much as possible. The continuous conversation between those two stakeholders' groups also allows hidden assumptions or possible bias to be tackled as soon as possible. The alignment between two key stakeholders' groups (business and IT) is thus a key success factor in analytics projects. However, as there is little research about how this challenge takes place in practice (Research Gap 2), we investigate it in depth in Section 4.2.

3.3 Methodology

To formalise the fraud analytics process (Research Gap 1), we first identify in the literature how the big data analytics domains have been applied to fraud detection. This step is summarised in Section 2. This allows for the development of a theoretical lens through which to examine two case studies within the Belgian Federal government: the detection of tax fraud by the Federal Public Service (FPS) Finances, and the detection of social security infringements by the Social Security Institutions (SSIs). By analysing these case studies, the organisational challenge of business-IT alignment can be clearly illustrated as well (Research Gap 2). Yin (2014) mentions that case study research is relevant in order to examine a current phenomenon when 'how' questions are raised and when the researcher has no intervention in the case. Furthermore, two case studies were selected to improve the external validity of the research and draw general conclusions about the fraud analytic process in Belgium.

Data from the cases were extracted through semi-structured interviews. The main insights from the literature were used to devise an interview guide. Indeed, this qualitative method is effective when covering a complex topic in detail (Baarda *et al.*, 1996; Boyce and Neale, 2006). Moreover, this technique is relevant for our research questions, as it centres around the expertise of the practitioners, and not on the validation of the knowledge of the researchers. The complete interview guide can be found in Appendix 3.A. The goal of this interview guide is to better understand the fraud analytics process, and its related challenges, implemented in the Belgian Federal governments to detect fraud. It is structured around the three big data analytics phase described in (Klievink *et al.*, 2017). The data collected through this guide was used to understand the impact of legal requirements for the fraud analytics process and to understand the organisational alignment challenge in fraud analytics.

In total, 21 interviews were performed with stakeholders from the two backgrounds of interest: IT and business profiles. This enabled us to create a complete understanding of the fraud analytics process and of the alignment challenge. The full list of interviewees can be found in Table 3.1.

Following the recommendations of Yin (2014), we analysed the data related to the two cases with four quality criteria in mind: construct validity, internal validity, external validity and reliability. We ensured construct validity by validating, with the interviewees, the main findings from the study. Regarding internal validity, the causal relationship between the business-IT alignment challenge

Table 3.1. List of interviewees.

Case	Function	Organisation
Tax Fraud	Business	FPS Finances
Tax Fraud	IT	FPS Finances
Tax Fraud	Business	FPS Finances
Tax Fraud	IT	FPS Finances
Tax Fraud	IT	FPS Finances
Tax Fraud	IT	FPS Finances
Tax Fraud	Business	FPS Finances
Tax Fraud	Business	FPS Finances
Social Security	IT	Smals (Private company)
Social Security	IT	Smals (Private company)
Social Security	IT	Smals (Private company)
Social Security	Business	CBSS (Social Security Database)
Social Security	Business	CBSS (Social Security Data)
Social Security	Business	ONEM (Job allocations)
Social Security	Business	INAMI (Health allocations)
Social Security	IT	INAMI (Health allocations)
Social Security	Business	INAMI (Health allocations)
Social Security	Business	INAMI (Health allocations)
Social Security	Business	INAMI (Health allocations)
Social Security	IT	ONSS (Social Security coordination)
Social Security	IT	ONSS (Social Security coordination)

and the fraud analytics process was once again ensured by the confirmation interviews. Although interviews could have been performed with complementary organisations (e.g. other SSIs or administrations at different government level) to increase this validity, this was not done here as we had reached code saturation. Regarding external validity, we ensured a replication logic by taking two case studies to study the fraud analytics process. Finally, to ensure reliability, we have stored all the interview reports in a case study database in the research data repository of our university.

The method used to analyse the findings extracted from the research data was *Thematic Content Analysis* (Anderson, 2007). The analysis started with a summary of the interviews in a data memo. In order to code the data, we skimmed the transcripts and highlight relevant sentences based on the research questions. Then, we inserted the codes into a table divided by the main stages of big data analytics. This method enabled us to link similar themes from every interview to each other, making it easier to analyse what was expressed and how it compares with other findings. We categorised the results from our interviews (textual data) to concrete steps of the fraud analytics process as well as the related challenges. We performed interviews until we reached code saturation, meaning that no new codes were identified after five consecutive interviews (Guest *et al.*, 2006).

3.4. Results

3.4.1 Formalising the fraud analytics process: insights from the case studies

From the 21 interviews we performed and analysed, we can formalise the fraud analytics processes of the two studied case studies (i.e. the detection of tax frauds and the detection of social security infringements).

Regarding the tax fraud detection process, data is first extracted from several sources and prepared for analysis. Then, data mining is used to signal potentially fraudulent cases that need to be further examined. These two tasks (in grey) are performed by data miners. Then, at the pre-investigation stage, the signals derived from the data mining tasks are enriched with data from other sources, and it is decided whether a proper investigation should be started. Finally, in the investigation stage, some of the potentially fraudulent cases are examined in-depth, with the support of analytics (e.g. text mining) to explore a large quantity of unstructured data. This stage is also referred to as e-auditing. Inspectors perform these inspection tasks and then give feedback to data miners about the relevance of the signals. It must be noted that cases to be investigated are sometimes also suggested by 'input services' that manually detect cases to be further investigated. The fraud analytics process, applied to tax fraud detection, is presented in a simplified version in Figure 3.2.

For the social security infringement detection process, it is important to understand that a 'Social Security Network' was created by the law of 15 January 1990,[3] in which all the Belgian Federal public SSIs are structured around the 'Crossroads Bank for Social Security' (CBSS) (Degrave,

[3] Loi du 15 janvier 1990 relative à l'institution et à l'organisation d'une Banque-carrefour de la sécurité sociale, *M.B.*, 22 février 1990.

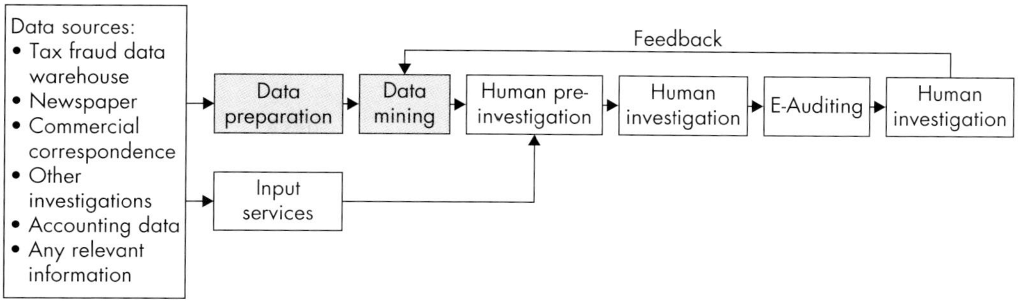

Figure 3.2. Fraud analytics process – tax fraud detection.

2020). The CBSS acts as the core of the network, and the SSIs are the nodes.[4] While these SSIs remain in control of their authoritative sources of personal social data, the CBSS acts as the central actor for the data-sharing between them.[5] The CBSS itself therefore does not store any data, but rather acts as a 'gatekeeper' to check that an SSI has the right to access data stored on one of the nodes of the network (another SSI). Regarding social security fraud, there is a difference between the types of techniques used to detect fraud committed by beneficiaries of social allocations, on the one hand, and employers, health institutions, independent workers, etc., on the other. For the former, SSIs mainly rely on data-matching techniques via bilateral crosschecks from other SSIs' databases, aimed at identifying incompatibilities in terms of allocations. These are performed either before or after the payment of the allocation. For the latter, social security institutions mainly rely on data mining techniques, using the OASIS data warehouse, where larger quantities of pseudonymised data are compiled. Moreover, one SSI is currently developing a Big Data

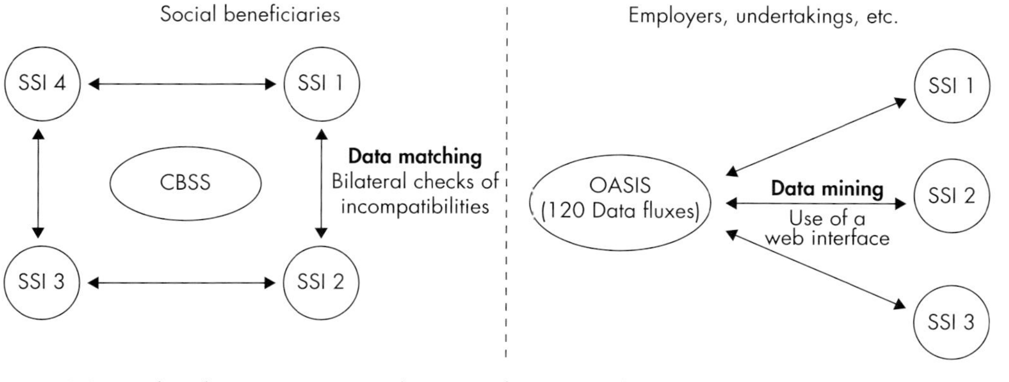

Figure 3.3. Fraud analytics process – social security infringements detection.

[4] https://www.ksz-bcss.fgov.be.
[5] Art. 3 of the Law of 15 January 1990.

Analytics Platform to improve the data governance mechanisms between SSIs, notably to tackle social fraud. The social security infringement detection process is presented in a simplified version in Figure 3.3.

3.4.2 Business-IT alignment challenges in the fraud analytics process

In Figure 3.4, a simplified fraud analytics process, abstracting the two above studied processes, is presented. This simplified process highlights the moments where alignment between the business and IT sides is necessary to ensure an efficient fraud detection. As identified in the interviews, alignment is particularly necessary for three key moments:

1. Requirements identification: how can the business team translate their needs into technical requirements for the IT team?
2. Presentation: how can the IT team present the output of the data analytics phase to the business team?
3. Feedback: how can the business team give feedback about the output of analytics to the IT team?

In the two cases studied above, several interviewees highlighted the lack of alignment for these three key interaction moments:

1. Business teams sometimes lack understanding about what the IT team needs. Furthermore, the data miners are often not fraud specialists and the needs of the business are sometimes difficult to understand and feel fragmented. As one interviewee nicely puts it: '*Business talks fraud and IT talks statistics*'. The IT teams can interpret this lack of communication in several ways: business experts may feel that the IT teams are a threat to their job; business experts

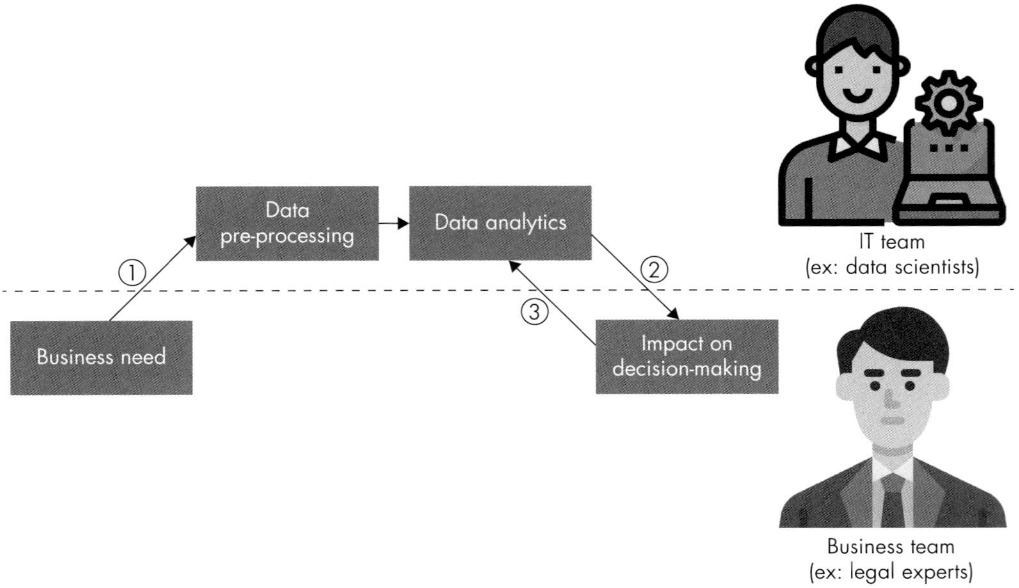

Figure 3.4. Simplified fraud analytics process highlighting the need for alignment.

do not understand the relevance of the miners; business experts do not want to change their traditional way of working. Furthermore, one investigator mentioned that it is essential that '*data scientists receive insights from the field due to the changing nature of fraud, as it impacts what the data mining needs to look for*'.
2. There are no standard ways to present the output of the analysis to business domain experts. It varies widely, depending on the IT team. Some IT teams try to make the output by '*spending a lot of time to make the analysis as readable as possible*'. For instance, they provide an excel sheet with clear headings or point separators. The goal is that the business experts immediately understand the output of the analytics phase.
3. Once this output is sent, the IT teams do not receive follow-up from the business experts. As one data miner puts it, she can '*only can assume that what she sent out is good*'. They have no idea what happens with the data, which is frustrating, as they do not know if the business experts appreciate the output of the analytics phase. They would welcome feedback to improve their output. In the current situation, IT teams can only assume to what extent the output is satisfying. However, it is possible that the business experts waste time adding data manually afterwards, even though the IT teams could easily do it.

3.5 Recommendations for alignment in fraud analytics

Based on the identified fraud analytics processes of the previous section, we better understand the organisational challenge related to collaboration between domain experts and data scientists. Therefore, in this section, we investigate the leads for solution from three research areas. Based on these leads, we suggest a tentative methodology to align data analytics and business domain in fraud analytics.

Even though the key questions raised in Section 4.2 might have specific answers, the literature about business-IT alignment offers promising ways of tackling the alignment challenges in an overarching manner. As shown in Figure 3.5, Avison *et al.* (2004) emphasise that this alignment can take place at two levels: the alignment between the business strategy and the IT strategy, and

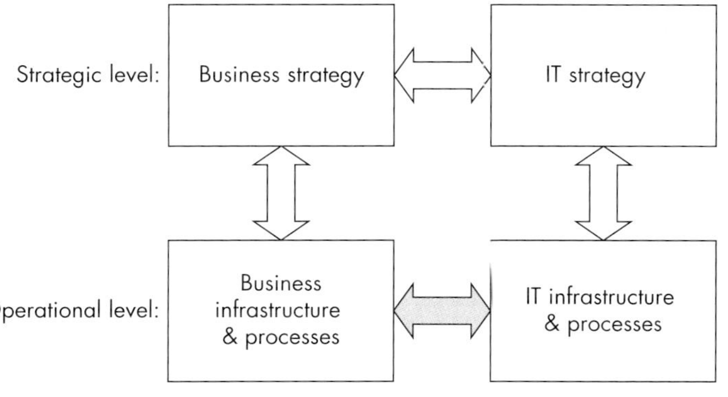

Figure 3.5. Simplified strategic alignment model by Avison *et al.* (2004).

the alignment between the operational business processes and IT processes. In this chapter, we focus on fostering the operational alignment between the business (e.g. controllers, investigators) and IT processes (e.g. data scientists, data miners), in yellow in Figure 3.5. In the following subsections, we present three ways of fostering this alignment for fraud analytics in the future.

3.5.1 Agile methods

Traditional systems development approaches, such as the Waterfall model, seemed to prevail for a long time in governments. No complete study has been found on current software development practices in governments but authors have underlined the predominance of the waterfall model (Følstad *et al.*, 2004; Pardo and Scholl, 2002). Such methods rely heavily on thorough planning and process standardisation and assume that the requirements remain static throughout the development process. They prevent public organisations from quickly adapting their processes to foster collaboration and participation of all stakeholders in the process. Nonetheless, over the last decade, some governmental organisations have become interested in a number of new techniques and approaches, such as agile development, to stimulate a more collaborative work environment in governments. Agile software development refers to a group of flexible and lightweight methodologies that rely on a set of principles and practices for the development of software (e.g. time-boxed iterations, customer involvement, daily meetings, continuous process improvement ...) (Beck *et al.*, 2001). Agile methods share a number of principles that drive the development process of practitioners. These 12 Agile Principles (AP) are described in the Agile manifesto (Beck *et al.*, 2001) and are listed below for the sake of clarity:

AP1: Customer satisfaction by early and continuous delivery of valuable software.
AP2: Welcome changing requirements, even in late development.
AP3: Working software is delivered frequently, weeks rather than months.
AP4: Close, daily cooperation between businesspeople and developers.
AP5: Projects are built around motivated individuals, who should be trusted.
AP6: Face-to-face conversation is the best form of communication.
AP7: Working software is the primary measure of progress.
AP8: Sustainable development, able to maintain a constant pace.
AP9: Continuous attention to technical excellence and good design.
AP10: Simplicity – the art of maximising the amount of work not done – is essential.
AP11: Best architectures, requirements, and designs emerge from self-organising teams.
AP12: Regularly, the team reflects on how to become more effective, and adjusts accordingly.

These general principles have been implemented through more formalised methods, such as Extreme Programming (XP), SCRUM, Feature Driven Development, Dynamic Systems Development Method (DSDM), Lean Development/Management (Cohen *et al.*, 2004). These methods can deliver high value for stimulating alignment between business and IT in fraud analytics instead of the sequential process shown in Figure 3.4. This process, which can be referred to as 'agile analytics', is an emerging research area (Collier, 2015).

3.5.2 Participation methods

Another solution for fostering alignment between business and IT in fraud analytics resides in the user participation in information systems literature. The information system research field has long proven that increased user satisfaction and early involvement in the development process (e.g. requirements engineering activities) improves system quality (Hartwick and Barki, 1994). The importance of user participation and customer involvement has been underlined by the evolution of the traditional Waterfall software development methods towards agile methods. These methods advocate more customer involvement in software development as shown in the agile principles AP1, AP2 and AP4 above. Indeed, the data analytics performed by the IT team can be considered as a software development project with a key phase of requirements identification. This identification can greatly be improved thanks to participation methods with the business team members as users.

This increased user participation has historically been implemented in the form of three main practices, giving different power of decision to the users: participatory design, user-centred design and user-innovation. Based on Holgersson *et al.* (2012), we provide the following definitions for these three concepts. *Participatory design* (PD) advocates an approach where good ideas are as likely to come from the user groups as from the decision-makers (Schuler and Namioka, 1993). In that regard, users and system developers are considered partners in the development process. In the context of this approach, the users can contribute as advisors (by assessing prototypes), as representatives to represent a particular user group, or as all-inclusive participants where all users contribute to the development work. *User-centred design* (UCD) emerged in the human-computer interaction field and underlines the important impact of user needs on the design of the interface (Abras *et al.*, 2004). Contrary to the previous approach, users and developers are not seen as equal because users only provide knowledge to the developers who, consequently, take into account this business domain knowledge. For instance, the developers could organise focus groups to gather this knowledge but still have the power to take all decision unilaterally. *User innovation* (UI) is the extreme counterpart of non-participation where the problem identification and design solutions emerge directly from the user, or more specifically from the 'lead users' group. This sub-group refers to users that have strong needs that will become more general in the marketplaces in the future (von Hippel, 1986).

These three main practices can be implemented through several participation methods, directly relevant in an analytics project:

- *Interviews*: interviews constitute a direct interaction method often used in the context of requirements engineering. Through this method, the IT team can ask specific questions to understand the needs of the business team for the analytics phase.
- *Representation in the project team*: in order to give more influence to users, Chan and Pan (2008) advocate the identification of salient intermediaries representing the users in all development stages. With this method, a representative from the business team (e.g. the lead fraud investigator) could be involved in every step of the analytics project for feedback.

- *User workshops*: the organisation of workshops to interact with a selected group of representative users is a method often used in the requirements identification stage to elicit innovation solutions (Følstad *et al.*, 2004; Oostveen *et al.*, 2004). In more recent research, these workshop are organised thanks to creativity techniques such as visualisation tools or improvisation principles (Mahaux and Maiden, 2008). Design thinking also represents a creative technique that can enhance user workshops.
- *Answer to surveys*: surveys are used for a number of purposes (market evaluation, research, etc.) but also for the large-scale participation of users, mainly in the evaluation phase of software development. De Róiste (2013) provides insights about this evaluation by users through online surveys, phone or in person surveys. Large-scale surveys can be useful to help the IT team received feedback about the output of the analytics phase.
- *Dedicated software*: in order to facilitate the large-scale participation of users, practitioners can develop dedicated software (in the form of platforms, applications, etc.) to gather users' ideas and needs (Berntzen and Johannessen, 2016). Crowd-Centric Requirements Engineering (CCRE) platforms apply the crowdsourcing paradigm in all phases of requirements engineering such as elicitation, negotiation and prioritisation (Snijders *et al.*, 2015).
- *Prototyping*: prototyping is a method often used to present a non-finished product to its potential users. Van Velsen *et al.* (2009) suggested a user-centric requirement engineering method for the design of online services with a rapid prototyping tested through focus groups, interviews or citizen walkthrough. These iterative developments of the output of analytics can deliver value to foster alignment.

3.5.3 Design thinking

In this section, we expand on the use of design thinking in analytics to stimulate alignment in fraud analytics. Design thinking can be considered a specific combination of the user innovation practice as well as the interview, prototyping and workshop participation methods. Indeed, a promising lead to stimulate the collaboration between IT (the developers) and business domain experts (the users) resides in the use of the design thinking methods as reported in Chongwatpol (2020). Design thinking is a creative process to develop solutions aligned with users' needs, combining several participation methods. It is applied in many contexts and recently investigated for analytics. These steps are generic but have recently shown potential for improving analytics projects (Chongwatpol, 2020):

- *Empathise*: understanding the business perspective. In this step, interviews with a representative set of stakeholders are performed to understand their needs, experience, and motivations.
- *Define*: unpacking and structuring the key findings from the interviews. In this step, the main challenges of the stakeholders are identified and features (or specific data analysis techniques) are mapped to solve them. Furthermore, the challenges are formulated as point-of-view statements to make them meaningful. If necessary, personas (fictional users with clear needs to be addressed by the solution) can be used.

- *Ideate*: brainstorming about which techniques or combination of techniques are relevant. In this step, workshops are used to design alternatives analytics solutions by thinking creatively through brainstorming, storyboards, mind maps, and idea clustering and selection. There must be a large quantity and diversity of ideas so that unexpected techniques are uncovered.
- *Prototype*: finalising the ideas for the solution space to evaluate practicability. In this step, low-fidelity analytics prototype (using mock-ups of visualisations, for instance) allow potential solutions to be demonstrated, leaving the details for later.
- *Test*: receiving feedback on the intermediary solution. In this step, feedback is collected through interviews about the prototype to better understand users' opinions, experiences and feedback.

3.5.4 Towards a methodology for business-IT alignment in fraud analytics

Drawing from the insights gathered from the agile methods, participation methods and design thinking literatures, we suggest a tentative methodology for business-IT alignment in fraud analytics. Figure 3.6 represents this methodology visually.

Drawing from the SCRUM agile methodology (Schwaber and Sutherland, 2017), the business team (fraud investigators) can be considered the product owners of the analytics process and the IT team (the data scientists) can be considered the development team in charge of iterative provision of analytics output. The methodology is composed of the following steps:

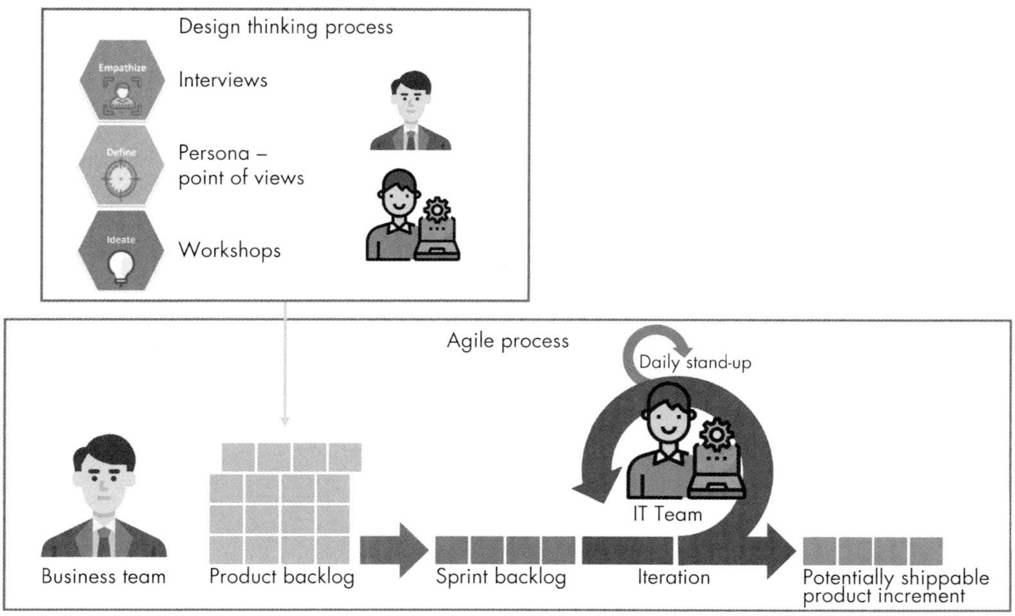

Figure 3.6. Methodology for business-IT alignment in fraud analytics.

1. The methodology starts with the design thinking process where business and IT work together. The business-side is interviewed by the IT-side (*'Empathise' stage, 'Interview' participation method*) so that clear requirements are defined ('Define' stage). Furthermore, the organisation of ideation workshops allows for creative thinking about analytics solutions to address the requirements (*'Ideate' stage, 'Workshop' participation method*). The refined leads for solutions are bundled into a product backlog that the business team is in charge of. This backlog summarises the requirements to be addressed by the IT team in a hierarchical manner.
2. The IT team self-organises and selects which elements from the backlog they will tackle in a sprint (time-boxed period of work, often 2-4 weeks). The team then works iteratively (*'Prototype' stage, 'Prototyping' participation method*), with daily stand-ups to reflect on the work done.
3. After the sprint, the IT team produce potentially shippable product increments submitted for feedback to the business team (*'Test' stage*). Based on the feedback received through several iterations, the analytics will be more aligned to the needs of the fraud investigators and this will foster collaboration.

Step 1 allows business and IT to formalise clear requirements through relevant participation methods as well as providing a usable method to manage them (the product backlog). In that sense, it helps to address the 'Requirements identification' alignment moment. Step 2 and the prototyping approach allow iteratively improving the way the IT team presents the output of analytics to the business team. In that sense, it helps to address the 'Presentation' alignment moment. Step 3 formalises feedback within the analytics process so that there is communication between the two teams and self-reflection about the overall process. This helps to address the 'Feedback' alignment moment.

Despite the promises of this generic methodology, particular attention should be paid to its customisation depending on the context. Indeed, the implementation of the methodology in a government context may raise challenges because of the intrinsic characteristics of government: need for regulatory compliance, lack of strategic or operational support, reluctance to change, etc. Indeed, the same challenge was met for the implementation of agile methods. As a consequence, a growing line of research has identified that practitioners also use tailored methods that fit the specificities of their organisations (Campanelli and Parreiras, 2015). The same customisation approach should be followed to iteratively improve and adapt the alignment methodology to each government organisation.

3.6 Conclusions

Business-IT alignment has for long been considered a key success factor for the digital governance of public administrations as it allows the IT operations and strategies to support the business operations and strategies (Avison *et al.*, 2004). However, the introduction of more advanced technologies (such as big data, artificial intelligence or machine learning) makes the need for alignment increasingly important. Indeed, as we have seen in this chapter, the introduction of analytics in the fraud detection domain does not simply affect the teams on the IT-side of the organisation but consists of a more global process where collaboration between business and IT teams is essential. Ensuring a dialogue between data scientists, data miners, legal specialists and tax investigators in public administrations becomes increasingly challenging as fraud becomes more and more difficult to detect, while the analytics used also become increasingly complex. The organisational alignment between diverse profiles is a pre-condition to ensure proper data-driven governance in public administrations.

In order to contribute to this need for alignment, this chapter formalises the fraud analytics process, illustrates one key organisational challenge (the alignment between business and IT teams) and provides leads for solutions to foster this alignment in fraud analytics. For this, we examine two case studies from the Belgian Federal government through 21 interviews: the detection of tax fraud and of social security infringements. This chapter is relevant for researchers as it formalises the fraud analytics process for both cases and then suggests three leads for solution, under-investigated for the fraud analytics domain, to foster business-IT alignment: agile methods, participation methods and design thinking. Furthermore, it is relevant for practitioners as the leads for solutions are bundled into a usable methodology, directly actionable to improve the effectiveness of fraud analytics. This managerial solution can be considered an effective governance tool in public administrations to increase organisational alignment in data-driven public administrations.

The research performed in this chapter also has two main inherent limitations that introduce leads for further research. First, only two cases were examined to formalise the fraud analytics process in government. These cases come from Belgium and an international comparison with other countries would be necessary to increase external validity. Second, three specific leads for solutions were selected for discussion in this chapter. Other leads are promising and deserve the attention of interested future researchers: the use of data logistics platforms to promote collaboration between business and IT, the coaching of stakeholders for cultural change to increase collaboration or the introduction of business analyst profiles to make the link between business and IT.

References

Abras, C., D. Maloney-Krichmar, and J. Preece. 2004. User-centered design. In W.S. Bainbridge (ed.) *Encyclopedia of human-computer interaction.* Berkshire Publishing Group, Great Barrington, MA, USA.

Anderson, R. 2007. Thematic content analysis (TCA): descriptive presentation of qualitative data. Unpublished manuscript. Available at: http://rosemarieanderson.com/wp-content/uploads/2014/08/ThematicContentAnalysis.pdf.

Archenaa, J., and E.A.M. Anita. 2015. A survey of big data analytics in healthcare and government. *Procedia Computer Science* 50: 408-413. https://doi.org/10.1016/j.procs.2015.04.021

Avison, D., J. Jones, P. Powell, and D. Wilson. 2004. Using and validating the strategic alignment model. *Journal of Strategic Information Systems* 13(3): 223-246. https://doi.org/10.1016/j.jsis.2004.08.002

Baarda, D.B., M.P.M. de Goede, and A.G.E. van der Meer-Middelburg. 1996. *Basisboek open interviewen: praktische handleiding voor het voorbereiden en afnemen van open interviews [Book about basics open interviewing: a practical guidance for preparing and conducting open interviews].* Stenfert Kroese, Groningen, the Netherlands.

Baesens, B., V. Van Vlasselaer, and W. Verbeke. 2015. *Fraud analytics using descriptive, predictive, and social network techniques*. Wiley, Hoboken, NJ, USA. https://doi.org/10.1002/9781119146841

Beck, K., M. Beedle, A. Van Bennekum, A. Cockburn, W. Cunningham, M. Fowler, J. Grenning, J. Highsmith, A. Hunt, R. Jeffries, J. Kern, B. Marick, R. C. Martin, S. Mellor, K. Schwaber, J. Sutherland, and D. Thomas. 2001. Agile manifesto. *Software Development* 9: 28-35. https://doi.org/10.1177/004057368303900411

Berntzen, L., and M.R. Johannessen. 2016. The role of citizen participation in municipal smart city projects: lessons learned from Norway. In *Smarter as the new urban agenda*. 299-314. Springer International Publishing, Cham, Switzerland. https://doi.org/10.1007/978-3-319-17620-8

Bertot, J.C., U. Gorham, P.T. Jaeger, L.C. Sarin, and H. Choi. 2014. Big data, open government and e-government: issues, policies and recommendations. *Information Polity* 19(1/2): 5-16. https://doi.org/10.3233/IP-140328

Bishop, C.M. 2006. *Pattern recoginiton and machine learning*. Springer Science+Business Media, LLC, New York, NY, USA.

Boyce, C., and P. Neale. 2006. Conducting in-depth interviews: a guide for designing and conducting in-depth interviews. *Evaluation* 2 (May): 1-16. https://doi.org/10.1080/14616730210154225

Campanelli, A.S., and F.S. Parreiras. 2015. Agile methods tailoring – a systematic literature review. *Journal of Systems and Software* 110: 85-100. https://doi.org/10.1016/j.jss.2015.08.035

Castellón González, P., and J.D. Velásquez. 2013. Characterization and detection of taxpayers with false invoices using data mining techniques. *Expert Systems with Applications* 40: 1427-1436. https://doi.org/10.1016/j.eswa.2012.08.051

Chan, C.M.L., and S.L. Pan. 2008. User engagement in e-government systems implementation: A comparative case study of two Singaporean e-government initiatives. *Journal of Strategic Information Systems* 17(2): 124-139. https://doi.org/10.1016/j.jsis.2007.12.003

Chantillon, M., A. Simonosfki, T. Tombal, R. Kruk, J. Crompvoets, and M. Snoeck. 2020. Analysing e-government through the multi-level governance lens – An exploratory study in Belgium. *CEUR Workshop Proceedings*, volume 2797, paper 17. Available at: http://ceur-ws.org/Vol-2797/paper17.pdf.

Chongwatpol, J. 2020. Operationalizing design thinking in business intelligence and analytics projects. *Decision Sciences Journal of Innovative Education* 18(3): 409-433. https://doi.org/10.1111/dsji.12217

Chung, W., and D. Zeng. 2018. Social-media-based policy informatics: cyber-surveillance for homeland security and public health informatics. In *Public administration and information technology*. Vol. 25. 363-385. Springer, Berlin, Germany. https://doi.org/10.1007/978-3-319-61762-6_16

Cohen, D., M. Lindval, and P. Costa. 2004. Agile software development. *Advances in Computers* 62: 1-66. https://doi.org/10.1016/S0065-2458(03)62001-2

Collier, K. 2015. *Agile analytics: a value-driven approach to business intelligence and data warehousing*. Addison-Wesley Professional, Boston, MA, USA.

Daniell, K.A., A. Morton, and D. Ríos Insua. 2016. Policy analysis and policy analytics. *Annals of Operations Research* 236(1): 1-13. https://doi.org/10.1007/s10479-015-1902-9

Davenport, B.T.H., and S L. Jarvenpaa. 2008. *The strategic use of analytics in government*. IBM Center for The Business of Government, Washington, DC, USA.

De Marchi, G., G. Lucertini, and A. Tsoukiàs. 2016. From evidence-based policy making to policy analytics. *Annals of Operations Research* 236(1): 15-38. https://doi.org/10.1007/s10479-014-1578-6

De Róiste, M. 2013. Bringing in the users: the role for usability evaluation in eGovernment. *Government Information Quarterly* 30(4): 441-449. https://doi.org/10.1016/j.giq.2013.05.007

De Roux, D., B. Pérez, A. Moreno, M. Del Pilar Villamil, and C. Figueroa. 2018. Tax fraud detection for under-reporting declarations using an unsupervised machine learning approach. *Proceedings of the ACM SIGKDD International Conference on Knowledge Discovery and Data Mining*. https://doi.org/10.1145/3219819.3219878

De Sousa, W.G., E.R.P. de Melo, P.H.D.S. Bermejo, R.A.S. Farias, and A.O. Gomes. 2019. How and where is artificial intelligence in the public sector going? A literature review and research agenda. *Government Information Quarterly* 36 (4): 101392. https://doi.org/10.1016/j.giq.2019.07.004

Degrave, E. 2020. The use of secret algorithms to combat social fraud in Belgium. *European Review of Digital Administration & Law* 1(1-2): 167-177.

Desouza, K.C., and B. Jacob. 2017. Big data in the public sector: lessons for practitioners and scholars. *Administration and Society* 49(7): 1043-1064. https://doi.org/10.1177/0095399714555751

EdChain. 2018. POW vs. PoS. A comparison of two blockchain consensus algorithms. Medium. Available at: https://medium.com/@EdChain/pow-vs-pos-a-comparison-of-two-blockchain-consensus-algorithms-f3effdae55f5.

Erickson, J.S., A. Viswanathan, J. Shinavier, Y. Shi, and J.A. Hendler. 2013. Open government data: a data analytics approach. *IEEE Intelligent Systems* 28(5): 19-23. https://doi.org/10.1109/MIS.2013.134

Følstad, A., H. D. Jørgensen, and J. Krogstie. 2004. User involvement in e-government development projects. *Third Nordic Conference on Human-Computer Interaction* 82: 217-224. https://doi.org/10.4018/978-1-59904-027-1.ch016

Gil-Garcia, J.R., T.A. Pardo, and L.F. Luna-Reyes. 2018. Policy analytics: definitions, components, methods, and illustrative examples. In *Policy Analytics, Modelling, and Informatics*. 1-16. Springer, Berlin, Germany. https://doi.org/10.1007/978-3-319-61762-6_1

Guest, G., A. Bunce, and L. Johnson. 2006. How many interviews are enough? An experiment with data saturation and variability. *Field Methods* 18(1): 59-82. https://doi.org/10.1177/1525822x05279903

Hamza, K., and S. Mellouli. 2018. Background on frameworks for policy analytics. In *Public Administration and Information Technology*. Vol. 25. 19-37. Springer, Berlin, Germany. https://doi.org/10.1007/978-3-319-61762-6_2

Han, J., M. Kamber, and J. Pei. 2012. *Data mining: concepts and techniques*. Morgan Kaufmann Publishers, Waltham, MA, USA. https://doi.org/10.1016/C2009-0-61819-5

Hartwick, J., and H. Barki. 1994. Explaining the role of user participation in information system use. *Management Science* 40(4): 440-465. https://doi.org/10.1287/mnsc.40.4.440

Henderson, J.C., and N. Venkatraman. 1999. Strategic alignment: leveraging information technology for transforming organizations. *IBM Systems Journal* 38: 472-484. https://doi.org/10.1147/SJ.1999.5387096

Holgersson, J., F. Karlsson, J. Holgersson, E. Söderström, and K. Hedström. 2012. Exploring user participation approaches in public e-service development. *Government Information Quarterly* 29(2): 158-168. https://doi.org/10.1016/j.giq.2011.07.009

Janssen, M., and G. Kuk. 2016. The challenges and limits of big data algorithms in technocratic governance. *Government Information Quarterly* 33(3): 371-377. https://doi.org/http://dx.doi.org/10.1016/j.giq.2016.08.011

Joseph, R.C., and N.A. Johnson. 2013. Big data and transformational government. *IT Professional* 15(6): 43-48. https://doi.org/10.1109/MITP.2013.61

Kalampokis, E., E. Tambouris, and K. Tarabanis. 2013. Linked open government data analytics. *Electronic Government: Proceedings of the 12th IFIP WG 8.5 International Conference, EGOV 2013*, volume 8074: 99-110.

Karjalainen, R. 2020. Governance in decentralised networks. *SSRN Electronic Journal*. http://dx.doi.org/10.2139/ssrn.3551099

Kim, G.H., S. Trimi, and J.H. Chung. 2014. Big-data applications in the government sector. *Communications of the ACM* 57(3): 78-85. https://doi.org/10.1145/2500873

Klievink, B., B.J. Romijn, S. Cunningham, and H. de Bruijn. 2017. Big data in the public sector: Uncertainties and readiness. *Information Systems Frontiers* 19(2):2 67-283. https://doi.org/10.1007/s10796-016-9686-2

Koutra, D., T.Y. Ke, U. Kang, D.H. Chau, H.K.K. Pao, and C. Faloutsos. 2011. Unifying guilt-by-association approaches: theorems and fast algorithms. *Lecture Notes in Computer Science* volume 6912. https://doi.org/10.1007/978-3-642-23783-6_16

Loukis, E.N. 2018. Citizen-sourcing for public policy making: Theoretical foundations, methods and evaluation. In *Public Administration and Information Technology*. Vol. 25. 179-203. Springer, Berlin, Germany. https://doi.org/10.1007/978-3-319-61762-6_8

Maciejewski, M. 2017. To do more, better, faster and more cheaply: using big data in public administration. *International Review of Administrative Sciences* 83 (Issue 1 Suppl): 120-135. https://doi.org/10.1177/0020852316640058

Mahaux, M., and N. Maiden. 2008. Theater improvisers know the requirements game. *IEEE Software* 25(5): 68-69. https://doi.org/10.1109/MS.2008.128

Mitchell, T. 1997. *Machine learning*. McGraw Hill, New York, NY, USA.

Negash, S. 2004. Communications of the association for information systems business intelligence. *Communications of the Association for Information Systems* 2: 175-193. https://doi.org/10.1007/978-3-540-48716-6_9

Oostveen, A.-M., P. van den Besselaar, and P. van den Besselaar. 2004. From small scale to large scale user participation: a case study of participatory design in e-government systems. *Eighth Conference on Participatory Design: Artful Integration: Interweaving Media, Materials and Practices*. 173-182. https://doi.org/10.1145/1011870.1011891

Pardo, T., and H.J. Scholl. 2002. Walking atop the cliffs – avoiding failure and reducing risk in large-scale e-government projects. *Proceedings of the 35th Hawaii International Conference on System Sciences (HICSS-35)*, 5: 124b (1-10).

Pencheva, I., M. Esteve, and S.J. Mikhaylov. 2018. Big data and AI – a transformational shift for government: so, what next for research? *Public Policy and Administration* 35(1): 4-44. https://doi.org/10.1177/0952076718780537

Schuler, D., and A. Namioka. 1993. *Participatory design: principles and practices*. CRC Press, Boca Raton, FL, USA.

Schwaber, K., and J. Sutherland. 2017. *The scrum guide: the definitive the rules of the game*. Available at: https://scrumguides.org/docs/scrumguide/v2020/2020-Scrum-Guide-US.pdf.

Snijders, R., A. Ozum, S. Brinkkemper, and F. Dalpiaz. 2015. *Crowd-centric requirements engineering : a method based on crowdsourcing and gamification*. Tech. Rep. UU-CS-2015-004. Department of Information and Computing Sciences, Utrecht University, the Netherlands.

Tsoukias, A., G. Montibeller, G. Lucertini, and V. Belton. 2013. Policy analytics: an agenda for research and practice. *EURO Journal on Decision Processes* 1(1-2): 115-134. https://doi.org/10.1007/s40070-013-0008-3

Ubaldi, B. 2013. *Open government data: towards empirical analysis of open government data initiatives.* OECD, Paris, France. https://doi.org/10.1787/19934351

Van Velsen, L., T. van der Geest, M. ter Hedde, and W. Derks. 2009. Requirements engineering for e-Government services: A citizen-centric approach and case study. *Government Information Quarterly* 26(3): 477-486. https://doi.org/10.1016/j.giq.2009.02.007

Van Vlasselaer, V., T. Eliassi-Rad, L. Akoglu, M. Snoeck, and B. Baesens. 2017. GOTCHA! Network-based fraud detection for social security fraud. *Management Science* 63(9): 2773-3145. https://doi.org/10.1287/mnsc.2016.2489

Viaene, S. 2013. Data scientists aren't domain experts. *IT Professional* 15(6): 12-17. https://doi.org/10.1109/MITP.2013.93

Von Hippel, E. 1986. Lead users: a source of novel product concepts. *Management Science* 32(7): 791-805. https://doi.org/10.1287/mnsc.32.7.791

Yin, R.K. 2014. *Case study research: design and methods.* Sage, Thousand Oaks, CA, USA.

Yu, F., Z. Qin, and X.L. Jia. 2003. Data mining application issues in fraudulent tax declaration detection. *International Conference on Machine Learning and Cybernetics.* https://doi.org/10.1109/icmlc.2003.1259872

Appendix 3.A. Interview guide

The goal of the following questions is to understand the fraud analytics process from the perspective of the interviewee. Furthermore, it allows eliciting the main organizational and legal challenges related to the process.

General questions:

- Can you describe your function within your organisation?
- What is your background ? What is your main expertise in the organization?
- What is your role related to Data Analytics in your organisation?
- What is your role related to Fraud Detection in your organisation?

Domain of fraud:

- What does your organisation consider as a Tax Fraud/Social Security Infringement?
- How do you detect these frauds? What are the different steps?
- What are the actors involved in the fraud investigation process?
- How is this policy implemented? Who are the main actors and what are they doing?

Pre-processing stage:

- Which data sources do you use for decision making? From which actors?
- Do you combine data from several sources? What are the challenges?
- How are the data stored?
- What is the ideal way you would like to see the data presented to you?
- Do you have procedures in place to answer data subjects' right request under the GDPR (access, rectification, erasure, ...)?
- How do you ensure the representativity of the datasets used to train the algorithms? How do you avoid biases?
- Could you tell us more about the data warehouse to fight tax fraud (Art. 5.1 of the Law of 3 August 2012)?
- Could you tell us more about the OASIS data warehouse (Art. 5bis of the Law of 15 January 1990)?
- Do you define the purposes before collecting and combining the data, or do you collect and combine the data first then define the purposes for which you'll use them?
- How do you determine these purposes? Is it always linked to a legal obligation?
- How do you determine the categories of data you will collect?
- How do you ensure that the data is kept up-to-date?
- Are the citizens informed about these data collections/combinations? If yes how?

Data analytics stage:

- How do you investigate a specific fraud?
- Which techniques do you apply to analyse the data related to fraud?
- How did you start to use these techniques?
- Are these technical tools developed internally or by private entities? Who holds the rights on the results?

Post-processing stage (use for fraud detection):

- Do you use different techniques depending on the type of fraud?
- How is the output of the analytics phase presented? How does it impact your fraud investigation process?
- Are some decisions solely based on automated processing, or is there always a human intervention in the fraud detection process (confirmation, investigation, etc.)? How significant is it?
- Is the citizen made aware of the logic involved behind the decision (relevant factors)?
- How can the citizen contest the decision?
- Are you able to explain the predictions/decisions made by the algorithm?
- What are the other main legal/organizational/technical challenges when using those techniques?
- To what degree do you take into account public opinion and the opinions of external stakeholders (such as businesses) on new fraud detection techniques and data-gathering methods?

Future directions and closing questions:

- What would be for you, personally, the future of fraud detection policy in 10 years? What would be the main factors to make it happen? What would be the main obstacles?
- Would you like to address another topic related to the research that we didn't highlight in the interview?
- Are there other stakeholders in your organisation that would be relevant for us to interview?

Chapter 4

Legal framework for the use of big data and blockchain in public governance

T. Tombal[1,2], P. Willem[3*] and C. De Terwangne[3]

[1]Faculty of Law, University of Namur, Rempart de la Vierge, 5, 5000 Namur, Belgium; [2]Tilburg Institute for Law and Technology, Tilburg University, Prof. Cobbenhagenlaan 221, 5037DE, Tilburg, the Netherlands; [3]Centre de recherche Information, Droit et Société, Namur Digital Institute, University of Namur, Rempart de la Vierge, 5, 5000 Namur, Belgium; pauline.willem@unamur.be

Abstract

By using technologies, such as big data and blockchain, public administrations are able to process a large amount of citizens' personal data. When processing these personal data, the administration must comply with the personal data protection rules contained in the General Data Protection Regulation. As the use of such technologies could have dramatic impacts on the lives of their citizens, it is fundamental to understand the limits that this legal framework puts on their use. This chapter analyses the interactions between personal data protection and big data and blockchain technologies. It explains the legal framework within which such technologies can be leveraged by public administrations for the provision of their public services. Focussing on data collection and combination (big data), this chapter concludes that the processing of personal should be based on a law meeting certain requirements. Data subjects' rights must also be respected. Regarding data storage (blockchain), several key and concrete takeaways are formulated for public administrations. While most of the following analysis will be equally applicable to any public administration within the European Union, this chapter will focus on two case studies within the Belgian public administration, namely the public policies and decision-making linked to social security infringements and tax fraud at the Belgian federal level. Indeed, these could have a significant impact on citizens' finances in particular, and their lives in general.

Keywords: big data, blockchain, GDPR, public administration, fraud

DOI 10.3920/978-90-8686-930-5_4,

4.1 Introduction

Public administrations consistently use more and more data, including their citizens' personal data,[1] to deliver their public services. Yet, when processing[2] these personal data, they have to comply with the personal data protection rules. For the public administrations of Members States of the European Union,[3] these rules are contained in the General Data Protection Regulation (hereafter 'GDPR').[4] The adoption of the GDPR presents several challenges for the administration as it is directly applicable to them and they might thus have to revise their former way of processing personal data. Indeed, the principles of accountability (Article 29 Working Party, 2010)[5] and of data protection by design and by default,[6] which are at the core of the GDPR, were not explicit in the former Directive 95/46,[7] and the administrations had to adapt their practices in order to meet the new standards set by the GDPR.

These privacy and personal data protection rules are especially important with the advent of new technologies, such as big data, artificial intelligence and blockchain, which bring new capabilities to public administration to process greater amounts of data in public service processes with unique implications in data privacy, transparency, security and governance. As the use of such technologies could have strong impacts on the lives of their citizens, it is fundamental to understand the limits that this legal framework puts on their use.

[1] Personal data is defined as 'any information relating to an identified or identifiable natural person ('data subject'); an identifiable natural person is one who can be identified, directly or indirectly, in particular by reference to an identifier such as a name, an identification number, location data, an online identifier or to one or more factors specific to the physical, physiological, genetic, mental, economic, cultural or social identity of that natural person' (Art. 4.1 of the GDPR).
[2] Processing 'means any operation or set of operations which is performed on personal data or on sets of personal data, whether or not by automated means, such as collection, recording, organisation, structuring, storage, adaptation or alteration, retrieval, consultation, use, disclosure by transmission, dissemination or otherwise making available, alignment or combination, restriction, erasure or destruction' (Art. 4.2 of the GDPR).
[3] For examples of data protection rules in other countries, see for instance the UK Data Protection Act (2018), available at https://www.legislation.gov.uk/ukpga/2018/12/contents/enacted; or the Australian Privacy Act (1988), available at https://www.legislation.gov.au/Details/C2021C00242. In the USA, there is no single Federal State legislation pertaining to data protection, but rather hundreds of (sector-specific) Federal and State legislations, some of which focus on a particular type of data (for more information and a brief overview, see https://iclg.com/practice-areas/data-protection-laws-and-regulations/usa).
[4] Regulation 2016/679 of the European Parliament and of the Council of 27 April 2016 on the protection of natural persons with regard to the processing of personal data and on the free movement of such data, and repealing Directive 95/46 (General Data Protection Regulation), OJ [2016] L 199/1. The GDPR was adopted on the 27 April 2016 and is applicable since the 25 May 2018.
[5] Art. 5.2 of the GDPR.
[6] Art. 25 of the GDPR.
[7] Directive 95/46/EC of the European Parliament and of the Council of 24 October 1995 on the protection of individuals with regard to the processing of personal data and on the free movement of such data, OJ [1995] L 281/31. In Belgium, which is the country that will be used as a case study (see *infra*), the provisions of the Directive had been included in an existing law of 1992 (Loi du 8 décembre 1992 relative à la protection de la vie privée à l'égard des traitements de données à caractère personnel, *M.B.*, 18 mars 1993).

Accordingly, this chapter and the following one will delve into the interactions between personal data protection and big data, artificial intelligence and blockchain technologies, in order to draw the legal framework within which such technologies can be leveraged by public administrations for the provision of their public services. To do so, the legal challenges will be structured around the data-processing lifecycle for decision-making, focussing first on data collection and data combination (big data – Section 4.2), and then on data storage (blockchain – Section 4.3). The legal considerations pertaining to the use of artificial intelligence to facilitate data analysis will, on the other hand, be dealt with in Chapter 5. Naturally, all of these technologies can potentially apply at all stages of the data-processing lifecycle, but for readability purposes, it has been decided to relate each technology to the stage of the lifecycle where it has the most impact. However, this does not mean that it does not have an impact on the other stages. Indeed, these technologies are intertwined and influence each other throughout the data-processing lifecycle.

While most of the following analysis will be equally applicable to any public administration within the European Union, this chapter will focus on one Member State, namely Belgium, in order to illustrate in more detail some of the legal challenges that public administrations face in practice when they wish to rely on these new technologies. More specifically, particular attention will be devoted, in this chapter, to two case studies within the Belgian public administration, namely the public policies and decision-making linked to social security infringements and tax fraud at the Belgian federal level, as these could have a significant impact on citizens' finances in particular, and their lives in general.

4.2 Data collection and combination – personal data protection and big data

As pointed out by Zarsky, 'when striving to define the 'big data' concept, the professional literature refers to the four V's: the Volume of data collected, the Variety of sources, the Velocity with which the analysis of the data can unfold, and the Veracity of the data which could (arguably) be achieved through the analytical process' (Zarsky, 2017). Big data represents 'a fundamental change in the way data is *collected*, *stored*, and subsequently *used* – all a result of recent technological developments' (emphasis in the original text) (Zarsky, 2017). Big data relies on the fact that large volumes of various data are gathered in order to extract information and draw inferences that would have otherwise not been possible to extract/draw with smaller volumes of data (Mayer-Schönberger and Padova, 2016). The more data is gathered, the more information can potentially be extracted from combining this data, in order, for example, to draw up public policies or to take decisions. For public administration, implementing big data technologies could notably derive from a greater sharing of data between administrations, thus breaking the existing silos that exist between these administrations (Chantillon *et al.*, 2017). It could also derive from the collection of data, held by private sector firms, for public interest purposes (High-Level Expert Group on Business-to-Government Data Sharing, 2020; Richter, 2020).[8]

---

[8] See also the French 'Loi Lemaire': Loi n° 2016-1321 pour une République numérique, 7 octobre 2016 (especially Articles 17 to 22 pertaining to 'data of general interest' ('données d'intérêt general')).

With the advent of today's technology, data can be collected more easily, notably through sensors and the Internet of Things (IoT); it can be stored in larger quantities, as the cost of storage is constantly decreasing; and it can be used in a wide variety of contexts, notably for decision-making (Zarsky, 2017). However, such an increase in the power of data analytics can have an impact on the data subjects' privacy and personal data protection. As pointed out by Zarsky, there is a 'double-sided tension' between big data analytics and personal data protection (Zarsky, 2017). As this author outlines, 'on the one hand, these advanced forms of data analyses can compromise the individuals' privacy rights and the control citizens have over their personal data. Thus, the availability of these tools might require stricter enforcement of privacy laws to so limit privacy-related harms. On the other hand, (...) stricter data protection and privacy laws compromise the growth of the big data industry and the benefits to be derived from it' (Zarsky, 2017).

Therefore, big data analytics, at least in some of its forms, is in tension with personal data protection. Indeed, the potential of big data relies on the fact that 'data needs to be gathered at an unprecedented scale whenever possible, and reused for different purposes over and over again (...) This puts big data on a direct collision course with the core principles of existing data protection laws' (Mayer-Schönberger and Padova, 2016). This is specifically apparent when focussing on five of the GDPR's provisions, namely the prohibition to process special categories of data,[9] the purpose limitation principle,[10] the data minimisation principle,[11] the data subject's right to information,[12] and the data subject's right not to be subject to automated individual decision-making[13] (Zarsky, 2017). Before delving into each of these provisions, it must first be outlined that any processing of personal data by the public administration must be lawful and fair.[14]

4.2.1 Lawfulness and fairness of processing and big data

The GDPR stipulates that the processing of personal data will be lawful 'only if and to the extent that at least one of the following applies: (a) the data subject has given consent to the processing of his or her personal data for one or more specific purposes; (b) processing is necessary for the performance of a contract to which the data subject is party or in order to take steps at the request of the data subject prior to entering into a contract; (c) processing is necessary for compliance with a legal obligation to which the controller is subject; (d) processing is necessary in order to protect the vital interests of the data subject or of another natural person; (e) processing is necessary for the performance of a task carried out in the public interest or in the exercise of official authority vested in the controller; (f) processing is necessary for the purposes of the legitimate interests

[9] Art. 9 of the GDPR.
[10] Art. 5.1.b) of the GDPR.
[11] Art. 5.1.c) of the GDPR.
[12] Arts. 12 to 14 of the GDPR.
[13] Art. 22 of the GDPR.
[14] Art. 5.1.a) of the GDPR.

pursued by the controller or by a third party, except where such interests are overridden by the interests or fundamental rights and freedoms of the data subject which require protection of personal data, in particular where the data subject is a child'.[15]

While a quick read of the GDPR might give the impression that the public authorities could thus potentially rely on six lawful bases of processing, in reality, two of those bases should be avoided by public authorities. On the one hand, the GDPR explicitly states that the 'legitimate interests' basis shall not apply to processing carried out by public authorities in the performance of their tasks.[16] In practice, this is not a major problem for public administrations as they will rely instead on the 'performance of a task carried out in the public interest' lawful basis.

On the other hand, public administrations should avoid relying on the data subject's consent as a lawful basis of processing. To understand why this is the case, the definition of consent should first be re-stated. According to the GDPR, the consent of the data subject 'means any *freely given*, specific, informed and unambiguous indication of the data subject's wishes by which he or she, by a statement or by a clear affirmative action, signifies agreement to the processing of personal data relating to him or her'[17] (emphasis added). The consent of the data subject thus needs to be freely given. In this regard, the GDPR outlines that 'consent should not be regarded as freely given if the data subject has no genuine or free choice or is unable to refuse or withdraw consent without detriment'.[18] This will especially be the case 'where there is a clear imbalance between the data subject and the controller, in particular *where the controller is a public authority and it is therefore unlikely that consent was freely given in all the circumstances of that specific situation*'[19] (emphasis added). This is confirmed by the European Data Protection Board (hereafter 'EDPB') in its revised guidelines on consent, though it outlines that it might be appropriate to use consent under certain circumstances, such as consenting to be included in a mailing list to receive information about the progress of road works (European Data Protection Board, 2020a).

Nevertheless, as suggested by the EDPB, other lawful bases are more appropriate to the activity of public authorities, namely either the processing necessary for compliance with a legal obligation or processing necessary for the performance of a task carried out in the public interest or in the exercise of official authority vested in the controller (European Data Protection Board, 2020a).[20] In practice, the lawful basis will often be a law. For instance, at the Belgian federal level, personal data processing to fight tax fraud is organised by the law of 3 August 2012,[21] while personal data

[15] Art. 6.1 of the GDPR.
[16] Art. 5.1 of the GDPR.
[17] Art. 4.11 of the GDPR.
[18] Recital 42 of the GDPR.
[19] Recital 43 of the GDPR.
[20] Respectively arts. 6.1.c) and 6.1.e) of the GDPR.
[21] Loi du 3 août 2012 portant dispositions relatives aux traitements de données à caractère personnel réalisés par le Service public fédéral Finances dans le cadre de ses missions, *M.B.*, 24 août 2012.

processing to tackle social security fraud is notably addressed by the law of 15 January 1990[22] and the 'Program Law (I)' of 29 March 2012.[23] It should be outlined from the outset that some of the provisions of the law of 3 August 2012 and of the law of 15 January 1990, which predate the adoption of the GDPR, have been modified or inserted in September 2018, in order to adapt them to the entry into force of the GDPR.[24]

Such laws, however, need to meet certain standards. Namely, they should specify 'the general conditions governing the lawfulness of processing by the controller; the types of data which are subject to the processing; the data subjects concerned; the entities to, and the purposes for which, the personal data may be disclosed; the purpose limitation; storage periods; and processing operations and processing procedures, including measures to ensure lawful and fair processing (...) [it shall also] meet an objective of public interest and be proportionate to the legitimate aim pursued'.[25] Moreover, as a rule of thumb, any personal data processing must be fair,[26] which implies that the laws on which this processing is based must be sufficiently explicit and understandable for the citizens. They cannot be taken by surprise and must be informed about this processing. This point, which is fundamental in terms of transparency, will be addressed in Section 4.2.5.

4.2.2 The prohibition to process special categories of data and big data

The processing of 'special categories of data', listed in Article 9.1 of the GDPR,[27] is, in principle, prohibited.[28] 'Data concerning health' are part of that list, and are defined as 'personal data related to the physical or mental health of a natural person, including the provision of health care services, which reveal information about his or her health status'.[29] These types of data are important to mention in the big data context, where more and more data will be collected by sensors, such as a connected watch calculating your number of daily steps. Indeed, according to the Article 29 Working Party (today the European Data Protection Board), such data could be considered as revealing health-related information, and should accordingly benefit from the reinforced protection for 'special categories of data' (Article 29 Working Party, 2015; Zarsky, 2017). This finding is important for public policies and decision-making linked to social security infringements, as they might rely on 'data concerning health'.

---

[22] Loi du 15 janvier 1990 relative à l'institution et à l'organisation d'une Banque-carrefour de la sécurité sociale, *M.B.*, 22 février 1990.

[23] Loi-programme (I) du 29 mars 2012, *M.B.*, 6 avril 2012. See in particular Article 101 of this Law.

[24] Loi du 5 septembre 2018 instituant le comité de sécurité de l'information et modifiant diverses lois concernant la mise en œuvre du Règlement (UE) 2016/679 du Parlement européen et du Conseil du 27 avril 2016 relatif à la protection des personnes physiques à l'égard du traitement des données à caractère personnel et à la libre circulation de ces données, et abrogeant la directive 95/46/CE, *M.B.*, 10 septembre 2018, arts. 9-40 and 70-85.

[25] Art. 6.3 of the GDPR.

[26] Art. 5.1.a) of the GDPR.

[27] 'Personal data revealing racial or ethnic origin, political opinions, religious or philosophical beliefs, or trade union membership, (...) genetic data, biometric data, data concerning health or data concerning a natural person's sex life or sexual orientation' (Art. 9.1 of the GDPR).

[28] Art. 9 of the GDPR.

[29] Art. 4.15 of the GDPR.

Moreover, some forms of big data analytics might blur the lines between special categories of data and 'regular' personal data. Indeed, 'an analysis merely relying on and addressing 'regular' categories can quite quickly end up pertaining to 'special categories.' For instance, health data can be deduced from a variety of datasets, such as shopping databases, and therefore this category has quickly and sharply expanded' (Zarsky, 2017). Once again, this is an important factor to consider for public administration relying on big data analytics for decision-making linked to social security infringements and tax fraud, as, even if they believe that they are relying on 'regular' data, the crossing of large numbers of such regular data could give rise to special categories of data, which can only be processed under stricter conditions.

Finally, it must be outlined that specific additional safeguards also apply to data relating to criminal convictions and offences (Art. 10 of the GDPR),[30] which might have to be considered for tax and social security fraud, for instance if the elaboration of algorithmic models to identify 'suspicious' profiles includes the analysis of historical criminal convictions and offences data.

4.2.3 Purpose limitation principle and big data

The purpose limitation principle provides that personal data shall be 'collected for specified, explicit and legitimate purposes and not further processed in a manner that is incompatible with those purposes' (Article 29 Working Party, 2013).[31] This means that the purpose of processing must be defined prior to the collection of the data, and that data that has been collected for a specific purpose cannot be further processed for a purpose that does not fit within this initial purpose. This principle, which is a core principle of personal data protection enshrined in Article 8.2 of the Charter of Fundamental Rights of the European Union, can be at odds with some forms of big data analytics, where a large number of data are collected 'for the sake of it', without a clearly defined purpose. Indeed, Mayer-Schönberger and Padova argue that 'there is a strong economic incentive to keep the data for as long as possible, and much beyond the initial use of it, to reuse it repeatedly as well as to combine it with other data' (Mayer-Schönberger and Padova, 2016).

One might think that resorting to broad definitions of purposes of processing could solve this incompatibility. However, as pointed out by Zarsky, 'trying to circumvent this limitation by initially defining a very broad and vague purpose for future uses would most likely not resolve this matter, as the stated purposes must also be 'specific.' Furthermore, stating an unnecessarily broad purpose might even be considered as 'illegitimate' and thus lead to unacceptable processing' (Article 29 Working Party, 2013; Hahn, 2021; Zarsky, 2017). In the same vein, obtaining consent from the data subjects for very broadly defined purposes would likely not

[30] 'Processing of personal data relating to criminal convictions and offences or related security measures based on Article 6(1) shall be carried out only under the control of official authority or when the processing is authorised by Union or Member State law providing for appropriate safeguards for the rights and freedoms of data subjects. Any comprehensive register of criminal convictions shall be kept only under the control of official authority'.

[31] Art. 5.1.b) of the GDPR.

amount to specific and informed consent (European Data Protection Board, 2020a; Mayer-Schönberger and Padova, 2016).[32] Indeed, the core idea of the purpose limitation principle is that the data subject should be aware of what is done with their data, and should be able to exercise control over this processing.

However, it should be remembered here that, according to Recital 50 and Article 6.4 of the GDPR, if data is processed for a new purpose that is compatible with the initial purpose of processing, no separate lawful basis is required. In this regard, it is important to outline that the GDPR provides that further processing for archiving purposes[33] in the public interest, scientific or historical research[34] purposes or statistical purposes[35] shall, in accordance with Article 89.1, not be considered incompatible with the initial purposes.[36] In such cases, appropriate technical and organisational safeguards, such as pseudonymisation,[37] would, however, have to be set.[38] As pointed out by Mayer-Schönberger and Padova, these exceptions could notably be used for some big data applications for statistical purposes (Mayer-Schönberger and Padova, 2016). Nevertheless, Recital 162 of the GDPR makes it explicit that pursuing statistical purposes implies that the personal data cannot be used in support of individual measures or decisions regarding any particular natural person. Accordingly, using big data statistical analysis 'to influence decision-making directly affecting a particular individual would be outside the meaning of 'statistical purposes,' and also violate the restrictions on automated individual decision-making, including profiling' (Mayer-Schönberger and Padova, 2016).

If one applies the above considerations to the specific topic of this contribution, the purpose limitation principle does not prevent the use of big data analytics for social security infringements and tax fraud per se, but only of big data analytics which do not have a specified, explicit and legitimate purpose. Accordingly, big data analytics can be applied for purposes that have been clearly defined in advance. Moreover, there must be a lawful basis for this processing. As a matter

[32] Art. 4.11 of the GDPR.

[33] According to Recital 158 of the GDPR, this should apply to bodies that have a 'legal obligation to acquire, preserve, appraise, arrange, describe, communicate, promote, disseminate and provide access to records of enduring value for general public interest'. This Recital also adds that the GDPR does not apply to deceased people.

[34] According to Recital 159 of the GDPR, scientific research purposes 'should be interpreted in a broad manner including for example technological development and demonstration, fundamental research, applied research and privately funded research (...) [and] should also include studies conducted in the public interest in the area of public health'. According to Recital 160 of the GDPR, historical research purposes 'should also include historical research and research for genealogical purposes, bearing in mind that this Regulation should not apply to deceased persons'.

[35] According to Recital 162 of the GDPR, statistical purposes mean 'any operation of collection and the processing of personal data necessary for statistical surveys or for the production of statistical results. Those statistical results may further be used for different purposes, including a scientific research purpose. The statistical purpose implies that (...) the personal data are not used in support of measures or decisions regarding any particular natural person'.

[36] Art. 5.1.b) and Recital 50 of the GDPR.

[37] 'Pseudonymisation means the processing of personal data in such a manner that the personal data can no longer be attributed to a specific data subject without the use of additional information, provided that such additional information is kept separately and is subject to technical and organisational measures to ensure that the personal data are not attributed to an identified or identifiable natural person' (Art. 4.5 of the GDPR).

[38] Art. 89.1 of the GDPR.

of principle, this should be a law (Art. 6.1.c) of the GDPR) because these personal data processing interfere with citizens' privacy (Degrave, 2015), and this law should be very specific regarding the purposes of processing it allows.[39]

In terms of tax fraud, Article 3 of the Law of 3 August 2012[40] states that the Belgian Federal Public Service (hereafter 'FPS') Finances can collect and process personal data to execute its legal missions, and that the data cannot be used for other purposes.[41] Within the FPS Finances, the various administrations and/or services of the FPS can exchange personal data, provided that they have the authorisation from the President of the Executive Committee.[42] The President can ask an opinion from the Information Security Committee in this regard.[43] Regarding, more specifically, the use of big data to fight tax fraud, Article 5.1 provides that the FPS Finances may aggregate data, collected to execute its legal missions, in a 'data warehouse' enabling 'data mining' and 'data matching' operations, including profiling. This can only be done to carry out, in the context of its legal missions, targeted controls on the basis of 'risk indicators' and of analyses on data coming from different administrations and/or services of the FPS Finances.

According to Degrave and Lachapelle, these purposes of data processing might be defined too broadly, as they simply refer to the execution of the FPS Finances' 'legal missions' (Degrave and Lachapelle, 2014). This might thus be problematic in terms of the purpose limitation principle. However, this concern is somewhat alleviated by the fact that the data miners have to fill in a Data Access Management (DAM) fiche, which has to be validated by the President of the Executive Committee of the SPF Finances.[44] In this DAM fiche, they have to state the objectives and purposes of the data mining and explain how it fits the organisation's mission. This is, however, not ideal from a democratic perspective (as Parliament does not define the concrete purposes of processing) and from a legal perspective (according to Article 8 of the European Convention on Human Rights[45] and Article 22 of the Belgian Constitution, the key elements of personal data processing by public administration must be clearly defined by law).

Personal data resulting from processing operations in the data warehouse shall be kept for no longer than is necessary for the purposes for which they are processed, with a maximal retention period of one year after the prescription of all actions falling within the competence of the controller.[46] In practice, a relevance check is performed every three months by the head of the data miners to see if the data are used in conformity with the DAM fiche and if the project is still relevant and advancing. If it is not, the data access ceases and the data must be deleted.

---

[39] Art. 6.3 of the GDPR.

[40] Loi du 3 août 2012 portant dispositions relatives aux traitements de données à caractère personnel réalisés par le Service public fédéral Finances dans le cadre de ses missions, *M.B.*, 24 août 2012.

[41] Art. 3, al.1 and 2 of the Law of 3 August 2012.

[42] Art. 4, al.1 of the Law of 3 August 2012.

[43] Art. 4, al.4 of the Law of 3 August 2012.

[44] Art. 4, al.1 of the Law of 3 August 2012.

[45] European Convention for the Protection of Human Rights and Fundamental Freedoms, Rome, 4 November 1950.

[46] Art. 5.1, al.3 of the Law of 3 August 2012.

In terms of social security fraud, the use of data-matching techniques, aimed at identifying incompatibilities in terms of social allocations, must be subject to a data transfer protocol, as provided in Article 20.1 of the Law of 30 July 2018,[47] unless provided otherwise in specific laws (e.g. in Article 15 of the Law of 15 January 1990 which requires, in some cases, a prior deliberation of the Information Security Committee). The protocol, which must notably contain the purposes of processing, must be submitted to the Data Protection Officers of the social security institutions (hereafter 'SSIs') involved in the sharing.[48] However, they are not subject to a prior validation by the Data Protection Authority, which would bring more certainty in terms of the legitimacy of the purpose of processing (Degrave, 2020a). Once this purpose is achieved, the data must be deleted.

SSIs also use data mining techniques. According to Article 5*bis* of the Law of 15 January 1990, they may aggregate and process data in a data warehouse, enabling them to carry out data-mining operations to prevent, establish, prosecute, and punish offences against social legislations that fall within their respective powers. This data warehouse is known as OASIS and has existed since 2005. According to Degrave, the purposes of processing in OASIS that are authorised by the law are not clearly defined, which could be problematic in terms of the purpose limitation principle (Belgian Privacy Commission, 2018; Degrave, 2020b). However, this concern is somewhat alleviated, although not optimally either from a democratic and legal perspective (see above), in the hypotheses contained in Articles 5*bis*, al.7 and 15 of the Law of 15 January 1990, as the authorisation to process data from the data warehouse must be subject to a prior deliberation by the Information Security Committee (ISC), which will evaluate the purposes of processing.[49] It must nevertheless be underlined here that the ISC should, in theory, be independent of the administrations (including the SSIs) to which it grants authorisations to process the data, which implies that its members should not also exercise mandates within these administrations.[50] This is currently not the case, which creates independence issues as some members of the CSI are both players and referees, and this has led to the launch of an infringement procedure by the European Commission against Belgium (Anonymous, 2021a,b; European Commission, 2021; Laloux, 2021). This situation will need to be remedied as soon as possible.

Moreover, if such a deliberation is not imposed, the data controllers taking part in the fraud detection processing will nevertheless have to conclude a protocol in this regard, notably specifying the desired processing purposes, as this is the standard for any exchange of personal data between administrations, in light of the accountability principle of the GDPR (Art. 5.2, GDPR).[51] In any case, personal data resulting from processing operations in the data warehouse shall be kept for no

[47] Loi du 30 juillet 2018 relative à la protection des personnes physiques à l'égard des traitements de données à caractère personnel, *M.B.*, 5 septembre 2018.
[48] Art. 20.2 of the Law of 30 July 2018.
[49] Art. 5*bis*, al. 1 of the Law of 15 January 1990.
[50] Art. 52 of the GDPR; Loi du 5 septembre 2018 instituant le comité de sécurité de l'information et modifiant diverses lois concernant la mise en œuvre du Règlement (UE) 2016/679 du Parlement européen et du Conseil du 27 avril 2016 relatif à la protection des personnes physiques à l'égard du traitement des données à caractère personnel et à la libre circulation de ces données, et abrogeant la directive 95/46/CE, *M.B.*, 10 septembre 2018, arts. 3 and 5.
[51] Article 20.1 of the Law of 30 July 2018.

longer than is necessary for the purposes for which they are processed, with a maximal retention period not exceeding one year after the prescription of all actions falling within the competence of the data controller.[52]

4.2.4 Data minimisation principle and big data

The data minimisation principle is another fundamental principle of the GDPR. It provides that only the adequate, relevant and necessary data for the fulfilment of the specific purpose of processing shall be processed.[53] This implies that, in combination with the purpose limitation principle, the categories and amount of data that can be processed should be limited to what is necessary to meet this purpose. Once again, this core principle of the GDPR is in stark contrast with some forms of big data analytics, which aim at collecting as much data as possible, in order to extract information and draw inferences that would have otherwise not been possible to extract/ draw with smaller volumes of data. Yet, there is a core tension between, on the one hand, collecting as much data as possible, without knowing which of it will actually be useful, and on the other hand collecting only the data that is necessary for a pre-defined specific purpose.

Moreover, the data minimisation principle can also be linked to the principle of storage limitation (Art. 5.1.e) of the GDPR), as the data can only be stored for the period during which it is useful for the specific purpose of processing, and has to be erased as soon as it is no longer necessary (Zarsky, 2017). Once again, this clashes with some forms of big data analytics that aim at retaining as much data as possible, for as long as possible, in order to discover potential new uses for the data (Zarsky, 2017).

Similar to the purpose limitation principle, the data minimisation principle can be more flexible for processing for archiving purposes in the public interest, scientific or historical research purposes or statistical purposes, provided that they are subject to appropriate technical and organisational safeguards.[54] Here, pseudonymisation is explicitly mentioned as a potential safeguard to ensure the respect of the data minimisation principle. Mayer-Schönberger and Padova argue that, since most of big data analytics are statistical in nature, this exception could allow data controllers to retain personal data for longer than is necessary for the original purpose of processing (Mayer-Schönberger and Padova, 2016). However, according to Zarsky, this might not be a viable option for some big data applications, as 'removing identifiers to achieve pseudonymity can potentially undermine the quality of the results derived, as the data would be purposefully altered and the aggregation of different datasets would be rendered difficult' (Zarsky, 2017). Alternatively, anonymisation[55] does not seem to be a viable option either, as truly effective anonymisation is

[52] Art. 5*bis*, al. 4 of the Law of 15 January 1990.

[53] Art. 5.1.c) of the GDPR.

[54] Art. 89.1 of the GDPR.

[55] The ISO 29100 standard defines anonymisation as the 'process by which personally identifiable information (PII) is irreversibly altered in such a way that a PII principal can no longer be identified directly or indirectly, either by the PII controller alone or in collaboration with any other party' (ISO 29100:2011, point 2.2, available at https://www.iso.org/obp/ui/#iso:std:iso-iec:29100:ed-1:v1:en).

difficult to achieve (Franceschi *et al.*, 2018; Graef *et al.*, 2018; Mayer-Schönberger and Padova, 2016; Wendehorst, 2017). This is especially true in light of the constant development of big data analytics, which increases the risk of re-identification of the data subjects. This failure to effectively anonymise personal data has been demonstrated several times in the literature (Rocher *et al.*, 2019; Sweeney, 2017), leading to the conclusion that what is often presented as anonymisation techniques are, in fact, merely pseudonymisation techniques.

Much like the purpose limitation principle, the data minimisation principle does not per se preclude the use of big data analytics by public administrations for decision-making linked to social security infringements and tax fraud. Rather, this principle frames the scope and scale of data that can be used. This therefore reinforces the need to clearly define in advance the specific purpose of the processing used to combat social security infringements and tax fraud, on the one hand, and the data that will be necessary to do so, on the other. The key is to be proportionate in the types of data collected and used. Even if public administrations could potentially have access to troves of data, a balance must be found not only with citizens' privacy and data protection, but also with commercial and professional secrecy requirements.

In terms of tax fraud, personal data exchange between the various administrations and/or services of the FPS Finances must be authorised by the President of the Executive Committee.[56] The President decides which types of personal data can be exchanged, on a systematic or ad hoc basis and for specific purposes, after having verified their adequacy, relevance and non-excessiveness.[57] Regarding data-mining operations in the data warehouse, Article 5 of the Law of 3 August 2012 provides that the FPS Finance can use 'data collected to execute its legal missions'. These are notably data collected from people's and undertakings' tax declarations, from the newspapers, from their own experience, from whistleblowers and from outputs of investigations.

According to Degrave and Lachapelle, the fact that Article 5 provides that the FPS Finance can use, via the data warehouse, any 'data collected in order to execute its legal missions' might be problematic from a data minimisation perspective, as this is a broad definition that does not define exactly which types of data can be used (Degrave and Lachapelle, 2014). However, this concern is somewhat alleviated by the fact that, as outlined above, a DAM fiche must be completed and submitted to the President of the Executive Committee. This constrains the data that data miners can access for a specific project. This is a pragmatic solution, as it would be very difficult for the legislator to predefine all the types of data that could be processed in this regard. Moreover, the technical access to the data warehouse is built in such a way that the agents of FPS Finances can only access the electronic records, data or applications that are adequate, relevant and non-excessive in light of the execution of the tasks that fall within their legal missions,[58] and this can

[56] Art. 4, al.1 of the Law of 3 August 2012.
[57] Art. 4, al.2 of the Law of 3 August 2012.
[58] Art. 10.1 of the Law of 3 August 2012.

be checked through access logs, which allows third party auditing (e.g. by the Data Protection Authority) of the process. This also materialises the data protection by design and by default principle.[59]

Regarding data-matching operations to fight social security infringements, they must be subject to a data transfer protocol or to a prior deliberation of the Information Security Committee (Section 4.2.3). In this regard, the SSIs must identify the data that are necessary and adequate for the data-matching purpose they pursue. Regarding the data-mining operations conducted in the OASIS database, Art 5*bis* of the Law of 15 January 1990 provides that 'all the necessary data for the purposes of applying the labour law and social security legislation' can be used. According to Degrave, this definition may be too broad as it does not allow citizens to know exactly which types of data are (or can be) processed (Degrave, 2020b). However, this concern is somewhat alleviated by the fact that access to data from the data warehouse must also be subject to a data transfer protocol or to a prior deliberation of the Information Security Committee, in which the necessary and proportionate nature of the accessed data will be controlled (Section 4.2.3).

Moreover, the data minimisation principle is enshrined in the fact that the data warehouse contains solely pseudonymised data and that it can only be accessed by a limited number of data miners/investigators. Importantly, the people who pseudonymise the data to be uploaded in the data warehouse are not the same as those who use the data warehouse in order to spot fraudulent patterns. In practice, the SSIs must draw up a list of the categories of persons with access to the personal data in the data warehouse, with a description of their role in relation to the data processing in question, and this list shall be kept at the disposal of the Data Protection Authority.[60] It is only once these data miners/investigators have identified a potential fraudulent case that the data at hand is de-pseudonymised, following a risk analysis, and extracted from the data warehouse, in order to start an investigation assessing whether there is indeed fraud.

4.2.5 The data subjects' right to information and big data

The right to information

Data has to be processed fairly and in a transparent manner.[61] This implies that citizens cannot be taken by surprise and that the public administration shall take appropriate measures to provide any information referred to in Articles 13 and 14 of the GDPR to the data subjects in a concise, transparent, intelligible and easily accessible form, using clear and plain language (Article 29 Working Party, 2018b).[62] If the public administration relies on big data to feed AI applications used to fight social security infringements and tax fraud, this means that the data subjects, whose data are collected, should notably receive information about:

[59] Art. 25 of the GDPR.
[60] Art. 5*bis*, al. 5 of the Law of 15 January 1990.
[61] Art. 5.1.a) of the GDPR.
[62] Art. 12.1 of the GDPR.

- The identity and contact details of the data controller and the contact details of its data protection officer;[63]
- The specific purposes of the processing for which the personal data are intended as well as the legal basis for the processing;[64]
- In cases where the data is acquired from a third party and thus not collected directly from the data subjects by the data controller, the categories of personal data that are processed and the source from which the personal data originate (notably whether it came from publicly accessible sources);[65]
- The recipients or categories of recipients of the personal data;[66]
- The period for which the personal data will be stored, or if that is not possible, the criteria used to determine that period;[67]
- The existence of the data subjects' rights and of the right to withdraw consent at any time if the processing is based on consent;[68] and
- The right to lodge a complaint with a supervisory authority.[69]

Moreover, in the specific case where the public administration intends to make use of this data for individual decision-making based 'solely' on automated processes, for instance to fight social security infringements and tax fraud, it will have to inform the data subject about the existence of such processes.[70]

Turning back to the examples of tax and social security fraud, citizens are generally informed about the existence of data-matching and data-mining operations through the laws mentioned in Section 4.2.3. Yet, according to some authors, these laws do not provide sufficiently clear information to the citizens, notably in terms of the concrete processing that will be conducted and in terms of the types of data that will be used (Degrave 2020b; Degrave and Lachapelle, 2014). To some extent, this lack of transparency is reduced by the fact that these concrete data processing will be subject to a DAM fiche, to an authorisation from the President of the Executive Committee of the FPS Finance, to a prior deliberation of the Information Security Committee or to the conclusion of a data transfer protocol, which will provide more specific information (Section 4.2.3). However, citizens do not have access to the DAM fiches or to the authorisations of the President of the Executive Committee. Moreover, while the deliberations of the Information Security Committee are published on the website of the CBSS,[71] it is hard to obtain information about a specific processing operation, as the search tool is quite basic. In a similar vein, while the

[63] Arts. 13.1.a), 13.1.b), 14.1.a) and 14.1.b) of the GDPR.
[64] Arts. 13.1.c) and 14.1.c) of the GDPR.
[65] Arts. 14.1.d) and 14.2.f) of the GDPR.
[66] Arts. 13.1.e) and 14.1.e) of the GDPR.
[67] Arts. 13.2.a) and 14.2.a) of the GDPR.
[68] Arts. 13.2.b), 13.2.c), 14.2.c) and 14.2.d) of the GDPR.
[69] Arts. 13.2.d) and 14.2.e) of the GDPR.
[70] Arts. 13.1.f) and 14.2.g) of the GDPR. For more details on this obligation, see Chapter 5, Section 5.2.2.
[71] https://www.ksz-bcss.fgov.be/fr/deliberations-csi-list?term_node_tid_depth=51.

data transfer protocols have to be published on the websites of the relevant data controllers,[72] the result is that they are published on a wide variety of websites, whose quality can vary greatly, making it almost impossible for citizens to have a good overview of the types of processing that are done with their data (Degrave, 2020b). For transparency purposes, it would be preferable to centralise the publication of all of these protocols and deliberations in a single source, such as the Data Protection Authority's website, where it should be possible to search through them on the basis of several criteria, such as the types of purposes or of data concerned (Degrave, 2020a). A good example of this is the city of Amsterdam's 'Algorithm register'.[73]

Restrictions to the right to information: the 'SyRI' case example

However, it must be pointed out that, according to Article 23.1 of the GDPR, the right to information, like any other data subject's right, can be restricted by a Member State law when such a restriction respects the essence of the fundamental rights and freedoms and is a necessary and proportionate measure in a democratic society to safeguard an important objective of general public interest (De Raedt, 2017; European Data Protection Board, 2020b; Scarcella, 2019). These objectives are listed in Article 23.1, and, in the context of big data collection to feed AI applications used to fight social security infringements and tax fraud, it could be resorted to 'an important economic or financial interest of the (...) Member State, including monetary, budgetary and taxation matters, public health and social security'.[74]

This provision is in line with Article 8.2 of the European Convention on Human Rights[75] and with Article 52.1 of the Charter of Fundamental Rights of the European Union.[76] Therefore, the interpretation of these provisions by the Court of Justice of the European Union and the European Court of Human Rights, respectively, is perfectly transposable to the interpretation of Article 23 of the GDPR. Accordingly, the restriction of data subjects' right must be provided by law, must respect the essence of the restricted fundamental rights and freedoms, must be necessary and proportionate in a democratic society, and must safeguard an important objective of general public interest (Article 29 Working Party, 2016; Tombal, 2018). For instance, in the context of the fight against tax fraud, the right to information, as well as the other data subject rights, can be delayed, limited or excluded, with regard to the processing of personal data for which the FPS Finances is the data controller, to guarantee public interest objectives in the budgetary, monetary

[72] Art. 20.3 of the Law of 30 July 2018.
[73] https://algoritmeregister.amsterdam.nl/en/ai-register/
[74] Art. 23.1.e) of the GDPR.
[75] 'There shall be no interference by a public authority with the exercise of this right except such as is in accordance with the law and is necessary in a democratic society in the interests of national security, public safety or the economic well-being of the country, for the prevention of disorder or crime, for the protection of health or morals, or for the protection of the rights and freedoms of others'.
[76] Charter of Fundamental Rights of the European Union, OJ [2012] C 326/391. Art. 52.1: 'Any limitation on the exercise of the rights and freedoms recognised by this Charter must be provided for by law and respect the essence of those rights and freedoms. Subject to the principle of proportionality, limitations may be made only if they are necessary and genuinely meet objectives of general interest recognised by the Union or the need to protect the rights and freedoms of others.

and fiscal field.[77] The goal is to prevent a citizen suspected of committing tax fraud from using this information in order to prejudice the investigation and to escape a sentence (Degrave and Lachapelle, 2014).[78]

In this regard, it is highly relevant to present the recent decision (5 February 2020) of the Rechtbank Den Haag,[79] on the compatibility, with Article 8 of the European Convention on Human Rights, of the Dutch government's 'Systeem Risico Indicatie ('SyRI' – system risk indication), which is a legal instrument used to detect various forms of fraud, including social benefits, allowances, and tax fraud. SyRI is a technical infrastructure in which data can be linked and analysed in a secure environment, in order to generate a risk report, which means that a legal or natural person is deemed worthy of investigating with regard to possible fraud, unlawful use and non-compliance with legislation.[80] The instrument is applied at the request of government bodies or other bodies with a public function, who decide to collaborate and exchange data.

The technique underlying the application of SyRI was first used between 2003 and 2013 without a legal basis.[81] On 1 January 2014, the SUWI Act[82] was amended in order to enshrine in law the application of SyRI, and the conditions for the application of SyRI are detailed in the SUWI Decree.[83,84] More precisely, the legal basis for the processing of the necessary information in SyRI for the purpose of carrying out risk analyses is Section 65 of the SUWI Act,[85] and Article 5a.1, §3 of the SUWI Decree lists the categories of data that qualify for processing in SyRI, such as work data, tax data, social assistance benefit data, pension data, etc.[86]

The legality of the SyRI instrument has been challenged, in the Court of The Hague, by a coalition of civil society interest groups and two natural persons.[87] In substance, they claimed that the SyRI instrument breached Article 8 of the European Convention on Human Rights because it constituted an infringement of people's privacy, and that the SyRI legislation did not provide

---

[77] Arts. 11.1, al.1; 11/1.1, al.1; 11/2.1, al.1; and 11/3.1, al.1 of the Law of 3 August 2012.

[78] Projet de loi portant dispositions relatives aux traitements de données à caractère personnel réalisés par le Service public fédéral Finances dans le cadre de ses missions, 6 juillet 2012, *Doc. parl.*, Chambre, sess. ord., 2011-2012, no 53-2343/001, p. 11.

[79] Rechtbank Den Haag, 5 februari 2020, Zaak n° C-09-550982-HA ZA 18-388, ECLI:NL:RBDHA:2020:865 (ECLI:NL:RBDHA:2020:1878 for the English version).

[80] *Ibid.*, points 3.1 and 3.2.

[81] *Ibid.*, points 3.5 to 3.10.

[82] Wet van 9 oktober 2013 tot wijziging van de Wet structuur uitvoeringsorganisatie werk en inkomen en enige andere wetten in verband met fraudeaanpak door gegevensuitwisselingen en het effectief gebruik van binnen de overheid bekende zijnde gegevens, *Stb.*, 2013, p. 405.

[83] Besluit van 1 september 2014 tot wijziging van het Besluit SUWI in verband met regels voor fraudeaanpak door gegevensuitwisselingen en het effectief gebruik van binnen de overheid bekend zijnde gegevens met inzet van SyRI, *Stb.*, 2014, p. 320.

[84] Rechtbank Den Haag, 5 februari 2020, Zaak n° C-09-550982-HA ZA 18-388, points 4.1 to 4.30.

[85] *Ibid.*, point 4.8.

[86] *Ibid.*, point 4.17.

[87] *Ibid.*, point 2.1.

sufficient safeguards.[88] The Court thus assessed whether the SyRI legislation complies with Article 8.2 of the European Convention on Human Rights. This particular provision requires striking a fair balance between the interests of the community as a whole, which the legislation serves, and the right to privacy of the individuals affected by the legislation.[89]

In this regard, the Court outlined that the State has a special responsibility when applying new technologies, and that it must strike the right balance between the benefits that technologies such as SyRI bring, and the violation of the right to a private life that they might entail.[90] In the case at hand, the Court concluded that, in its current form, the SyRI legislation failed to comply with Article 8.2 of the European Convention on Human Rights, because the legislation does not strike the 'fair balance' required under the ECHR between the social interest the legislation serves and the violation of private life to which the legislation gives rise.[91] Indeed, taking into account the GDPR principles of transparency, purpose limitation and data minimisation, the Court ruled that the application of the SyRI legislation was especially deemed to be unlawful because it was insufficiently transparent and verifiable.[92]

Without entering into too much detail here, we will simply outline that the following circumstances were deemed relevant by the Court in reaching this conclusion: (1) the principle of transparency is insufficiently respected in the SyRI legislation because it does not provide information on which objective factual data can justifiably lead to the conclusion that there is an increased risk, but only provides a few examples of indicators that can indicate an increased risk and a potential hit;[93] (2) the SyRI legislation does not provide information on the functioning of the risk model, for instance the type of algorithms used in the model, nor does it provide information on the risk analysis method as applied by the inspection services;[94] (3) a data subject whose data were processed in SyRI, but which did not result in a risk report, will not be informed about this processing, and therefore cannot verify that her data was processed on correct grounds;[95] (4) because it is impossible to verify how the risk report has been generated and which steps it comprises, a data subject will not be able to defend herself against the fact that a risk report has been submitted about her;[96] and (5) because large amounts of data qualify for processing in SyRI and because, in a concrete SyRI project, the test of necessity is carried out by the separate participants in the project, with no independent assessment, the SyRI legislation therefore contains insufficient safeguards and breaches the principles of purpose limitation and data minimisation.[97]

---

[88] *Ibid.*, point 6.1.
[89] *Ibid.*, executive summary.
[90] *Ibid.*, point 6.6. See also ECtHR, *S. and Marper v. the United Kingdom*, 4 December 2008, req. n°s 30562/04 and 30566/04, §112.
[91] Rechtbank Den Haag, 5 februari 2020, Zaak n° C-09-550982-HA ZA 18-388, point 6.7.
[92] *Ibidem.*
[93] *Ibid.*, point 6.87.
[94] *Ibid.*, point 6.89.
[95] *Ibid.*, point 6.90.
[96] *Ibidem.*
[97] *Ibid.*, point 6.106.

4.2.6 The data subjects' right not to be subject to automated individual decision-making and big data

According to Article 22.1 of the GDPR, the data subject shall have the right not to be subject to a decision based solely on automated processing, including profiling, which produces legal effects concerning her or similarly significantly affects her (Article 29 Working Party, 2018a). We will further address this right in our chapter dedicated to AI,[98] but we wish to point out here that, according to Rouvroy, achieving Article 22's aims in a big data world is 'both unrealistic and deeply paradoxical (...) especially when [big data analytics] involve self-learning algorithms' (Rouvroy, 2016). Indeed, as pointed out by Zarsky, there are several tensions between Article 22 of the GDPR and big data analytics:

> First, prohibiting automated analysis obviously undermines many of the big data practices (...). Second, even if one of the many exceptions to the prohibition on automation (...) is met, the specific disclosures (...) which call for enabling a human response to the machines' decisions are still required. To meet these disclosure obligations, big data processes must be conducted in a manner that would assure they are interpretable – i.e. they can be explained to the inquiring individual. Constantly meeting an 'interpretability' requirement might call upon those designing the automated processes to compromise some of the system's precision to enable the delivery of this form of detailed disclosure. Third, allowing human interjection would further encumber the automated process and slow down the innovative technologies they bring about (Zarsky, 2017).

4.3 Data storage – personal data protection and blockchain

Blockchain technology is a distributed ledger technology (DLT) 'that makes it possible to avoid the use of a trusted third party for transactions, and which is notably at the basis of Bitcoin' (Villani *et al.*, 2018). As pointed out by Finck, 'a distributed ledger can be described as a shared and synchronised digital database that is maintained by a consensus algorithm and stored on multiple nodes (computers)' (Finck, 2017). From a technical point of view, blockchains 'only designate the variants of DLT that record data in packages ('blocks') that are hashed ('chained') to another' (Finck, 2017). As Finck outlines, 'data is usually grouped into blocks that, upon reaching a certain size, are chained to the existing ledger through a hashing process. Through this process, data is chronologically ordered in a manner that makes it difficult to tamper with information without altering subsequent blocks' (Finck, 2017). Concretely, digital ledger technologies, such as blockchains 'rely on a two-step verification process with asymmetric encryption. Every user has a public key (...) that is shared with others to enable transactions. In addition, each user holds a private key (...) that must never be shared with others. Both keys have a mathematical relationship by virtue of which the private key can decrypt data that is encrypted through the public key. Public keys thus hide the identity of the individual unless they are linked to additional identifiers.

[98] See Chapter 5, Section 5.2.5.

The nodes are the computers on which the ledger is stored. (...) In public and permission-less blockchains, anyone can entertain a node by downloading and running the relevant software. Some (but not all) nodes also function as 'miners', which aggregate transactions into candidate blocks and hash a new block to the chain on the basis of a predetermined consensus protocol (such as proof-of-work or proof-of-stake)' (Finck, 2017).

Similarly to big data and AI, blockchain, as a new technology, creates challenges in terms of compatibility with the GDPR. This is notable because the GDPR was built for a world where data is collected, stored and processed centrally, whereas blockchain technology decentralises each of these processes (Finck, 2017; Lyons *et al.*, 2018). Accordingly, some types of blockchains (notably those that are public and permissionless) may be in tension with the GDPR as the decentralised method of data storage and protection of blockchains cannot be easily reconciled with data protection mechanisms developed for centralised data silos (Finck, 2017). However, 'GDPR compliance is not about the technology, it is about how the technology is used' (Lyons *et al.*, 2018). Therefore, a creative legal interpretation of the GDPR and the implementation of technical solutions could reconcile some of these tensions (Finck, 2017; Lyons *et al.*, 2018).

Before diving into these potential tensions between blockchain technology and personal data protection, a difference must be made between two types of blockchains, namely public and permissionless blockchains on the one hand, and private and permissioned blockchains on the other. Public and permissionless blockchains are open-source and open-access, which means that anyone can download or design software to run nodes (Finck, 2017). Private and permissioned blockchains run on a private network (an intranet or a VPN) and a person must have been granted permission by an administrator in order to maintain a node (Finck, 2017). To take an example, the European Blockchain Services Infrastructure (EBSI) will be a public but permissioned network, as only pre-validated participating nodes will have writing, storing, processing and transmitting of (personal and non-personal) data on the ledger (European Commission, 2020). For those types of blockchains, it will be easier to comply with the GDPR (Lyons *et al.*, 2018). Public and permissionless blockchains, on the other hand, are the ones that create the most issues from a personal data protection perspective, because of their extremely distributed nature (Lyons *et al.*, 2018). Therefore, those are the ones we will focus on in the remainder of this section.

4.3.1 Two preliminary questions

Two preliminary questions must first be outlined, namely whether data stored on the blockchain are personal data and, if this is the case, who should be responsible for the application of the GDPR.

Two sets of data could potentially be considered as personal data, namely the data stored in the blocks themselves (we will refer to these as the 'block data'), and the public keys that are used to link a specific operation to an individual (Finck, 2017). Regarding the block data, personal data can either be stored in plain text, in an encrypted form or hashed (Finck, 2017). For the plain text data, it is obvious that this remains personal data. For the encrypted and hashed data, the question is whether this constitutes pseudonymised data,[99] to which the GDPR applies, or anonymised data,[100] to which the GDPR does not apply. The difference between the two concepts is that it is impossible to link back to the data subject when their data has been anonymised, while this is possible with pseudonymised data. Yet, when data is encrypted, the data subject can still be reidentified through the use of the right encryption keys, so it must be considered as pseudonymised data subject to the GDPR (Finck, 2017; Lyons *et al.*, 2018). The same goes for hashed data (Article 29 Working Party, 2014; Lyons *et al.*, 2018), though some authors argue that this is debatable, and that it will, in fact, depend on the circumstances of the case and on the hashing technique that has been used (Lyons *et al.*, 2018). In any case, truly effective anonymisation is difficult to achieve at present (Franceschi *et al.*, 2018; Graef *et al.*, 2018; Wendehorst, 2017).

That being said, advanced cryptographic (such as zero-knowledge proofs) and data aggregation techniques could potentially lead to robust anonymisation in the future (Lyons *et al.*, 2018). For blockchain technology, this could notably be the case if 'personal data could be stored off-chain and merely linked to the blockchain through a hash pointer. In such a scenario, personal data is recorded in a referenced encrypted and modifiable database and not on the blockchain. Under this formula, no personal data is stored on-chain' (Finck, 2017). Storing personal data off-chain could indeed be an interesting avenue to explore, especially in light of the tensions that will be presented below, as a number of them could be solved through storing the personal data off-chain. However, this is an attempt to design GDPR compliant blockchains that limits the role of the chain, as it merely holds proof that the data, which is held on a private storage, is valid (Finck, 2017).

Regarding the public keys, as they are used for the pseudonymous identification of an individual's operations on the blockchain, they shall also be considered as personal data, as this individual can be reidentified by matching additional information to the key (Finck, 2017). Contrary to the block data, the public keys could not be stored off-chain, as they are necessary to the chain's functioning (Finck, 2017).

Since the block data and the public keys can be considered as personal data, this raises the question of who should be the data controller, i.e. who is accountable for the application of the GDPR. In the GDPR, the data controller is defined as 'the natural or legal person, public authority, agency or other body which, alone or jointly with others, determines the purposes and means of the processing of personal data'.[101] For private and permissioned blockchains, it might not be too complicated to identify a central entity that can qualify as the controller, for example the

[99] Pseudonymisation means 'the processing of personal data in such a manner that the personal data can no longer be attributed to a specific data subject without the use of additional information, provided that such additional information is kept separately and is subject to technical and organisational measures to ensure that the personal data are not attributed to an identified or identifiable natural person' (Art. 4.5 of the GDPR).
[100] The ISO 29100 standard defines anonymisation as the 'process by which personally identifiable information (PII) is irreversibly altered in such a way that a PII principal can no longer be identified directly or indirectly, either by the PII controller alone or in collaboration with any other party' (ISO 29100:2011, point 2.2, available at: https://www.iso.org/obp/ui/#iso:std:iso-iec:29100:ed-1:v1:en).
[101] Art. 4.7 of the GDPR.

administrator that grants permissions to maintain a node (Finck, 2017; Lyons *et al.*, 2018). Public and permissionless blockchains, on the other hand, are a decentralised peer-to-peer network of nodes, and consequently, this leads to uncertainties as to who should be considered as the data controller(s) (Lyons *et al.*, 2018). According to Finck, either none of the nodes can be defined as *the* data controller, as none of them determines the purposes and means of the processing, or, more likely, *every* node qualifies as a data controller, which means that each of them would be accountable for the application of the GDPR and would have to be able to answer to a data subject's request regarding one of its rights (Finck, 2017). The same conclusion is reached by the European Commission in its EBSI assessment, where it indicates that all the nodes should be qualified as 'joint data controllers for the transactional data that they to verify, store, and put on/off chain' (European Commission, 2020). Moreover, these decentralised models might make it difficult to draw clear boundaries between data controllers and mere data processors (European Commission, 2020). This clearly shows that the GDPR was built for a world where data is collected, stored and processed centrally, which clashes with the functioning of decentralised blockchains. This is especially problematic because 'nodes: (1) only see the encrypted or hashed version of the data; and (2) are unable to make any changes thereto. Nodes are thus decentralised entities that cannot respond to the tasks the GDPR requires of centralised agents' (Finck, 2017).

Another issue derives from the GDPR's territorial scope, as decentralised blockchains rely on nodes that are spread out across the world (Finck, 2017). Yet, the GDPR's scope of application is relatively wide, as it applies to the processing of personal data in the context of the activities of an establishment of a controller or a processor in the Union, regardless of whether the processing takes place in the Union or not;[102] and to the processing of personal data of data subjects who are in the Union by a controller or processor not established in the Union, where the processing activities are related to the offering of goods or services to these data subjects, or to the monitoring of their behaviour.[103] In light of this wide territorial scope of application, the GDPR obligations may be binding for a lot of blockchain-based applications throughout the world, even if they only have an indirect link with the EU (Finck, 2017).

4.3.2 Tensions between blockchain technology and personal data protection

We will now turn to the potential tensions between blockchain technology and personal data protection. In this regard, we should clarify from the outset that our analysis starts from the assumption that the data controller uses the blockchain for a specified, explicit and legitimate purpose,[104] relies on a lawful basis of processing to do so (Lyons *et al.*, 2018),[105] and that the data subject has been informed about this.[106] Accordingly, we start from the assumption that the blockchain is simply a technical means to achieving that purpose.

[102] Art. 3.1 of the GDPR.
[103] Art. 3.2 of the GDPR.
[104] Art. 5.1.b) of the GDPR. See Section 4.2.3. Purpose limitation principle and big data.
[105] Art. 6.1 of the GDPR. See Section 4.2.1. Lawfulness and fairness of processing and big data.
[106] Arts. 12 to 14 of the GDPR. See Section 4.2.5. The data subjects' right to information and big data.

Data minimisation and storage limitation principle

According to the data minimisation principle, only the adequate, relevant and necessary data for the fulfilment of the specific purpose of processing shall be processed.[107] Yet, integral copies of the chain are stored on each node, which multiplies, rather than minimises, the personal data that is used (Finck, 2017). Additionality, in light of the storage limitation principle, data shall be kept in a form which permits identification of data subjects for no longer than is necessary for the purposes for which the personal data are processed.[108] This principle is inherently at odds with blockchain technology, as personal data that has been added to the chain will perpetually remain part of it, as the chain is an append-only database that continuously grows and expands (Finck, 2017; Lyons *et al.*, 2018). Because data will never be deleted from the blockchain, the chain might contain data that are no longer adequate, relevant and necessary for the purpose of processing for which they were originally processed, and might contain data that is stored for longer than necessary, thus contradicting these two key principles of the GDPR. It should however be outlined that, in its EBSI assessment, the European Commission indicated that 'it can be argued that the existence of public keys on blockchains, combined with necessary privacy enhancing mechanisms (PEMs), will fulfil the data minimisation requirements of the GDPR' (European Commission, 2020). The efficiency of these PEMs however remains to be demonstrated (Lyons *et al.*, 2018).

One way to circumvent this tension would be to store the block data off-chain, as, in this case, the chain simply contains a link, via a hash pointer, towards an off-chain database, where the personal data can more easily be minimised or deleted, in full compliance with the GDPR (Finck, 2017). Indeed, if the data is deleted off-chain, the link will remain on the chain, but it will point to nothing. However, whether this will actually work in practice remains to be seen.

Data accuracy principle and right to rectification

According to the data accuracy principle, the data controller must ensure that the personal data that it processes is accurate and that it is kept up to date.[109] This principle is complemented by the right to rectification, which grants the data subject the right to obtain, without undue delay, the rectification of inaccurate personal data concerning them, and the right to have incomplete personal data completed.[110] Once again, these principles are at odds with blockchain technology, in light of the latter's immutability (Finck, 2017; Lyons *et al.*, 2018). Regarding the right to have incomplete data completed, the GDPR does outline that this could be done by 'providing a supplementary statement' (Finck, 2017). Adding a new block with supplementary information might thus potentially satisfy this requirement, though the former block containing the incomplete

---

[107] Art. 5.1.c) of the GDPR. See Section 4.2.4. Data minimisation principle and big data.
[108] Art. 5.1.e) of the GDPR.
[109] Art. 5.1.d) of the GDPR.
[110] Art. 16 of the GDPR.

information will remain. If the request is to rectify data rather than to complete it, then adding a new block might not be sufficient (Finck, 2017). Alternatively, the block data could be stored off-chain, in order to facilitate its rectification and update.

Right of access

The data subjects' right of access provides that the data subject has the right to obtain, from the controller, the confirmation as to whether or not it processes personal data concerning him or her.[111] If it is the case, the controller will have to provide access to the data, as well as to the information listed in points (a) to (h) of Article 15.1 of the GDPR.[112] Moreover, the right of access provides that the data subject has a right to obtain a copy of the personal data that is processed by the controller.[113] In the context of blockchain technology, this right might be extremely difficult to implement for the various nodes of the chain, as they will not know exactly which data is stored on the chain because they will likely only have access to encrypted or hashed data, and will thus not be able to tell a specific data subject whether their data is being processed in the chain or to provide them with a copy of her data (Finck, 2017; Lyons *et al.*, 2018). Nevertheless, if the block data is stored off-chain, the controller can more easily identify if data concerning the data subject is processed, and provide them with a copy of the said data.

Right to erasure

Article 17 of the GDPR stipulates that the data subject shall have the right to obtain from the controller the erasure of personal data concerning them, without undue delay, if: '(a) the personal data are no longer necessary in relation to the purposes for which they were collected or otherwise processed; (b) the data subject withdraws consent on which the processing is based (...) and where there is no other legal ground for the processing; (c) the data subject objects to the processing (...); (d) the personal data have been unlawfully processed; (e) the personal data have to be erased for compliance with a legal obligation in Union or Member State law to which the controller is subject; (f) the personal data have been collected in relation to the offer of information society services [to a child]'.[114] Once again, this right is at odds with blockchain technology, in light of the latter's immutability (Finck, 2017; Lyons *et al.*, 2018). By definition, it is almost impossible to delete the data from the chain. Nevertheless, as pointed out by Finck, 'the precise meaning of 'erasure' is not defined in the GDPR, opening the door to other interpretations than absolute deletion' (Finck, 2017). For instance, the French CNIL (CNIL, 2018), indicates that the combination of encryption techniques and key destruction could potentially be considered as an erasure (Lyons *et al.*, 2018). Moreover, the German legislator accepts that, if the specific technical mean of storage makes it impossible to delete

[111] Art. 15 of the GDPR.

[112] For example: the purposes of the processing, the categories of personal data concerned, the recipients to whom the data will be or has been disclosed, the storage period, etc.

[113] Art. 15.3 of the GDPR.

[114] Art. 17.1 of the GDPR.

the data, erasure can be achieved through other means, such as limiting the processing that are tolerated,[115] e.g. the data can only remain stored but cannot be used, and this example might open the door to an interpretation of 'erasure' that accounts for the blockchain's immutability (Finck, 2017). However, we do not find this alternative compelling, as the GDPR provides for a distinct right to the restriction of the processing, which precisely aims at limiting the processing that can be performed on the data to its sole storage.[116] Accordingly, if the GDPR was meant to make erasure possible through restriction rather than through deletion, even if only in specific cases, two distinct rights would probably not have been created. That being said, data erasure could be facilitated by storing the block data off-chain.

Data protection by design and by default

According to the data protection by design principle, the controller will have to implement appropriate technical and organisational measures, both at the time of the determination of the means for processing and at the time of the processing itself, which are designed to implement data protection principles in an effective manner and to integrate the necessary safeguards into the processing.[117] For instance, pseudonymisation could be used in order to ensure the respect of the data minimisation principle.[118] Here, the encryption and hashing of the block data before including it in the chain might actually be considered as a form of privacy by design, and the objectives of the GDPR and of blockchain technology are thus aligned in this regard (Finck, 2017; Lyons *et al.*, 2018). This will especially be the case if the data is stored off-chain.

According to the data protection by default principle, the controller shall implement appropriate technical and organisational measures in order to ensure that, by default, only personal data which are necessary for each specific purpose of the processing are processed.[119] This echoes the data minimisation principle that has been presented above. By default, the amount of personal data collected, the extent of its processing, the period of its storage and accessibility should be reduced to the minimum necessary for the purpose of processing.[120] In terms of accessibility, this means that, by default, the personal data should only be made accessible to a limited number of people.[121] Yet, as mentioned above, in decentralised blockchains, integral copies of the chain are stored on each node, which multiplies, rather than minimises, the number of people that have access to the data (Finck, 2017). Additionality, because data will never be deleted from the blockchain, the chain might contain data that are no longer adequate, relevant and necessary for the purpose of processing for which they were originally processed (Finck, 2017).

---

[115] Article 35 of the Gesetz zur Anpassung des Datenschutzrechts an die Verordnung (EU) 2016/679 und zur Umsetzung der Richtlinie (EU) 2016/680.
[116] Art. 18 of the GDPR.
[117] Art. 25.1 of the GDPR.
[118] Art. 25.1 of the GDPR.
[119] Art. 25.2 of the GDPR.
[120] Art. 25.2 of the GDPR.
[121] Art. 25.2 of the GDPR.

As outlined above, a way to address this issue would be to store the block data off-chain, where the personal data can more easily be minimised and where the access can be limited to specific people. Another solution would be to apply PETs, such as zero knowledge proofs (European Commission, 2020). The efficiency in practice of these PETs is however uncertain at the moment (Lyons *et al.*, 2018).

Transborder data flows

Finally, since the nodes of a chain are located all around the world, and each of them contains a full copy of the chain, the rules pertaining to the transfer of personal data to 'third countries', namely countries outside of the EU and the European Economic Space (EES), must be kept in mind (Finck, 2017; Lyons *et al.*, 2018). As a matter of principle, data can only be transferred to a third country if the European Commission has decided that it ensures an adequate level of protection.[122] As this list is quite limited, some of the nodes will very likely be located in third countries that are not deemed to grant adequate protection. In such cases, Article 46.1 of the GDPR provides that a controller may transfer personal data to a third country if it has provided appropriate safeguards, and on condition that enforceable data subject rights and effective legal remedies for data subjects are available. This could notably be done through binding corporate rules or a code of conduct.[123] In theory, the nodes could be requested to adhere to one of these mechanisms in order to join the blockchain, but in light of the other tensions that we have highlighted, such mechanisms may, in fact, not be deemed to grant appropriate safeguards (Finck, 2017). A final solution would be to rely on the data subject's explicit consent for such a transfer,[124] but this consent may be very hard to get in practice (Finck, 2017).

4.3.3 Use of blockchain, by the public administration, to fight social security and tax fraud

In light of the above, it stems that the use, by public administration, of blockchain technology to fight social security infringements and tax fraud raises numerous challenges and potential tensions. The starting point for the public administration should be to determine whether they actually need blockchain technology, or whether other technical solutions could be used, as it should not be assumed that using blockchain will automatically be more secure and cheaper (Lyons *et al.*, 2018).

If a public administration wishes to rely on blockchain technology, it is recommended that they use a private and permissioned blockchain with a clearly identified controller, who will grant authorisations to run nodes of the blockchain, in order to limit the number of people that can

---

[122] Art. 45.1 of the GDPR. For a list of these countries, see https://ec.europa.eu/info/law/law-topic/data-protection/international-dimension-data-protection/adequacy-decisions_en.
[123] Respectively arts. 46.2.b) and e) of the GDPR.
[124] Art. 49.1.a) of the GDPR.

get access to the block data (Lyons *et al.*, 2018). In this regard, authorisations should be solely granted to bodies, whether public or private, such as SMALS,[125] that have been authorised by a law to process such data in order to fight social security infringements and tax fraud (Section 4.2.1).

Moreover, the blockchain should, ideally, be designed in such a way that the block data, which qualifies as personal data,[126] should be encrypted, aggregated and stored off-chain (Lyons *et al.*, 2018). In this regard, 'it could be argued that blockchain networks should be used to store immutable proofs that certain data exists, rather than to store the data itself' (Lyons *et al.*, 2018). As we have seen above, this would facilitate the compliance with the data minimisation, storage limitation, data accuracy, and data protection by design and by default principles, and would facilitate the exercise of the data subjects' rights of rectification, access and erasure, all of which may not be possible to do if the block data was directly stored on the chain. However, it is uncertain whether this will actually be workable in practice.

Finally, it should be outlined that, depending on the circumstances of the case, blockchain technology might enhance transparency about the processing done by the data controllers, which are stored on the chain. Indeed, blocks in the chain contain not only 'block data' pertaining to the different processing that occurred (e.g. which data has been processed), but also headers that outline the identity of the data source and data recipient, as well as a timestamp (Finck, 2017). Accordingly, depending on how the blockchain is built (e.g. whether the data subject can have access to the block data and whether the data controller can link the block data to a specific data subject), the data subject could check the chain in order to obtain information about whether data concerning them has been processed, but also about the entities between which their data has been transferred, and at what moment this occurred. Thus, if information about the processing of citizens' data in order to fight social security infringement and tax fraud were to be stored on a blockchain, the citizen could check, depending on how the blockchain is built, which of their data has been processed in this regard, among which public administration the data has circulated and at what moment. Naturally, in light of the sensitive nature of the data at hand, and in light of the data protection by default principle, natural or legal persons other than the data subject and the relevant public administration should not be able to access the chain (private and permissioned blockchain).

An interesting example to mention in this regard is Estonia's use of blockchain technology to secure its residents' health records (Einaste, 2018).[127] Indeed, Estonia has set up a blockchain that stores all the log files that record the data-processing activities performed on its residents' health records (which are themselves not stored on the blockchain) (Einaste, 2018). Each access or each change to a patient's electronic records is recorded and timestamped, and this can be checked by the patients (Einaste, 2018).

---

[125] SMALS is a Belgian non-profit organisation whose objective is to support the Belgian Federal public administration regarding information management and related issues for integrated IT service provision (https://www.smals.be/fr). It is one of the Belgian nodes of the European Blockchain Services Infrastructure (EBSI) (see https://ebsi4be.eu/).

[126] See Section 4.3.1.

[127] See https://e-estonia.com/blockchain-healthcare-estonian-experience/.

4.4 Conclusions

As outlined throughout this chapter, public administrations are increasingly relying on new technologies, such as big data and blockchain, which increase their capability to process greater amounts of data, in order to provide public services and to support their decision-making. While this can lead to significant benefits, notably in terms of efficiency, for these administrations, it must not be overlooked that the use of such technologies could also have significant impacts on the lives of their citizens. Therefore, administrations need to comply with the legal framework that aims at limiting the uses they can make of such technologies, in order to circumscribe these impacts. In this regard, the aim of this chapter was precisely to outline how this could be done in practice.

In terms of key takeaways pertaining to the use of big data by public administrations, it should be remembered that the lawful basis underlying big data processing should ideally be a law.[128] Importantly, this law needs to meet certain standards in order to be sufficiently clear and predictable. Indeed, as a rule of thumb, any personal data processing must be fair, which implies that the laws on which this processing is based must be sufficiently explicit and understandable for citizens. Furthermore, the purpose of processing must be defined prior to the collection of the data, and data that has been collected for a specific purpose cannot be further processed for a purpose that does not fit within this initial purpose.[129] This requirement can be at odds with big data analytics if a large number of data are collected 'for the sake of it', without a clearly defined purpose. Accordingly, if public administrations wish to rely on big data analytics, they will need to clearly define the purposes of processing in advance. Linked to this is the data minimisation requirement, according to which only the adequate, relevant and necessary data for the fulfilment of the specific purpose of processing shall be processed.[130] This can also be at odds with big data analytics that rely on the processing of data coming from the largest possible number of sources, independently of their relevance. Therefore, public administrations willing to rely on big data analytics will also need to clarify, in advance, which data they will use and why these data are necessary to achieve the desired purpose. Finally, public administration relying on big data analytics to process their citizen's personal data must respect the latter's data subject rights, such as the right to information and the right not to be subject to automated individual decision-making.[131]

Regarding the use of blockchain, several key takeaways can also be formulated for public administration.[132] First, it is advised to use a private and permissioned blockchain with a clearly identified controller, who will grant authorisations to run nodes of the blockchain, in order to limit the number of people that can get access to the block data. Moreover, such authorisations should only be granted to bodies that have been authorised by a law to process such data. Second,

[128] See Section 4.2.1.
[129] See Section 4.2.3.
[130] See Section 4.2.4.
[131] See Sections 4.2.5 and 4.2.6.
[132] See Section 4.3.3.

the blockchain should, ideally, be designed in such a way that the personal data from the blocks are encrypted, aggregated and stored off-chain. Third, the blockchain should be designed in a way that enhances transparency about the processing done by the data controllers on the data stored on the chain. In this regard, the data subject should be able to check the chain in order to obtain information about whether data concerning them has been processed, but also about the entities between which their data has been transferred, and at what moment this occurred.

Finally, it must be outlined that while the analysis pertaining to big data has focussed on data collection and combination, and the analysis pertaining to blockchain has focussed on data storage, this does not mean that these technologies do not have an impact on the other stages of the data processing lifecycle. Indeed, these technologies are intertwined and influence each other. For instance, a blockchain based system can enable the sharing of personal data between public administration in order to complement other sources of data, which would allow these administrations to conduct big data analytics on a greater scale and scope of data. Furthermore, big data and blockchain technologies can contribute to the development of artificial intelligence techniques, by increasing the amount of available data on which the AI algorithms can be trained. Naturally, the use of artificial intelligence technologies by public administration must also comply with the rules of personal data protection, as will be analysed in Chapter 5.

References

Anonymous, 2021a. European Commission gives Belgium two months to restore independence to ODA. Available at: https://www.archyde.com/european-commission-gives-belgium-two-months-to-restore-independence-to-oda/.

Anonymous, 2021b. Mathieu Michel invites the federal parliament to assume its responsibilities in the APD file: 'A highly problematic situation'. Available at: https://www.archyde.com/mathieu-michel-invites-the-federal-parliament-to-assume-its-responsibilities-in-the-apd-file-a-highly-problematic-situation/.

Article 29 Working Party, 2010. Opinion 3/2010 on the principle of accountability. WP 173. Available at: https://ec.europa.eu/justice/article-29/documentation/opinion-recommendation/files/2010/wp173_en.pdf.

Article 29 Working Party, 2013. Opinion 03/2013 on purpose limitation. Available at: https://ec.europa.eu/justice/article-29/documentation/opinion-recommendation/files/2013/wp203_en.pdf.

Article 29 Working Party, 2014. Opinion 05/2014 on Anonymisation Techniques. Available at: https://ec.europa.eu/justice/article-29/documentation/opinion-recommendation/files/2014/wp216_en.pdf.

Article 29 Working Party, 2015. Annex – Health data an apps and devices (annex to a letter to the European Commission on the scope of health data in relation to lifestyle and well-being apps). Available at: https://ec.europa.eu/justice/article-29/documentation/other-document/files/2015/20150205_letter_art29wp_ec_health_data_after_plenary_annex_en.pdf.

Article 29 Working Party, 2016. Working Document 01/2016 on the justification of interferences with the fundamental rights to privacy and data protection through surveillance measures when transferring personal data (European Essential Guarantees). Available at: https://ec.europa.eu/justice/article-29/documentation/opinion-recommendation/files/2016/wp237_en.pdf.

Article 29 Working Party, 2018a. Guidelines on Automated individual decision-making and Profiling for the purposes of Regulation 2016/679. Available at: https://ec.europa.eu/newsroom/article29/item-detail.cfm?item_id=612053.

Article 29 Working Party, 2018b. Guidelines on transparency under Regulation 2016/679. Available at: http://ec.europa.eu/newsroom/article29/item-detail.cfm?item_id=622227.

Belgian Privacy Commission, 2018. Opinion n° 34/2018. Available at: https://www.autoriteprotectiondonnees.be/publications/avis-n-34-2018.pdf.

Chantillon, M., R. Kruk, A. Simonofski, T. Tombal, J. Crompvoets, C. de Terwangne, N. Habra, M. Snoeck, and B. Vanderose. 2017. FLEXPUB Public e-Service Strategy – Work package 2 – Baseline Measurement. KU Leuven Public Governance Institute, Leuven, Belgium.

CNIL, 2018. Blockchain et RGPD: quelles solutions pour un usage responsable en présence de données personnelles? Available at: https://www.cnil.fr/fr/blockchain-et-rgpd-quelles-solutions-pour-un-usage-responsable-en-presence-de-donnees-personnelles.

De Raedt, S. 2017. The impact of the GDPR for tax authorities. *Revue du droit des technologies de l'information* 66-67: 129-143.

Degrave, E. 2015. La protection de la vie privée dans l'e-gouvernement. In C. De Terwangne, and E. Degrave (eds.) *Vie privée et données à caractère personnel*. 963-1003. Politeia, Brussels, Belgium.

Degrave, E. 2020a. Le R.G.P.D., les lois belges et le secteur public: Les traitements de données dans l'administration en réseaux et l'Autorité de protection des données. *Anthémis* 195: 281-317.

Degrave, E. 2020b. The use of secret algorithms to combat social fraud in Belgium. *European Review of Digital Administration & Law* 1: 167-177.

Degrave, E. and A. Lachapelle. 2014. Le droit d'accès du contribuable à ses données à caractère personnel et la lutte contre la fraude fiscal. *Revue Générale du Contentieux Fiscal* 28: 322-335.

Einaste, T. 2018. Blockchain and healthcare: the Estonian experience. Available at: https://nortal.com/blog/blockchain-healthcare-estonia/.

European Commission, 2020. EBSI GDPR Assessment: Report on data protection within the EBSI version 1.0 infrastructure. Available at: https://ec.europa.eu/cefdigital/wiki/display/CEFDIGITALEBSI/Legal+Assessment+Reports.

European Commission, 2021. June infringements package: key decisions. Available at https://ec.europa.eu/commission/presscorner/detail/en/inf_21_2743.

European Data Protection Board, 2020a. Guidelines 05/2020 on consent under Regulation 2016/679 (V.1.1). Available at: https://edpb.europa.eu/sites/edpb/files/files/file1/edpb_guidelines_202005_consent_en.pdf.

European Data Protection Board, 2020b. Guidelines 10/2020 on restrictions under Article 23 GDPR, Version 1.0. Available at: https://edpb.europa.eu/sites/default/files/consultation/edpb_guidelines_202010_article23_en.pdf.

Finck, M. 2017. Blockchains and data protection in the European Union. Max Planck Institute for Innovation & Competition Research Paper No. 18-01. https://doi.org/10.2139/ssrn.3080322

Franceschi, A., R. Schulze, M. Graziadei, O. Pollicino, F. Riente, S. Sica, and P. Sirena. 2018. Legal challenges of the changing role of personal and non-personal data in the data economy. In: De Franceschi, A. and Schulze, R. (eds.) *Digital revolution: data protection, smart products, blockchain technology and bitcoins challenges for law in practice*. 19-41. C.H. Beck Verlag, München, Germany. https://doi.org/10.17104/9783406759048-19

Graef, I., R. Gellert, and M. Husovec. 2018. Towards a holistic regulatory approach for the european data economy: why the illusive notion of non-personal data is counterproductive to data innovation. TILEC Discussion Paper No. 2018-028. https://doi.org/10.2139/ssrn.3256189

Hahn, I. 2021. Purpose limitation in the time of data power: is there a way forward? *European Data Protection Law Review* 7: 31-44.

High-Level Expert Group on Business-to-Government Data Sharing, 2020. Towards a European strategy on business-to-government data sharing for the public interest – Final Report. Available at: https://ec.europa.eu/digital-single-market/en/news/experts-say-privately-held-data-available-european-union-should-be-used-better-and-more.

Laloux, P. 2021. Vie privée: la Commission lance une procédure d'infraction au RGPD contre la Belgique. Available at https://plus.lesoir.be/376968/article/2021-06-08/vie-privee-la-commission-lance-une-procedure-dinfraction-au-rgpd-contre-la.

Lyons, T., L. Courcelas, and K. Timsit, 2018. Blockchain and the GDPR: report prepared by the European Union Blockchain Observatory and Forum. Available at: https://www.eublockchainforum.eu/sites/default/files/reports/20181016_report_gdpr.pdf.

Mayer-Schönberger, V., and Y. Padova, 2016. Regime change? Enabling big data through Europe's new Data Protection Regulation. *Columbia Science & Technology Law Review* XVII: 315-335.

Richter, H. 2020. The law and policy of government access to private sector data ('B2G data sharing'). Max Planck Institute for Innovation and Competition Research Paper No. 20-06. https://doi.org/10.2139/ssrn.3594109

Rocher, L., J. Hendrickx, and Y.-A. de Montjoye. 2019. Estimating the success of re-identifications in incomplete datasets using generative models. *Nature Communications* 10: 3069. https://doi.org/10.1038/s41467-019-10933-3

Rouvroy, A. 2016. 'Of data and men': fundamental rights and freedoms in a world of big data. Bureau of the Consultative committee of the Convention for the protection of individuals with regard to automatic processing of personal data. Available at: https://rm.coe.int/CoERMPublicCommonSearchServices/DisplayDCTMContent?documentId=09000016806a6020.

Scarcella, L. 2019. Tax compliance and privacy rights in profiling and automated decision making. *Internet Policy Review* 8: 1-19.

Sweeney, L. 1997. Weaving technology and policy together to maintain confidentiality. *Journal of Law, Medicine & Ethics* 25: 98-110.

Tombal, T. 2018. Les droits de la personne concernée dans le RGPD. In C. de Terwangne, and K. Rosier (eds.) *Le Règlement Général sur la Protection des Données (RGPD/GDPR): analyse approfondie*. 407-557. Larcier, Brussels, Belgium.

Villani, C., M. Schoenauer, Y. Bonnet, C. Berthet, A-C. Cornut, F. Levin, and B. Rondepierre. 2018. Donner un sens à l'intelligence artificielle: pour une stratégie nationale et européenne. Available at: https://hal.inria.fr/hal-01967551/document.

Wendehorst, C. 2017. Of elephants in the room and paper tigers: how to reconcile data protection and the data economy. In S. Lohsse, R. Schulze, and D. Staudenmayer (eds.) *Trading data in the digital economy: legal concepts and tools*. 327-355. Hart Publishing, Baden Baden, Germany.

Zarsky, T., 2017. Incompatible: the GDPR in the age of big data. *Seton Hall Law Review* 47: 995-1020.

Chapter 5

Legal framework for the use of artificial intelligence and automated decision-making in public governance

T. Tombal[1,2], P. Willem[3*] and C. De Terwangne[3]

[1]*Faculty of Law, University of Namur, Rempart de la Vierge, 5, 5000 Namur, Belgium;* [2]*Tilburg Institute for Law and Technology, Tilburg University, Prof. Cobbenhagenlaan 221, 5037DE, Tilburg, the Netherlands;* [3]*Centre de recherche Information, Droit et Société, Namur Digital Institute, University of Namur, Rempart de la Vierge, 5, 5000 Namur, Belgium; pauline.willem@unamur.be*

Abstract

European public administration increasingly relies on personal data to deliver their public services, which implies that they must comply with the rules contained in the General Data Protection Regulation. These rules are especially important in the advent of new artificial intelligence (AI) technologies, which increase the public administration's capability to make informed decisions in order to provide public services and to support their decision-making. This chapter presents the legal framework within which these AI technologies can be used, namely data protection rules and core principles of administrative law. Several key takeaways are given. First, data subjects' rights must be respected (right to information, access, erasure, not to be subject to decisions based solely on automated processing). Those rights impact the design of the technologies used by public administrations and have concrete implications. Moreover, the legal framework aims to give more control to the citizens and keep them at the centre. In this regard, it is highly reassuring to see that the European Commission has proposed the adoption of an AI Act. The chapter analyses this proposal, keeping in mind the use of new technologies by public administrations in the fight against fraud.

Keywords: artificial intelligence, GDPR, right not to be subject to automated individual decision-making, administrative law, AI Act Proposal

DOI 10.3920/978-90-8686-930-5_5,

5.1 Introduction

As outlined in Chapter 4, European public administration increasingly relies on personal data[1] to deliver their public services, which implies that they must comply with the rules contained in the General Data Protection Regulation (hereafter 'GDPR').[2] These rules are especially important with the advent of new artificial intelligence technologies, which increase the public administration's capability to take informed decisions in order to provide public services and to support their decision-making. Indeed, while artificial intelligence (AI) technologies can potentially increase the efficiency of the public administration's decision-making, they can also generate significant impacts on the lives and fundamental rights of the citizens to which automated decision-making systems are applied.

Accordingly, this chapter will aim at presenting the legal framework within which these AI technologies can be used. This legal framework is composed of two categories of rules. First, public administrations willing to rely on AI techniques to improve the delivery of their public services must respect personal data protection rules (Section 5.2). Then, since these AI technologies are used by the administration in the context of the pursuit of their public service missions, they will also have to comply with the core principles of administrative law (Section 5.3). Indeed, these additional legal challenges, distinct from personal data protection challenges, must also be considered when reflecting on the development of algorithmic decision-making tools by the public administration.

Similar to Chapter 4, while most of the following analysis will be equally applicable to any public administration in the European Union, the chapter will focus on the Belgian case, to detail the legal challenges that public administrations face in practice. More specifically, the accent will be put on the use of AI for social security and tax fraud. This is because, in both cases, the use of algorithmic and automated decision-making processes could have a significant impact on citizens' finances and their general well-being.

5.2 Personal data protection and artificial intelligence

AI 'refers to systems that display intelligent behaviour by analysing their environment and taking actions – with some degree of autonomy – to achieve specific goals' (European Commission, 2018). AI thus aims at understanding how human cognition works, in order to reproduce it and to create computer cognition resembling that of humans (Villani *et al.*, 2018). AI methods are, in

[1] Personal data are defined as 'any information relating to an identified or identifiable natural person ('data subject'); an identifiable natural person is one who can be identified, directly or indirectly, in particular by reference to an identifier such as a name, an identification number, location data, an online identifier or to one or more factors specific to the physical, physiological, genetic, mental, economic, cultural or social identity of that natural person' (Art. 4.1 of the GDPR).
[2] Regulation 2016/679 of the European Parliament and of the Council of 27 April 2016 on the protection of natural persons with regard to the processing of personal data and on the free movement of such data, and repealing Directive 95/46 (General Data Protection Regulation), OJ [2016] L 199/1.

fact, numerous and diverse (expert systems, neural networks, machine learning, reinforcement learning, deep learning...) and are not new, as many have been developed and refined since the infamous Dartmouth conference of 1956 (Villani *et al.*, 2018).

For the purpose of this contribution, we will simply further elaborate here on two categories of AI, namely *expert systems* and *artificial neural networks*, as the distinction between the two will be relevant for our analysis below. An expert system can be defined as 'a computer system that emulates the decision-making ability of a human expert. Expert systems are designed to solve complex problems by reasoning through bodies of knowledge, represented mainly as 'if-then rules'.[3] In short, expert systems are rules-based AI systems. Based on rules that have been programmed, the AI system will draw inferences, following a form of decision-tree (If A then C, but if B then D, etc.).

On the other hand, artificial neural networks 'are computing systems (..) [that] 'learn' to perform tasks by considering examples, generally without being programmed with task-specific rules. (...) Instead, they automatically generate identifying characteristics from the examples that they process'.[4] Here the rules are not provided in advance to the program. Rather, the program has to understand the rules for itself, through trial and error, and through self-improvement (machine learning, reinforcement learning, deep learning...). The big difference between these two categories of AI is that, while the human programmer knows the rules it has given to an expert system, it will not necessarily understand which rules have been applied by the neural network, because the program has created its own rules, through trial and error. This is sometimes referred to as the 'Black box' (De Streel *et al.*, 2020; Pasquale, 2015).

From a personal data protection point of view, it should be outlined from the start that, as the use of AI algorithms requires a large amount of data, all the developments in Chapter 4 pertaining to big data and personal data protection[5] need to be kept in mind here. Indeed, the collection, linking and processing of large volumes of data to train an algorithm will need to rely on a lawful basis of processing[6] and to respect the purpose limitation and data minimisation principles.[7] That analysis will not be repeated here. Rather, we will focus, in this section, on the specific challenges posed by AI in terms of the GDPR, and these mostly revolve around the need to respect the data subjects' rights. More particularly, we will analyse the interactions between AI and the data subjects' right to information,[8] the data subjects' right of access,[9] the data subjects' right to erasure[10] and the

[3] https://en.wikipedia.org/wiki/Expert_system.
[4] https://en.wikipedia.org/wiki/Artificial_neural_network.
[5] See Chapter 4, Section 4.2.
[6] Art. 6.1 of the GDPR. See Chapter 4, Section 4.2.1.
[7] Arts. 5.1.b) and 5.1.c) of the GDPR. See Chapter 4, Sections 4.2.3 and 4.2.4.
[8] Arts. 12 to 14 of the GDPR.
[9] Art. 15 of the GDPR.
[10] Art. 17 of the GDPR.

data subjects' right not to be subject to automated individual decision-making.[11] However, prior to delving into the analysis of these data subject rights, a word must be said about the importance of considering the risks of AI before taking the decision to rely on such a technology.

5.2.1 A preliminary step: considering the risks of AI use (impact assessments and the proposal for a regulation on AI)

Like any technology, while AI can offer significant benefits, it also presents significant risks, notably in terms of safety and liability, security, bias, discrimination, but also personal data processing (European Commission, 2018).[12] In light of this latter risk, it is important to have Article 35 of the GDPR in mind, which pertains to personal data impact assessments. According to Article 35.1 of the GDPR:

> Where a type of processing in particular using new technologies, and taking into account the nature, scope, context and purposes of the processing, *is likely to result in a high risk to the rights and freedoms of natural persons*, the controller shall, prior to the processing, carry out an assessment of the impact of the envisaged processing operations on the protection of personal data. A single assessment may address a set of similar processing operations that present similar high risks (emphasis added).

Any public administration wishing to rely on AI techniques to support its policy- or decision-making, notably in the field of tax and social security fraud, should thus question whether this will imply a processing of personal data and, if this is the case, whether this is likely to result in a high risk to the rights and freedoms of natural persons (Edwards and Veale, 2017).

In assessing the likelihood of this high risk, public administration should be particularly attentive to Article 35.3 of the GDPR, as, in light of this provision, a public administration will have to perform a data protection impact assessment prior to the use of AI techniques to support its policy- or decision-making, if it implies a systemic and extensive evaluation of personal aspects relating to natural persons, which is based on automated processing[13] and produces legal effects on the person or similarly significantly affects them (Art. 35.3.a). Such an impact assessment would thus be required if a public administration intends to rely on automated processing to detect tax or social security fraud. In fact, the need to conduct such an impact assessment might also be justified, in the context of tax and social security fraud, in the case where a large scale of special categories of personal data (e.g. health data) or of personal data relating to criminal convictions and offences (e.g. data about prior convictions to establish 'suspicious' profiles) are used (Art. 35.3.b) (Edwards and Veale, 2017).[14]

---

[11] Art. 22 of the GDPR.

[12] For a depiction of some of these risks, see Alston, 2019. Extreme poverty and human rights. Note by the Secretary-General of the United Nations. Available at: https://undocs.org/pdf?symbol=en/A/74/493.

[13] On this point, see Section 5.2.5.

[14] See Chapter 4, Section 4.2.2.

If such a data protection impact assessment has to be performed, Article 35.7 provides that it shall contain, at least:

> (a) a systematic description of the envisaged processing operations and the purposes of the processing, including, where applicable, the legitimate interest pursued by the controller; (b) an assessment of the necessity and proportionality of the processing operations in relation to the purposes; (c) an assessment of the risks to the rights and freedoms of data subjects referred to in paragraph 1; and (d) the measures envisaged to address the risks, including safeguards, security measures and mechanisms to ensure the protection of personal data and to demonstrate compliance with this Regulation taking into account the rights and legitimate interests of data subjects and other persons concerned.

Moreover, the Data Protection Authority will have to be consulted before the process is launched, and will have the power to ban, temporarily or permanently, the use of the system if it is not satisfied with the risk identification and/or the guarantees offered by the public administration to mitigate the risks for the data subjects (Edwards and Veale, 2017).[15]

Finally, it is worth adding that according to a law proposal of 6 April 2021, these impact assessments of the algorithmic tools put in place by the administrations would have to be published in order to increase transparency.[16]

Importantly, the fact that public administration must consider whether an AI application entails 'high risk' to the rights and freedoms of natural persons is not only relevant for their right to personal data protection, but more broadly for all of their human rights (see, for example, Committee of Ministers of the Council of Europe, 2020). In this regard, it should be outlined that the European Commission's White Paper on AI also suggested the adoption of a risk-based approach when it comes to AI applications (European Commission, 2020). For the Commission, the extent of regulatory intervention on AI should be proportionate and should differentiate between categories of AI applications, focussing on 'high risk' applications (European Commission, 2020[17]). In fact, this approach has been confirmed by the Commission in its recent proposal for

[15] Art. 36.1 and 36.2 of the GDPR.

[16] Proposition de loi modifiant la loi relative à la publicité de l'administration du 11 avril 1994 afin d'introduire une plus grande transparence dans l'usage des algorithmes par les administrations, 6 avril 2021, *Doc. parl.*, Chambre, sess. ord., 2020-2021, no 55-1904/001, p. 6 and 8.

[17] In that White Paper, the Commission had provided that an AI application should be considered 'high risk' when it meets the two following cumulative criteria (p. 17):

'First, the AI application is employed in a sector where, given the characteristics of the activities typically undertaken, significant risks can be expected to occur. [These are] areas where, generally speaking, risks are deemed most likely to occur. (...) For instance, healthcare; transport; energy and *parts of the public sector.*

Second, the AI application in the sector in question is, in addition, used in such a manner that significant risks are likely to arise. This second criterion reflects the acknowledgment that not every use of AI in the selected sectors necessarily involves significant risks. (...) The assessment of the level of risk of a given use could be based on the impact on the affected parties. For instance, uses of *AI applications that produce legal or similarly significant effects for the rights of an individual or a company*; that pose risk of injury, death or significant material or immaterial damage; [or] that produce effects that cannot reasonably be avoided by individuals or legal entities' (emphasis added).

an Artificial Intelligence Act (European Commission, 2021).[18] Indeed, the proposal differentiates between 'uses of AI that create: (1) an unacceptable risk; (2) a high risk; and (3) low or minimal risk' (European Commission, 2021).

First, the proposal suggests prohibiting AI uses that create an unacceptable risk, notably because they would contravene EU values and/or would violate fundamental rights (European Commission, 2021).[19] These are:

- AI systems that deploy subliminal techniques beyond a person's consciousness in order to materially distort a person's behaviour in a manner that causes or is likely to cause that person or another person physical or psychological harm (Art. 5.1.a);
- AI systems that exploit any of the vulnerabilities of a specific group of persons due to their age, physical or mental disability, in order to materially distort the behaviour of a person pertaining to that group in a manner that causes or is likely to cause that person or another person physical or psychological harm (Art. 5.1.b);
- AI systems used by public authorities, or on their behalf, for the evaluation or classification of the trustworthiness of natural persons over a certain period of time based on their social behaviour or known or predicted personal or personality characteristics, with the social score leading to either or both of the following: (1) detrimental or unfavourable treatment of certain natural persons or whole groups thereof in social contexts which are unrelated to the contexts in which the data was originally generated or collected; (2) detrimental or unfavourable treatment of certain natural persons or whole groups thereof that is unjustified or disproportionate to their social behaviour or its gravity (Art. 5.1.c);
- 'Real-time' remote biometric identification systems in publicly accessible spaces[20] for the purpose of law enforcement, unless and in as far as such use is strictly necessary for one of the following objectives: (1) the targeted search for specific potential victims of crime, including missing children; (2) the prevention of a specific, substantial and imminent threat to the life or physical safety of natural persons or of a terrorist attack; or (3) the detection, localisation, identification or prosecution of a perpetrator or suspect of a criminal offence referred to in

[18] European Commission, Proposal for a Regulation of the European Parliament and of the Council laying down harmonised rules on artificial intelligence (Artificial Intelligence Act) and amending certain Union legislative acts, Brussels, 21 April 2021, COM(2021) 206 final, available at https://digital-strategy.ec.europa.eu/en/library/proposal-regulation-european-approach-artificial-intelligence.

[19] See also Article 5.

[20] A 'real-time' remote biometric identification system is 'a remote biometric identification system [i.e. an AI system for the purpose of identifying natural persons at a distance through the comparison of a person's biometric data with the biometric data contained in a reference database, and without prior knowledge of the user of the AI system whether the person will be present and can be identified (Art. 3.36 of the proposal)] whereby the capturing of biometric data, the comparison and the identification all occur without a significant delay. This comprises not only instant identification, but also limited short delays in order to avoid circumvention' (Art. 3.37 of the proposal). Publicly accessible space 'means any physical place accessible to the public, regardless of whether certain conditions for access may apply' (Art. 3.39 of the proposal).

On AI applications for the purposes of remote biometric identification (facial recognition), see also Consultative Committee of the Convention for the protection of individuals with regard to automatic processing of personal data, 2021. Moreover, it is worth underlining that Art. 35.3.c of the GDPR provides that a data protection impact assessment will be required in the case of systematic monitoring of a publicly accessible area on a large scale.

Article 2(2) of Council Framework Decision 2002/584/JHA (Council, 2002) and punishable in the Member State concerned by a custodial sentence or a detention order for a maximum period of at least three years (Art. 5.1.d).[21]

Second, the proposal lays down a risk methodology to define 'high-risk' AI systems that pose significant risks to the health and safety or fundamental rights of persons. The classification of an AI system as high-risk is based on the 'intended purpose'[22] of the AI system, which implies that this classification 'does not only depend on the function performed by the AI system, but also on the specific purpose and modalities for which that system is used'.[23] More precisely, the proposal identifies two main categories of high-risk AI systems:[24]

- AI systems, listed in Annex 2 of the proposal, that are intended to be used as safety component of products, or as products themselves, and that are subject to a third-party *ex ante* conformity assessment in order to be placed on the market (Art. 6.1); and Stand-alone AI systems listed in Annex III (Art. 6.2).

The list of Annex III contains a limited number of AI systems whose risks (mainly in terms of fundamental rights) have already materialised or are likely to materialise in the near future.[25] These high-risk AI systems are divided into eight areas, namely: (1) biometric identification and categorisation of natural persons; (2) management and operation of critical infrastructure; (3) education and vocational training; (4) employment, workers management and access to self-employment; (5) access to and enjoyment of essential private services and public services and benefits; (6) law enforcement; (7) migration, asylum and border control management; and (8) administration of justice and democratic processes.

In the context of tax and social security fraud, it is worth highlighting that 'AI systems intended to be used by public authorities or on behalf of public authorities to evaluate the eligibility of natural persons for public assistance benefits and services, as well as to grant, reduce, revoke, or reclaim such benefits and services'[26] are considered as high-risk AI systems. This is because 'natural persons applying for or receiving public assistance benefits and services from public

[21] In order to apply one of the exceptions of Article 5.1.d), account shall be taken of 'the nature of the situation giving rise to the possible use, in particular the seriousness, probability and scale of the harm caused in the absence of the use of the system' and of 'the consequences of the use of the system for the rights and freedoms of all persons concerned, in particular the seriousness, probability and scale of those consequences', and these uses 'shall comply with necessary and proportionate safeguards and conditions in relation to the use, in particular as regards the temporal, geographic and personal limitations' (Art. 5.2 of the proposal). Moreover, this will be subject to a prior authorisation from a judicial or independent administrative authority, except in duly justified situations of urgency, where the authorisation can be requested during or after the use (Art. 5.3 of the proposal).

[22] 'The use for which an AI system is intended by the provider, including the specific context and conditions of use, as specified in the information supplied by the provider in the instructions for use, promotional or sales materials and statements, as well as in the technical documentation' (Art. 3.12 of the proposal).

[23] Proposal for an Artificial Intelligence Act, p. 13.

[24] See Articles 6 and 7 of the proposal for an Artificial Intelligence Act.

[25] Proposal for an Artificial Intelligence Act, p. 16.

[26] Annex III, point 5.a) of the proposal for an Artificial Intelligence Act.

authorities are typically dependent on those benefits and services and in a vulnerable position in relation to the responsible authorities. If AI systems are used for determining whether such benefits and services should be denied, reduced, revoked or reclaimed by authorities, they may have a significant impact on persons' livelihood and may infringe their fundamental rights, such as the right to social protection, non-discrimination, human dignity or an effective remedy'.[27] This would arguably cover AI systems used to allocate social benefits or to fight against social security fraud.

Moreover, the following are also considered high-risk: 'AI systems intended to be used by law enforcement authorities for predicting the occurrence or reoccurrence of an actual or potential criminal offence based on profiling of natural persons or assessing personality traits and characteristics or past criminal behaviour of natural persons or groups';[28] 'AI systems intended to be used by law enforcement authorities for profiling of natural persons in the course of detection, investigation or prosecution of criminal offences';[29] and 'AI systems intended to be used for crime analytics regarding natural persons, allowing law enforcement authorities to search complex related and unrelated large data sets available in different data sources or in different data formats in order to identify unknown patterns or discover hidden relationships in the data'.[30] Arguably, AI systems used to fight tax fraud, through data matching and data mining, could be considered as falling within these types of high-risk AI systems. However, it should be underlined that Recital 38 of the proposal provides that 'AI systems specifically intended to be used for administrative proceedings by tax and customs authorities should not be considered high-risk AI systems used by law enforcement authorities for the purposes of prevention, detection, investigation and prosecution of criminal offences'. This seems to suggest that AI systems used to fight tax fraud would not be considered as 'high-risk'. Yet, as Recitals are not binding, this creates some uncertainty in this regard.

The Commission will have to conduct an annual assessment of the list of high-risk AI systems,[31] which it can update according to the methodology described in Article 7 of the proposal. This list can notably be expanded to AI systems intended for use in any of the areas listed in Annex III, which 'pose a risk of harm to the health and safety, or a risk of adverse impact on fundamental rights, that is, in respect of its severity and probability of occurrence, equivalent to or greater than the risk of harm or of adverse impact posed by the high-risk AI systems already referred to in Annex III'.[32]

[27] Recital 37 of the proposal for an Artificial Intelligence Act.
[28] Annex III, point 6.e) of the proposal for an Artificial Intelligence Act.
[29] Annex III, point 6.f) of the proposal for an Artificial Intelligence Act.
[30] Annex III, point 6.g) of the proposal for an Artificial Intelligence Act.
[31] Art. 84.1 of the proposal for an Artificial Intelligence Act.
[32] Art. 7.1 of the proposal for an Artificial Intelligence Act.

These high-risk AI systems will have to undergo *ex ante* conformity assessment procedures,[33] and they will have to be registered in a public EU-wide database, operated by the European Commission, to increase public transparency and oversight and strengthen *ex post* supervision by competent authorities.[34] Furthermore, any AI system meeting this 'high-risk' threshold will have to comply with the legal requirements set out in the proposal.[35] According to the Commission, these requirements 'are strictly necessary to mitigate the risks to fundamental rights and safety posed by AI and that are not covered by other existing legal frameworks'.[36] These requirements pertain to the need: (1) to establish, implement, document and maintain a risk management system; (2) to produce technical documentation; (3) to ensure that high-quality data is used; (4) to adopt an appropriate data governance allowing record-keeping – i.e. logs –, transparency and the provision of information to users; (5) to ensure human oversight; and (6) to ensure a certain level of accuracy, robustness, and cybersecurity (European Commission, 2020).[37] Interestingly, several of these requirements contribute to the minimisation of the risks of algorithmic discrimination, namely those pertaining to the design and the quality of datasets, and to the obligations for testing, risk management, documentation and human oversight throughout the AI systems' lifecycle.[38] Finally, the Commission outlined that 'harmonised standards and supporting guidance and compliance tools will assist providers and users in complying with the requirements laid down by the proposal and minimise their costs'.[39]

Third, the proposal suggests imposing transparency obligations for certain low-risk systems, namely those that: '(1) interact with humans; (2) are used to detect emotions[40] or determine association with (social) categories based on biometric data;[41] or (3) generate or manipulate content ('deep fakes')'.[42] Moreover, the proposal establishes a framework for the creation of codes of conduct, in order to encourage low-risk AI systems' providers to voluntarily apply the above-mentioned mandatory requirements for high-risk AI systems.[43]

Finally, it is worth noting that, in order to enforce the above-mentioned rules, the proposal establishes, at the EU level, a 'European Artificial Intelligence Board', composed of representatives from the Member States and the Commission, which will 'facilitate a smooth, effective and harmonised implementation of this regulation by contributing to the effective cooperation of the

---

33 Proposal for an Artificial Intelligence Act, p. 3. See also Articles 40 to 50, and Annexes VI and VII.
34 Proposal for an Artificial Intelligence Act, p. 12-14. See also Articles 51 and 60, and Annex VIII.
35 Article 8 of the proposal for an Artificial Intelligence Act.
36 Proposal for an Artificial Intelligence Act, p. 13.
37 See Articles 9 to 15 and Annex IV of the proposal for an Artificial Intelligence Act.
38 Proposal for an Artificial Intelligence Act, p. 4.
39 *Ibid.*, p. 7. See also Articles 40 to 42.
40 Emotion recognition system: 'an AI system for the purpose of identifying or inferring emotions or intentions of natural persons on the basis of their biometric data' (Art. 3.34 of the proposal).
41 Biometric categorisation system: 'an AI system for the purpose of assigning natural persons to specific categories, such as sex, age, hair colour, eye colour, tattoos, ethnic origin or sexual or political orientation, on the basis of their biometric data' (Art. 3.35 of the proposal).
42 Proposal for an Artificial Intelligence Act, p. 14. See Article 52 for more details.
43 *Ibid.*, p. 16. See also Article 69.

national supervisory authorities and the Commission and providing advice and expertise to the Commission. It will also collect and share best practices among the Member States'.[44] Furthermore, 'Member States will have to designate one or more national competent authorities and, among them, the national supervisory authority, for the purpose of supervising the application and implementation of the regulation'.[45]

5.2.2 AI and the data subject's right to information

As mentioned in Chapter 4, data has to be processed fairly and in a transparent manner.[46] Moreover, if the public administration makes use of individual decision-making, based 'solely' on automated processes, for instance to fight social security infringements and tax fraud, it will have to inform the data subject about the existence of such processes.[47] In that case, the data subject should, at least, receive meaningful information about the logic involved, as well as the significance and the envisaged consequences of such processing for the data subject.[48] This is key for the data subjects' right to informational self-determination, as it allows them to understand what is being done with their data (De Terwangne, 2015). Indeed, the data subject can legitimately wish to know the criteria that have been used, and their respective weight, in the automated decision (De Terwangne, 2015).

In this regard, it is interesting to mention the SyRI decision of the Court of The Hague,[49] which has been presented in Chapter 4.[50] As a reminder, the Court was asked to assess the compatibility, with Article 8 of the European Convention on Human Rights, of the Dutch government's 'Systeem Risico Indicatie ('SyRI' – System Risk Indication), which is a legal instrument used to detect various forms of fraud, including social benefits, allowances, and tax fraud.[51] In that case, the Court seems to have addressed the obligation to inform the data subject about the use of individual decision-making based 'solely' on automated processes, although not explicitly. Indeed, the Court pointed out that there is a clear transparency problem, because:

> [T]he SyRI legislation does not provide for a duty of disclosure to those whose data are processed in SyRI so that these data subjects can be reasonably assumed to know that their data are or have been used for that processing. The SyRI legislation also does not provide for an obligation to notify the data subjects individually, as appropriate, that a risk report has

[44] *Ibid.*, p. 15. See also Articles 56 to 58.
[45] *Ibid.*, p. 15. See also Article 59.
[46] Arts. 5.1.a) and 12.1 of the GDPR. See Chapter 4, Section 4.2.5.
[47] Arts. 13.1.f) and 14.2.g) of the GDPR.
[48] Art. 15.1.h) of the GDPR.
[49] Rechtbank Den Haag, 5 februari 2020, Zaak n° C-09-550982-HA ZA 18-388, ECLI:NL:RBDHA:2020:865 (ECLI:NL:RBDHA:2020:1878 for the English version).
[50] See Chapter 4, Section 4.2.5.
[51] Wet van 9 oktober 2013 tot wijziging van de Wet structuur uitvoeringsorganisatie werk en inkomen en enige andere wetten in verband met fraudeaanpak door gegevensuitwisselingen en het effectief gebruik van binnen de overheid bekende zijnde gegevens, *Stb.*, 2013, p. 405; Besluit van 1 september 2014 tot wijziging van het Besluit SUWI in verband met regels voor fraudeaanpak door gegevensuitwisselingen en het effectief gebruik van binnen de overheid bekend zijnde gegevens met inzet van SyRI, *Stb.*, 2014, p. 320.

> been submitted. There is only a statutory obligation to announce the start of a SyRI project beforehand by way of publication in the Government Gazette and after the processing access to the register of risk reports upon request. (…). Data subjects are also not informed automatically afterwards. This only occurs if there is a control and investigation in response to a risk report. This does not happen as a matter of course.[52]

Although the Court did not rule specifically on the application of the right to be informed about the logic involved,[53] this right seems not to have been respected in this case, because the data subjects are unaware of the existence of a risk report, while the submission of a risk report has a significant effect on them.[54] Indeed, a data subject whose data were processed in SyRI, but which did not result in a risk report, will not be informed about this processing, and therefore cannot verify that their data was processed on correct grounds.[55]

Additionally, it should be underlined that, depending on the type of AI that is used and on the legal entity that holds the rights to the algorithm, it might be complicated for the public administration to comply with this requirement to provide the data subject with meaningful information about the logic involved.

On the one hand, a difference must be made between expert systems and neural networks. As outlined above, expert systems are rules-based AI systems. Accordingly, the public administration is more likely to be aware of the rules that have been applied by the algorithm and of the logic involved in the decision, as these have been explicitly programmed and dictated to the AI, and should be able, in such cases, to provide the data subject with meaningful information about the logic involved.[56] On the contrary, neural networks create their own rules, through trial and error (e.g. machine learning). Accordingly, the public administration might not necessarily understand which rules have been applied by the neural network, and thus might not be able to provide the data subject with meaningful information about the logic involved.[57]

On the other hand, a difference will have to be made depending on whether the rights to the algorithm are held by the public administration itself, or by a private entity that has been tasked, by means of a public procurement, with developing the algorithm for the public administration. Indeed, in the vast majority of cases, the algorithm will benefit from the copyright protection

---

52 Rechtbank Den Haag, 5 februari 2020, Zaak n° C-09-550982-HA ZA 18-388, point 6.54.
53 Arts. 13.1.f) and 14.2.g) of the GDPR.
54 Rechtbank Den Haag, 5 februari 2020, Zaak n° C-09-550982-HA ZA 18-388, point 6.82.
55 *Ibid.*, point 6.90.
56 Art. 15.1.h) of the GDPR.
57 Art. 15.1.h) of the GDPR.

granted to computer programs.[58] Moreover, it might also be protected as a trade secret.[59] In cases where these rights are held by the private entity, this may prevent the public authority from understanding the logic involved behind the algorithm, and therefore from providing this information to their citizens.[60] Indeed, the private entity could invoke its rights to the algorithm in order to refuse to disclose this commercially sensible information to the public administration, which in turn will not be able to communicate it to citizens. It is thus of the utmost importance for public administration that call upon private parties to develop an algorithm, to specify clearly in the public procurement either that it will hold the rights to the algorithm, or that it has the right to receive information about the logic involved behind the algorithm, in order to be able to provide it to the data subjects whose data are being processed.[61] However, an important caveat here is that the private entity itself may not necessarily be able to explain the logic involved in the decision taken by a neural network AI. This is sometimes referred to as the 'black box' (De Streel *et al.*, 2020; Pasquale, 2015).

Finally, it should be added that there have been proposals to include *algorithmic impact assessments* (AIAs) as part of a public administration's procurement procedures pertaining to automated decision-making systems (Misuraca and van Noordt, 2020). The potential benefits of rolling out such AIAs would notably include: 'better communication with the general public; increase of in-house expertise of public agencies; higher levels of accountability of automated decision-making systems and a meaningful way for the public to question them' (Misuraca and van Noordt, 2020).

5.2.3 AI and the data subject's right of access

The data subject's right of access has a double impact for public administrations. On the one hand, it stipulates that the data subject has the right to obtain, from the public administration, the confirmation as to whether or not it processes personal data concerning them.[62] The goal of this right is for the data subject to be aware of, and to verify, the lawfulness of the processing.[63] In the context of AI applications used to fight social security infringements and tax fraud, this means that the data subject has the right to obtain, from the public administration, the confirmation as to whether or not the algorithm processes personal data concerning them. If this is the case, the public administration will have to provide access to the data, as well as to the information

[58] Directive 2009/24/EC of the European Parliament and of the Council of 23 April 2009 on the legal protection of computer programs, OJ [2009] L 111/16.
[59] Directive (EU) 2016/943 of the European Parliament and of the Council of 8 June 2016 on the protection of undisclosed know-how and business information (trade secrets) against their unlawful acquisition, use and disclosure, OJ [2016] L 157/1. Art. 2.1 defines a trade secret as any 'information which meets all of the following requirements: (a) it is secret in the sense that it is not, as a body or in the precise configuration and assembly of its components, generally known among or readily accessible to persons within the circles that normally deal with the kind of information in question; (b) it has commercial value because it is secret; (c) it has been subject to reasonable steps under the circumstances, by the person lawfully in control of the information, to keep it secret'.
[60] Art. 15.1.h) of the GDPR.
[61] Art. 15.1.h) of the GDPR.
[62] Art. 15.1 of the GDPR.
[63] Recital 63 of the GDPR.

listed in points (a) to (h) of Article 15.1 of the GDPR. Since, these elements of information are, in substance, the same as those contained in Articles 13 and 14 of the GDPR, we will simply refer here to what has been said in Chapter 4, Section 4.2.5.

On the other hand, the right of access provides the data subject with the right to obtain a copy of the personal data that is processed by the public administration (Tombal, 2018).[64] In the context of AI applications used to fight social security infringements and tax fraud, this means that the data subject has the right to obtain a copy of the personal data concerning them that is processed by the algorithm. In this regard, simply providing a mass of incomprehensible data for any human being would not be sufficient, as any communication under Articles 15 to 22 of the GDPR must be made in a concise, transparent, intelligible and easily accessible form, using clear and plain language.[65] This means that the data subject should be able to understand the data that she receives. Moreover, and similarly to all of the data subjects' rights, the public administration will have to answer without undue delay and, in any case, within one month of the receipt of the request, and the exercise of the right of access should be free.[66] However, a reasonable fee based on administrative costs may be charged for any further copies requested by the data subject.[67]

Like the right to information, this right of access can be restricted by a Member State law when such a restriction respects the essence of the fundamental rights and freedoms and is a necessary and proportionate measure in a democratic society to safeguard an important objective of general public interest.[68]

5.2.4 AI and the data subject's right to erasure

Article 17 of the GDPR provides that the data subject shall have the right to obtain from the controller the erasure of personal data concerning them, without undue delay, in certain hypotheses.[69] It should, however, be outlined from the outset that this right to erasure shall not apply to the extent that the processing is necessary for the compliance with a legal obligation to which the controller is subject, or for the performance of a task carried out in the public interest or in the exercise of official authority vested in the controller.[70]

That being said, even in cases where such right should apply, doing so in an AI world can turn out to be much more complicated that it seems, as the GDPR does not define the notion of erasure (Fosch Villaronga *et al.*, 2017). As pointed out by Fosch Villaronga *et al.* in their seminal paper,

---

[64] Art. 15.3 of the GDPR. This clarification is important and puts an end to the debate that existed, prior to the GDPR, when Directive 95/46 was applicable, as it was uncertain whether the right of access, as set out in Article 12 of the Directive, implicitly allowed the data subject to obtain a copy of their personal data.
[65] Art. 12.1 of the GDPR.
[66] Arts. 12.3 and 12.5 of the GDPR.
[67] Art. 15.3 of the GDPR.
[68] Art. 23.1 of the GDPR. See Chapter 4, Section 4.2.5.
[69] Art. 17.1 of the GDPR.
[70] Art. 17.3.b) of the GDPR.

Article 17 'seems to push toward the simple deletion of the personal data or the folder containing the personal data from the data controller's system, as if data on a computer was like a physical file that can simply be destroyed' (Fosch Villaronga *et al.*, 2017). Yet, data deletion requirements pose crucial challenges in AI environments, and might actually be practically impossible to satisfy (Fosch Villaronga *et al.*, 2017).

To understand why this is the case, it must be outlined that AI 'minds' do not function exactly as human minds and that, as a consequence, an AI cannot forget data the way humans do, making it much more complex to delete data (Fosch Villaronga *et al.*, 2017). Indeed, it is questionable whether data deletion is, in fact, actually possible in modern AI environments relying on relational database management system (DBMS) (Fosch Villaronga *et al.*, 2017). As outlined by Fosch Villaronga *et al.*:

> [E]very data record added to the database might not only reside at one specific point in the file system, but might be stored at various locations inside internal database mechanisms, as well as across different replicated databases, in logfiles and backups (...) When asking for deletion in a strict sense, these spaces must be identified and overwritten with random information. In several internal mechanisms like the database transaction log, the latter is especially impossible without seriously endangering the consistency of the database, or even simply breaking it altogether (Fosch Villaronga *et al.*, 2017).

Moreover, in most AI environments, when data is 'deleted', it is not directly overwritten with other data, but is only marked as deleted and removed from the search indexes, and it can take a very long time before the space marked as 'deleted' is effectively reused (effectively destroying the old data) (Fosch Villaronga *et al.*, 2017). Accordingly, determining if and when a deletion has occurred, and thus whether the data controller has complied with the data subject's right to erasure, will depend on the interpretation of the words erasure and deletion. Is it the removal from the search index, the overwriting in the file system, the deletion from the log-files and backups or is it the removal from all internal mechanisms? (Fosch Villaronga *et al.*, 2017)

In any case, even if it is assumed that data can be erased from AI systems in a way that complies with Article 17 of the GDPR, such deletion might have an impact on the quality of the AI's results (Fosch Villaronga *et al.*, 2017). As outlined by Fosch Villaronga *et al.*:

> This is especially interesting considering algorithms that use a so-called 'knowledgebase' for calibration, i.e. the algorithm takes the knowledgebase with pre-calculated results as reference data and extracts the common artifacts. It then uses these 'learned' rules on new data, which have to be very close to the training data in terms of data structure and statistical properties. Furthermore, the resulting categorisations are again fed into the knowledge base in order to get even better training data for the next run, thus iteratively extending the knowledge base. (Fosch Villaronga *et al.*, 2017)

In other words, once an algorithm has trained on data to produce a result, it will use these 'learned' results to train on the next batch of data. Accordingly, even if we could assume that the data subject's data can be deleted from the training data, traces of this data will probably still be found in the 'learned' results, which will be used in the training iterations. This may be problematic if it is impossible to erase these traces. In fact, it will likely be impossible to do so, and doing so might actually have large-scale effects on the algorithm's efficiency (Fosch Villaronga *et al.*, 2017). While several approaches have been suggested to solve this issue, none has so far offered satisfactory results (Fosch Villaronga *et al.*, 2017).

In light of the above, a public administration wishing to use AI applications in order to fight social security infringements and tax fraud should anticipate these potential erasure requests in the way it builds its AI system. More specifically, it should define, in advance, the moment at which data will be considered as being deleted (removal from the search index, overwriting in the file system, deletion from the log-files or from all internal mechanisms) and justify, in light of the accountability principle,[71] why this complies with Article 17 of the GDPR. Moreover, in light of the iterative way of working of AI systems, which rely on the previous 'learned' results for the next training iterations, it should reflect, from the outset, on the impact that the right of erasure will have on these 'learned' results: is it possible to erase the data subject's data not only from the training data, but also any trace of it in the 'learned' results?

5.2.5 AI and the data subject's right not to be subject to automated individual decision-making

According to Article 22.1 of the GDPR, the data subject shall have the right not to be subject to a decision based solely on automated processing, including profiling, which produces legal effects concerning him or her or similarly significantly affects him or her. This right echoes the strong will of the human being not to be entirely subject to the machine, as the former does not accept the idea that a decision may be imposed on it solely on the basis of conclusions reached by that machine (de Terwangne, 2015). Accordingly, it is fundamental for a public administration that wishes to rely on AI to fight social security infringements and tax fraud to comply with this provision of the GDPR.

Automated decision-making based *solely* on automated processing

It is important to highlight from the outset that Article 22 only applies to decisions based *solely* on automated processing. This means that Article 22 could potentially 'be sidestepped relatively easily by inserting human intervention into the process. In other words, once the process is not 'solely' automated, this provision will not apply' (Scarcella, 2019; Zarsky, 2017). Nevertheless, it remains to be seen how courts and regulators would react to the introduction, by data controllers, of fictitious or negligible human intervention in the automated decision process, simply in order to avoid (potentially in bad faith) the application of Article 22 (Zarsky, 2017).

[71] Art. 5.2 of the GDPR.

For instance, if the whole decision-making process is automated, but has to be validated by a human before being effectively applied (e.g. the result of the decision-making appears on the human's screen and she has to validate it by clicking 'Ok' or 'Validate'), can it be said that there has been human intervention in the decision-making process? As coined by Edwards and Veale: 'When does 'nominal' human involvement become no involvement?' (Edwards and Veale, 2017).

The answer will likely differ depending on the leeway that the human has in the automated decision: does he have to follow it or can he divert from it? Moreover, the human's capability to interpret the data and to be sceptical about the result might also have an impact. Indeed, 'human involvement can also be rendered nominal by 'automation bias,' a psychological phenomenon where humans either over or under-rely on decision support systems' (Edwards and Veale, 2017). If the human, whose intervention has been added at the very end of the process, always simply trusts the algorithm's 'suggestion' without ever questioning it (either because it is not able to or because it doesn't want to), this human intervention will have no concrete impact on the decision-making process. For instance, in the specific field of customs fraud, while some fraud indicators result from human knowledge, there is also an automated model that analyses all of the feedback from the controllers on a continuous basis and updates itself every day. Based on these updates, it will produce hundreds of updated selection rules every day to determine which goods/undertakings should be controlled. Therefore, only the feedbacks are provided by humans, not the rules inferred from them. In such cases, it is fundamental to ensure that the inspectors keep collaborating by giving feedback on those newly suggested indicators, rather than simply following what the AI suggests, without any critical thinking. Yet, looking towards the future, it is possible that, in light of the constant budget cuts and reduction in personnel, there is a risk that the few inspectors left will simply end up trusting the machine without any critical thinking, because they have to meet their quotas of controls that they must do, and no longer have time to check the relevance of the indicators suggested by the machine. In such situations, we believe that this should be considered, *de facto*, as a decision based *solely* on automated processing. In this regard, the Article 29 Working Party outlines that:

> The controller cannot avoid the Article 22 provisions by fabricating human involvement. For example, if someone routinely applies automatically generated profiles to individuals without any actual influence on the result, this would still be a decision based solely on automated processing. To qualify as human involvement, the controller must ensure that any oversight of the decision is meaningful, rather than just a token gesture. It should be carried out by someone who has the authority and competence to change the decision. As part of the analysis, they should consider all the relevant data (Article 29 Working Party, 2018).

The key is thus whether a solely automated decision is taken, which produces legal effects concerning the data subject or similarly significantly affects them. In the case of AI applied to social security infringements or tax fraud, there will be a clear difference in whether the AI decides itself that a person has committed a social security infringement or tax fraud, or whether the AI merely warns the civil servants working for the relevant public administration that a specific person has a 'suspicious profile', and that this will lead to a human intervention, by the civil servant, who will look into the case. There is indeed a major difference between an AI 'deciding' itself, and an AI doing a preliminary analysis in the large quantity of data available to the public administration, in order to prioritise the cases on which the civil servants should focus because, by essence, they do not have the time to check every single case.

In this regard, it should be outlined that, at the Belgian FPS Finances, humans play an important role in the pre-investigation and investigation stages. For instance, human controllers have first established a set of typologies (types of suspicious profiles they want to detect) and use analytics to support detection. Indeed, the indicators used to identify these typologies are either proposed by humans or by the machine, which will propose a predictive shortlist of profiles that closely correspond to the typology that the investigators are looking for. To this effect, the machine will identify the most effective factors to detect these typologies, but the final decision to investigate (or not) a profile remains in the hand of the human controller. In fact, the investigators will often not investigate all suspicious cases identified in the pre-investigation. Rather, they will test some of these and will provide feedback on the usefulness of the signals at the end of the investigation. If the signals are relevant, they will investigate more cases from the suggested list. Additionally, for some types of fraud such as those linked to 'direct income taxes', the cases that are investigated following a data-mining recommendation are relatively small compared to the cases that controllers investigate on their own initiative (about 20%). Data mining is however, extremely important for, and well-suited to, other specific types of fraud, such as VAT fraud where about 80% of the cases derive from data mining.

Regarding the social security infringements, a distinction must be made between data-matching operations and the data-mining operations conducted in the OASIS data warehouse. For the former, some forms of *ex ante* data-matching cross-checks (e.g. checks before the social allocation is paid) do not need human intervention and are fully automated. This is because they are used to identifying objective obstacles to the payment of the allowances (e.g. no unemployment benefit if a person has a professional income). It is thus not a matter of interpretation, as there is no flexibility for the machine. This could easily be reviewed by a human, if requested by a data subject. *Ex post* data-matching cross-checks (e.g. checks after the social allocation is paid), on the other hand, always require human verification. Indeed, it is necessary for them to hear the person and ensure the rights of defence before taking a decision. This makes it possible to find cases that have escaped the *ex ante* cross-checks. All these *ex post* cross-checks are justified by a decision in due form with legal and factual justification, which can give rise to complaints to the ombudsman, and appeals. Data-mining operations, on the other hand, merely suggest cases to investigate, while the concrete investigation will always be done by a human. Moreover, the indicators integrated in OASIS have, in fact, been suggested by humans, namely inspectors in the field, who translate their experience of the cases they investigated into indicators. The machine simply looks for those indicators in the large amount of data.

However, even if the machine does not decide on its own that a person is a fraudster, this pre-selection could, in and of itself, be considered as a solely automated decision that significantly affects the data subject, as they are placed on the 'suspect list' and this leads to the opening of an investigation (De Raedt, 2017; Degrave, 2020). Indeed, this may be considered as having an effect on them, as it will entail additional scrutiny from the public administration of their behaviour (see, by analogy, Edwards and Veale, 2017). If this interpretation is followed, this would require implementing appropriate safeguards, such as the right to obtain a human intervention (Section 5.2.5 – Exceptions to the right ...).

Finally, it is interesting to outline that, in the SyRI decision of the Court of the Hague mentioned above,[72] the Court also assessed whether an automated individual decision-making occurred when SyRI was applied.[73] This is because the claimants argued that the submission of a risk report by the Social Affairs and Employment Inspectorate can be considered a decision with legal effect, or at least a decision that affects the data subjects significantly in another way, and that this decision is taken on the basis of automated individual decision-making within the meaning of Article 22 of the GDPR.[74] These claimants added that there is no meaningful human intervention prior to the submission of a risk report, as the mere removal of 'false positives' cannot qualify as such, nor can the assessment of the participating parties after receipt of a risk report.[75] The Dutch State, on the other hand, contested the fact that automated individual decision-making occurred and added that, in any case, the exceptions of Article 22.2 of the GDPR were met and that the amended legislation contained sufficient safeguards to protect privacy, as provided in Article 22.3.[76]

As a starting point, the Court outlined that the SyRI legislation leaves the option open whether predictive analyses, 'deep learning' and data mining can be used in the SyRI infrastructure, but that, at that point in time, no use was made of deep learning and data mining in the implementation of the SyRI legislation.[77] Regarding the potential effects of SyRI on the data subjects, the Court ruled that while the use of SyRI in and of itself is not aimed at having legal effect, a risk report nevertheless does have a similarly significant effect on the private life of the person to whom the risk report pertains.[78] Indeed, according to the Court, '[t]he fact that a risk report does not necessarily always lead to further investigation, or to an administrative or criminal-law sanction, and may also not be used as the sole basis for an enforcement decision, does not alter the significant effect on the private life of the data subject'.[79]

However, and quite surprisingly, the Court decided not to rule on whether this constituted an automated individual decision-making in the GDPR and, insofar as this is the case, on whether one or more of the exceptions to the prohibition in the GDPR had been met, because it deemed this to be irrelevant in the context of the assessment of whether the SyRI legislation meets

---

72 See Section 5.2.2.
73 Rechtbank Den Haag, 5 februari 2020, Zaak n° C-09-550982-HA ZA 18-388, ECLI:NL:RBDHA:2020:865 (ECLI:NL:RBDHA:2020:1878 for the English version), points 6.55 to 6.60.
74 *Ibid.*, point 6.57.
75 *Ibid.*
76 *Ibid.*, point 6.58.
77 *Ibid.*, point 6.51.
78 *Ibid.*, point 6.59.
79 *Ibid.*

the requirements of Article 8 of the European Convention on Human Rights.[80] This is quite disappointing, as such an assessment would have provided more clarity for administrations on what constitutes a decision based 'solely' on automated processing, and could also have shed more light on the safeguards that need to be put in place in such cases. Public administration will thus have to be careful when using AI in order to combat social security and tax fraud, as there remains a significant level of uncertainty about what can and cannot be done in this regard.

Exceptions to the right not to be subject to automated individual decision-making, and appropriate safeguards

According to Article 22.2 of the GDPR, individual decision-making based solely on automated processing can nevertheless be used if the decision: '(a) is necessary to enter into, or for the performance of, a contract between the data subject and a data controller; (b) is authorised by a law to which the controller is subject, provided that this law lays down suitable measures to safeguard the data subject's rights and freedoms and legitimate interests; or (c) is based on the data subject's explicit consent'.

In the case where a public administration wishes to rely on AI to fight social security infringements and tax fraud, its only viable option will be to rely on a law. Indeed, it is very unlikely that a contract can be concluded with all the citizens who would likely not be willing to allow the public administration to link their data in order to identify fraudulent behaviour. Moreover, as outlined in Chapter 4,[81] public administrations should avoid relying on the data subject's consent, as there will likely be a clear imbalance between the data subject and the controller, leading to the conclusion that the consent is not freely given (European Data Protection Board, 2020).[82] Accordingly, individual decision-making based solely on automated processing, in order to fight social security infringements and tax fraud, should be authorised by a law.[83] In fact, Recital 71 of the GDPR explicitly states that such automated individual decision-making 'should be allowed where expressly authorised by Union or Member State law to which the controller is subject, *including for fraud and tax-evasion monitoring*' (emphasis added).

This law will, however, have to lay down suitable measures to safeguard the data subject's rights and freedoms and legitimate interests.[84] While Article 22 does not explicitly indicate what such safeguards should be when the automated processing is authorised by a law, it does provide, in its Article 22.3, that in the case of automated processing based on consent or a contract, the data controller should, at least, provide the data subject with the right to obtain human intervention on the part of the controller, to express his or her point of view and to contest the decision. Logically, a law authorising individual decision-making based solely on automated processing should also,

---

[80] *Ibid.*, point 6.60.
[81] See Chapter 4, Section 4.2.1.
[82] Recital 43 of the GDPR.
[83] Art. 22.2.b) of the GDPR.
[84] Art. 22.2.b) of the GDPR.

at least, provide for these three safeguards. This is supporting by the wording of Recital 71 of the GDPR, which provides that '*in any case*, such processing should be subject to suitable safeguards, which should include specific information to the data subject and the right to obtain human intervention, to express his or her point of view, to obtain an explanation of the decision reached after such assessment and to challenge the decision' (emphasis added). Recital 71 indeed targets all three of the exceptions.

Regarding Recital 71, it is also interesting to point out that it invites, in its second paragraph, the data controller to take into account the specific circumstances and context in which the personal data are processed, in order to ensure fair and transparent processing in respect of the data subject. In this regard, Recital 71 contains the following recommendations:

> [T]he controller should use appropriate mathematical or statistical procedures for the profiling; implement technical and organisational measures appropriate to ensure, in particular, that factors which result in inaccuracies in personal data are corrected and the risk of errors is minimised; [and] secure personal data in a manner that takes account of the potential risks involved for the interests and rights of the data subject and that prevents, inter alia, discriminatory effects on natural persons on the basis of racial or ethnic origin, political opinion, religion or beliefs, trade union membership, genetic or health status or sexual orientation, or that result in measures having such an effect.

At first glance, these recommendations appear to have limited binding force, as they are not included in the text of Article 22 of the GDPR. In reality, these recommendations are only the formulation of binding obligations formulated elsewhere in the GDPR, namely the principle of data protection by design[85] (appropriate technical and organisational measures), the requirement of data accuracy[86] (reduction of the risk of errors and correction of errors), and the requirement of data security[87] (secure personal data by taking into account the risks for the data subjects).

Finally, it should be outlined that the use of these exceptions is limited when they lead to the processing of special categories of data.[88] Indeed, Article 22.4 of the GDPR provides that decisions based solely on automated processing shall not be based on special categories of personal data referred to in Article 9.1 of the GDPR, unless point (a) or (g) of Article 9.2 applies and suitable measures to safeguard the data subject's rights and freedoms and legitimate interests are in place. This can have a substantial impact for public administrations wishing to use automated individual decision-making in order to fight social security infringements, as they might want to rely on

[85] Art. 25 of the GDPR.
[86] Art. 5.1.d) of the GDPR.
[87] Art. 5.1.f) of the GDPR.
[88] 'Personal data revealing racial or ethnic origin, political opinions, religious or philosophical beliefs, or trade union membership, (...) genetic data, biometric data, data concerning health or data concerning a natural person's sex life or sexual orientation' (Art. 9.1 of the GDPR).

'data concerning health',[89] which are listed in the special categories of data.[90] Indeed, such data could be processed if the data subject has given explicit consent to their processing (Art. 9.2.a) of the GDPR). Yet, as we have seen in Chapter 4,[91] public administration should avoid relying on the data subject's consent, which will likely not be deemed as being freely given. Accordingly, the processing of health data by an AI in order to fight social security infringements will have to be necessary for reasons of substantial public interest and will have to be based on a law, which shall be proportionate to the aim pursued, respect the essence of the right to data protection and provide for suitable and specific measures to safeguard the fundamental rights and the interests of the data subject (Art. 9.2.g) of the GDPR).[92] As has been outlined in the SyRI case mentioned above,[93] fighting social security infringements can be considered of substantial public interest, as the Court in The Hague ruled that combatting fraud is key to maintaining citizen support in the social security system, which is one of the pillars of society, that new technological possibilities to prevent and combat fraud should therefore be used, and that the SyRI legislation thus pursues an important objective of general public interest.[94]

The right to obtain an explanation of the decision and AI explainability

An attentive reader of the GDPR will have noticed that Recital 71 goes further than Article 22.3 in terms of the appropriate safeguards that must be implemented, as it also mentions the right to obtain an explanation of the decision reached after the automated assessment, which does not appear in Article 22 of the GDPR.

The right to obtain an explanation of the automated decision

According to some authors, this implies that Article 22 of the GDPR does not, in fact, provide a right to obtain an explanation about how the automated decision was reached because, contrary to the Articles of the GDPR, the Recitals are not legally binding (Edwards and Veale, 2017; Wachter *et al.*, 2017). For these authors, this is not a mere omission, but a genuine desire not to include this right in the text of Article 22.3 of the GDPR (Wachter *et al.*, 2017). The European Parliament had indeed proposed to include this right in the article of the GDPR (Committee on Civil Liberties, Justice and Home Affairs, 2013), whereas the Council was of the opinion that this right should be mentioned only in the recitals (Presidency of the Council of the European Union, 2015), which clearly shows, according to them, that the final text of the GDPR is the result of a deliberate choice made during the trialogue negotiations, and not of a mere drafting error (Wachter *et al.*, 2017). However, other authors express more reservations on whether this implies that such a right to explanation does not exist, as 'many issues too controversial for agreement in the main text have

[89] Art. 4.15 of the GDPR.
[90] Art. 9.1 of the GDPR.
[91] See Chapter 4, Section 4.2.1.
[92] Art. 22.4 of the GDPR
[93] See Sections 5.2.2 and 5.2.5 – Automated decision making based solely on automated processing.
[94] Rechtbank Den Haag, 5 februari 2020, Zaak n° C-09-550982-HA ZA 18-388, ECLI:NL:RBDHA:2020:865 (ECLI:NL:RBDHA:2020:1878 for the English version), points 6.3 and 6.4.

been kicked into the long grass of the recitals, throwing up problems of just how binding they are' (Edwards and Veale, 2017). In fact, even the former group of authors admits that it could be argued that 'although it is certainly not explicit in the phrasing of Article 22.3, the right to obtain human intervention, express views or contest a decision is meaningless if the data subject cannot understand how the contested decision was taken' (Wachter *et al.*, 2017).

Thus, this right to obtain an explanation about the decision based solely on automated processing could be implicitly encapsulated in Article 22.3 of the GDPR. This point of view seems to be supported by the explanatory report of the Modernised Convention 108, which outlines that:

> Data subjects should be entitled to know the reasoning underlying the processing of data, including the consequences of such a reasoning, which led to any resulting conclusions, in particular in cases involving the use of algorithms for automated decision-making including profiling. (...) [T]hey should be entitled to know the logic underpinning the processing of their data and resulting (...) decision, and not simply information on the decision itself. Having an understanding of these elements contributes to the effective exercise of other essential safeguards such as the right to object and the right to complain to a competent authority. (Committee of Ministers of the Council of Europe, 2018)

The Article 29 Working Party seems to embrace the same view, as it indicates that: 'the controller should find simple ways to tell the data subject about the rationale behind, or the criteria relied on in reaching the decision' (Article 29 Working Party, 2018) and that the information that is provided should be 'sufficiently comprehensive for the data subject to understand the reasons for the decision' (Article 29 Working Party, 2018). Indeed, according to the Article 29 Working Party, 'the data subject will only be able to challenge a decision or express their view if they fully understand how it has been made and on what basis' (Article 29 Working Party, 2018).

Moreover, we believe that it could be argued that this right to obtain an explanation is also intrinsically included in the data subject's right to information,[95] as well as their right of access.[96] These two rights, which have been presented above,[97] enable them to receive useful information concerning the underlying logic of the processing operation, which should not only enable the data subject to know what is being done with their data, but also to understand the underlying logic of the processing (De Terwangne, 2015). That being said, it would have been preferable, for the sake of clarity, to explicitly include the reference to this right to obtain an explanation in Article 22.3 of the GDPR, rather than only in Recital 71.

---

[95] Arts. 13.2.f) and 14.2.g) of the GDPR. These articles both stipulate that the data subject should receive information about 'the existence of automated decision-making, including profiling, referred to in Article 22.1 and 22.4 and, at least in those cases, meaningful information about the logic involved, as well as the significance and the envisaged consequences of such processing for the data subject'.

[96] Art. 15.1.h) of the GDPR, which uses the same phrasing as Arts. 13.2.f) and 14.2.g).

[97] See Sections 5.2.2 and 5.2.3.

However, even if, on the basis of the above, it was to be deemed that the data subject has a right to an explanation of the decision, which is controversial, the ability of the public administration to provide explanations about the decisions taken by an AI used to fight social security infringements and tax fraud will depend on the type of AI that is used (expert systems vs. neural networks) and on the legal entity that holds the rights on the algorithm (the public administration itself vs a private entity). Indeed, the above analysis of these factors in the context of the obligation to inform the data subjects about the logic involved in the AI decision-making equally applies to the provision of explanations about the decision.[98]

Yet, it should be underlined that, at this point, public administrations seem to be aware of the importance of explainability. For instance, the Belgian FPS Finances is conscious that it needs to be able to explain why a certain person or company is suspected of tax fraud. For each case, the data miners are able to explain the reasoning behind the detection (indicators, techniques applied, etc.). Even more advanced techniques, such as social network analysis, used to detect more complex fraud types (e.g. 'domino bankruptcies'), are designed by the data miners. However, it should be mentioned that, when investigating a specific case, controllers can rely on AI techniques delivered by private software companies. It can therefore not be excluded that these private companies might hide behind commercial secrecy to refuse to provide explanations about the functioning of their algorithm, and this should be a key point of attention when dealing with those software providers.

Regarding social security infringements, even if the *ex ante* data-matching cross-checks are fully automated (Section 5.2.5 – Automated decision-making based solely on automated processing), they remain explainable because they are used to identify objective obstacles to the payment of the allowances. The machine thus does not have any margin of interpretation. Regarding bilateral *ex post* data-matching cross-checks, their results are also explainable, since they always imply a human verification. Similarly, the results of the data-mining operations conducted in the data warehouse are also explainable, since the indicators that are used to pinpoint suspicious cases have, in fact, been suggested by humans (the data miners). Yet, in the future, as fraud becomes more and more complex, the use of simpler algorithms with explainable business rules may become an issue, especially if public administrations increasingly need to resort to private sector providers.

However, it should be outlined that a person or an undertaking will not be informed that it has been flagged as being a potential (tax or social security) fraudster following data-mining operations conducted in the data warehouse, if the follow-up investigation did not result in the finding of fraud. Consequently, this person/undertaking might be repeatedly flagged as 'suspicious', although erroneously, without being aware of it and without being able to request explanations about why this is the case. This highlights that it is complex for public administrations to find a balance between being fully transparent and explaining the data-mining processes and models used, and the need to protect the confidentiality of their fraud analytics processes, as otherwise the fraudsters will adapt and avoid being detected.

---

[98] See Section 5.2.2.

To conclude, it is fundamental to emphasise that all of the above depends on how the notion of 'explainability' should be understood. What does it mean for a public administration to explain an algorithmic decision? Arguably, explaining 'the decision' requires more precision than simply explaining the 'logic involved', which could be more general. If we revert to the SyRI case, explaining the 'logic involved' would probably amount to explaining which indicators and which risk model are used,[99] while explaining the 'decision' may require an explanation of how the weight of each indicator has, in the specific case, led to the decision whether the person has a 'suspicious profile' or not.[100] We now briefly turn to the analysis of this notion in the AI context.

AI explainability

In the computer science literature, there are essentially two models of AI explainability (Bibal and Frenay, 2016). On the one hand, there are *interpretable models*, which are 'understandable either because their mathematical expressions are easy to understand (as it is the case with linear models) or can be represented in an easily understandable manner (as it is the case with decision trees)' (De Streel *et al.*, 2020). On the other hand, there are *black box models*, which 'are not easy to understand because their mathematical expression is neither straightforward nor easily representable in an understandable manner. For those models, understanding can be improved through explanations by using methods which are external to the models such as visualisation or approximation with interpretable models (Mittelstadt *et al.*, 2019)' (De Streel *et al.*, 2020).

In law and ethics, however, there is no commonly-agreed definition of AI explainability, although this has become a major concern for policy-makers across the world (De Streel *et al.*, 2020; Pasquale, 2015). At the European level, the requirement of AI explainability is directly highlighted in the 'Ethics Guidelines for Trustworthy AI' of the High-Level Expert Group on AI (High-Level Expert Group on Artificial Intelligence, 2019), and indirectly highlighted in the European Commission's White Paper on AI (European Commission, 2020). (De Streel *et al.*, 2020). Indeed the former lists four ethical principles in the context of AI systems, among which the 'Explicability' principle, which can be understood as a synonym of explainability (High-Level Expert Group on Artificial Intelligence, 2019), and lists seven requirements for trustworthy AI, among which the 'Transparency' requirement which explicitly includes explainability (De Streel *et al.*, 2020; High-Level Expert Group on Artificial Intelligence, 2019). The latter lists six requirements for high-risk AI applications, among which the need for 'Information provision', which does not refer explicitly to explainability, but which refers to the need for transparency and to the need to clearly inform about the functioning of the AI systems (European Commission, 2020).

---

[99] Rechtbank Den Haag, 5 februari 2020, Zaak n° C-09-550982-HA ZA 18-388, ECLI:NL:RBDHA:2020:865 (ECLI:NL:RBDHA:2020:1878 for the English version), point 4.22. See Sections 5.2.2 and 5.2.5 – Automated decision-making based solely on automated processing.
[100] *Ibid.*, points 4.29 to 4.30.

The requirement of AI explainability can also be observed indirectly in the Commission's proposal for an Artificial Intelligence Act, where it is provided that, in order to address 'the opacity that may make certain AI systems incomprehensible to or too complex for natural persons' (European Commission, 2021), high-risk AI systems 'shall be designed and developed in such a way to ensure that their operation is sufficiently transparent to enable users to interpret the system's output and use it appropriately'.[101] Furthermore, Annex IV of that proposal stipulates that the technical documentation pertaining to the AI system should contain: 'the design specifications of the system, namely the general logic of the AI system and of the algorithms; the key design choices including the rationale and assumptions made, also with regard to persons or groups of persons on which the system is intended to be used; the main classification choices; [and] what the system is designed to optimise for and the relevance of the different parameters'.[102]

Regarding the principle of 'explicability' of AI, the 'Ethics Guidelines for Trustworthy AI' of the High-Level Expert Group on AI stipulate that:

> Explicability is crucial for building and maintaining users' trust in AI systems. This means that processes need to be transparent, the capabilities and purpose of AI systems openly communicated, and decisions – to the extent possible – explainable to those directly and indirectly affected. Without such information, a decision cannot be duly contested. An explanation as to why a model has generated a particular output or decision (and what combination of input factors contributed to that) is not always possible. These cases are referred to as 'black box' algorithms and require special attention. In those circumstances, other explicability measures (e.g. traceability, auditability and transparent communication on system capabilities) may be required, provided that the system as a whole respects fundamental rights. The degree to which explicability is needed is highly dependent on the context and the severity of the consequences if that output is erroneous or otherwise inaccurate (High-Level Expert Group on Artificial Intelligence, 2019).

Regarding the 'explainability' of AI, the 'Ethics Guidelines for Trustworthy AI' of the High-Level Expert Group on AI stipulate that:

> Explainability concerns the ability to explain both the technical processes of an AI system and the related human decisions (e.g. application areas of a system). Technical explainability requires that the decisions made by an AI system can be understood and traced by human beings. Moreover, trade-offs might have to be made between enhancing a system's explainability (which may reduce its accuracy) or increasing its accuracy (at the cost of explainability). Whenever an AI system has a significant impact on people's lives, it should be possible to demand a suitable explanation of the AI system's decision-making process. Such explanation should be timely and adapted to the expertise of the stakeholder concerned (e.g. layperson, regulator or researcher). In addition, explanations of the degree to which

[101] Article 13.1 of the proposal for an Artificial Intelligence Act.

[102] Article 11.1 and Annex IV, point 2.b) of the proposal for an Artificial Intelligence Act.

an AI system influences and shapes the organisational decision-making process, design choices of the system, and the rationale for deploying it, should be available (hence ensuring business model transparency) (High-Level Expert Group on Artificial Intelligence, 2019).

Moreover, additional guidance about the notion of explainability can be found in the 'Explaining decisions made with AI' report of the UK's Information Commissioner's Office and the Alan Turing Institute (De Streel *et al.*, 2020; Information Commissioner's Office and the Alan Turing Institute, 2020). This report first outlines that there are two subcategories of explanation, namely '*process-based explanations* which give you information on the governance of your AI system across its design and deployment; and *outcome-based explanations* which tell you what happened in the case of a particular decision' (Information Commissioner's Office and the Alan Turing Institute, 2020) (emphasis added). If we apply this to our previous developments, 'process-based explanations' would pertain to the explanation of the 'logic involved' behind the algorithm used to fight social security infringements or tax fraud, while 'outcome-based explanations' would pertain to the explanation of the concrete reasons why the algorithm considers that an individual has a 'suspicious profile'. These two subcategories of explanation are sometimes also referred to as 'model-centric' and 'subject-centric' explanations (Edwards and Veale, 2017). While the former focusses more information on the AI model, the training metadata, the performance metrics and the estimated global logics, the latter aims at providing information about a specific query, output or decision (Edwards and Veale, 2017).

Additionally, the 'Explaining decisions made with AI' report identifies six main types of explanation:
1. '*Rationale explanation*: the reasons that led to a decision, delivered in an accessible and non-technical way.
2. *Responsibility explanation*: who is involved in the development, management and implementation of an AI system, and who to contact for a human review of a decision.
3. *Data explanation*: what data has been used in a particular decision and how.
4. *Fairness explanation*: steps taken across the design and implementation of an AI system to ensure that the decisions it supports are generally unbiased and fair, and whether or not an individual has been treated equitably.
5. *Safety and performance explanation*: steps taken across the design and implementation of an AI system to maximise the accuracy, reliability, security and robustness of its decisions and behaviours.
6. *Impact explanation*: steps taken across the design and implementation of an AI system to consider and monitor the impacts that the use of an AI system and its decisions has or may have on an individual, and on wider society' (emphasis in the original text) (Information Commissioner's Office and the Alan Turing Institute, 2020).

Two reservations must, however, be made regarding AI explanations. Firstly, AI systems that rely on a low number of variables are easier to 'explain', yet 'systems with more variables will typically perform better than simpler systems, so we may end up with a trade-off between performance and explicability' (Edwards and Veale, 2017). Moreover, these complex models may be especially difficult to explain because 'the features that are being fed in might lack any convenient or clear human interpretation in the first place, even if we are creative about it. LinkedIn, for example, claims to have over 100,000 variables held on every user that feed into ML modelling. Many of these will not be clear variables like 'age,' but more abstract ways you interact with the webpage, such as how long you take to click, the time you spend reading, or even text you type in a text box but later delete without posting' (Edwards and Veale, 2017).

Secondly, because AI systems do not function like human minds, it might be impossible, in some cases, to 'explain' in a way that is satisfactory for humans how a complex AI system relying on thousands of variables and correlations has taken a decision, i.e. to provide a 'humanly-understandable decision' (Busuioc, 2020). Accordingly, some authors argue that the focus should instead be set on the human interpretability of a decision (i.e. it can be understood and interpreted by a layman) rather than on its 'explanation' per se (Busuioc, 2020). In this regard, "pedagogical' systems which create explanations around a model rather than from decomposing it may be useful and benefit from not relying on disclosure of proprietary secrets or IP' (Edwards and Veale, 2017).

In light of the above, the public administration using an AI to fight social security infringements and tax fraud must thus be able to explain how the algorithm has, technically, reached its decision in a way that can be understood and verified by humans (High-Level Expert Group on Artificial Intelligence, 2019). This implies the ability to explain not only the logic involved behind the algorithm (process-based explanation) but also the concrete decision of why the algorithm considers that a person has a 'suspicious profile' (outcome-based processing) (Information Commissioner's Office and the Alan Turing Institute, 2020). Similarly, it must be able to explain how human decisions, based on the algorithm's decision, have been taken. This explanation should be adapted to the person to whom it is given, as the person must be able to understand it (a layperson will not have the same level of understanding as a researcher in AI).

In order to be compliant with Article 22 of the GDPR, we believe that – provided that the data subject has a right to an explanation of the AI decision, which is controversial –,[103] the public administration should, at least, be able to explain in an accessible and non-technical way, the reasons that led to the decision (rationale explanation), to identify who is involved in the development, management and implementation of an AI system, and to identify who to contact for a human review of the decision (responsibility explanation) and what data has been used, and how, in the particular decision (data explanation) (Information Commissioner's Office and the Alan Turing Institute, 2020). Indeed, the other three types of explanation outlined by the 'Explaining decisions made with AI' report (fairness explanation, safety and performance explanation, and impact explanation) (Information Commissioner's Office and the Alan Turing Institute, 2020) do not so much focus on the explanation of the decision as such, but more on how the AI system that has taken the decision is built. It is, however, recommended to also provide these types of explanation in order to reinforce the citizens' trust in the use of AI by public administrations.

---

[103] See previous Section 5.2.5 – The right to obtain an explanation of the automated decision.

As mentioned above,[104] the ability of the public administration to provide such explanations about the decisions taken by an AI used to fight social security infringements and tax fraud will depend on the type of AI that is used and on the legal entity that holds the rights to the algorithm. In those circumstances, 'other explicability measures (e.g. traceability, auditability and transparent communication on system capabilities) may be required, provided that the system, as a whole, respects fundamental rights' (High-Level Expert Group on Artificial Intelligence, 2019). To ensure that it meets these explanation requirements, the public administration could use the 'Explainability checklist' provided in the 'Ethics Guidelines for Trustworthy AI' of the High-Level Expert Group on AI:

> 'Did you assess:
> - to what extent the decisions and hence the outcome made by the AI system can be understood?
> - to what degree the system's decision influences the organisation's decision-making processes?
> - why this particular system was deployed in this specific area?
> - what the system's business model is (for example, how does it create value for the organisation)?
>
> Did you ensure an explanation as to why the system took a certain choice resulting in a certain outcome that all users can understand?
> Did you design the AI system with interpretability in mind from the start?
> Did you research and try to use the simplest and most interpretable model possible for the application in question?
> Did you assess whether you can analyse your training and testing data? Can you change and update this over time?
> Did you assess whether you can examine interpretability after the model's training and development, or whether you have access to the internal workflow of the model?' (High-Level Expert Group on Artificial Intelligence, 2019, 2020).

5.3 Additional legal challenges for the development of algorithmic decision-making tools by the public administration

In Section 5.2, we outlined the main personal data protection rules that must be factored in by public administrations willing to rely on AI techniques to improve the delivery of their public services. In this regard, particular attention was devoted to public policies and decision-making linked to social security infringements and tax fraud. With the same objective in mind, this section will be dedicated to the analysis of additional legal challenges that shall be considered when reflecting on the development of algorithmic decision-making tools by the public administration. These legal challenges revolve around four topics, namely the citizens' right to a human public service, their right to an equal access to public services, the transparency of the public administration and the administrative decision-making through AI systems (Gérard, 2017).

[104] See Section 5.2.2.

5.3.1 Risk of 'dehumanised' public services

Like any new technology, resorting to AI in public services presents both risks and opportunities. The use of AI could first make public services more available to citizens, as they could access it 24 hours a day, 7 days out of 7, and it would arguably speed up services (Gérard, 2017). Moreover, delegating the repetitive and non-complex tasks to an AI would grant more time to the civil servants to focus on more complex cases, which would ultimately benefit the citizens (Gérard, 2017). However, resorting to AI rather than humans might 'dehumanise' the public service (Gérard, 2017). Although there is, as such, no 'right to a human public service', the right to human dignity could come into play (Gérard, 2017). Indeed, this right is recognised as 'the very essence' of the European Convention on Human Rights[105] and is explicitly enshrined in Article 1 of the Charter of Fundamental Rights of the European Union (Gérard, 2017). Accordingly, it should be ensured that machines (AI) do not become the norm, and people the exception (Nevejans, 2016).

5.3.2 Equal access to public services

Resorting to AI, like resorting to any other technology, may lead to a digital divide among citizens, as some of them will not be able to use these technologies, either because they lack the skills to do so, or because they have physical inabilities that prevent them from doing so (Chantillon *et al.*, 2017). In the specific case of AI, the complexity of the technical knowledge required for the effective use of this technology is such as to make it totally or partially inaccessible to part of the population (Gérard, 2017). This may have an impact on two key principles of the public service, namely the principle of equality and the principle of accessibility.

The principle of equality

This principle of equality is enshrined in several supranational Conventions[106] and in Articles 10 and 11 of the Belgian Constitution, and it prevents the legislator from treating differently two equal situations or categories of people, or conversely, from treating equally two different situations or categories of people (Gérard, 2017).[107] Applied to the public service, the principle of equality is defined as the 'law of equality', and constitutes, together with the 'law of mutability' and the 'law of continuity', the three 'laws of the public service' (Gérard, 2017). According to this law, all the users of the public service should be treated equally and should benefit from the same services and advantages (Gérard, 2017).

---

[105] ECtHR, *Christine Goodwin v. United Kingdom*, 11 July 2002, req. n° 28957/97, point 90; *S.W. v. United Kingdom*, 22 November 1995, req. n° 20166/91, point 44.

[106] Article 14 of the European Convention on Human Rights and Articles 20, 21 and 23 of the Charter of Fundamental Rights of the European Union.

[107] See Belgian Constitutional Court, case n° 21/89, 13 July 1989, point B.4.5.b; and case n° 16/92, 12 March 1992, point B.3.3.

If applied to AI, this means that all the users should be treated equally by the algorithm, which shall not be biased and shall not entail discriminations against some categories of the population. Indeed, even if AI systems are often presented as being deprived of any bias, they might reflect biases that have been integrated, intentionally or not, by their human creators, but also biases that derive from the training data that has been selected to build them (Défenseurs des droits et CNIL, 2020). One such bias is the lack of representative data, which, in the field of facial recognition technologies, led to the finding that the chances of 'false positives' were much greater for woman and for people of colour, because the algorithm had, to a large extent, been trained with images of white men (Buolamwini and Gebru, 2018; Défenseurs des droits et CNIL, 2020). Indeed, even if, at first sight, the AI systems rely on 'neutral criteria', the combination of several of these criteria can lead to biases and, as a matter of consequence, to automatic and systemic discriminations (Défenseurs des droits et CNIL, 2020).

Moreover, these discrimination risks are enhanced by the fact that AI systems often tend to target and control minorities and marginalised social groups (Défenseurs des droits et CNIL, 2020; Eubanks, 2018). This is especially relevant to keep in mind for AI systems aiming at fighting social security infringements and tax fraud. For instance, in the SyRI case brought before the Court in The Hague,[108] the claimants argued that the use of SyRI had a discriminatory and stigmatising effect, because it was allegedly used to 'further investigate neighbourhoods that are known as 'problem areas'. This increases the chances of discovering irregularities in such areas as compared to other neighbourhoods, which in turn confirms the image of a neighbourhood as a problem area, contributes to stereotyping and reinforces a negative image of the occupants of such neighbourhoods, even if no risk reports have been generated about them'.[109] The Court agreed that SyRI had only been applied to 'problem districts' and that there was a risk that this could lead to biases towards people with a lower socio-economic status or an immigration background,[110] but outlined that, because of the lack of transparency of the SyRI legislation, it was unable to assess whether this risk of discrimination had been sufficiently neutralised by the State.[111] The court thus acknowledged the risk of the discrimination but was unable to verify whether this risk materialised *in casu*.

In light of the above, public administrations wishing to rely on AI to fight social security infringements and tax fraud will have to audit their AI systems on a regular basis in order to ensure that they are not biased and that they do not discriminate against some categories of citizens (Défenseurs des droits et CNIL, 2020). Naturally, this implies that the public authorities need to be able to both understand and explain how the AI system works.[112] In this regard, it is relevant to mention that, in the context of tax fraud, data quality checks are performed in the data warehouse, in order to ensure that the data is not biased at the application level and does not

[108] Rechtbank Den Haag, 5 februari 2020, Zaak n° C-09-550982-HA ZA 18-388, ECLI:NL:RBDHA:2020:865 (ECLI:NL:RBDHA:2020:1878 for the English version). See Sections 5.2.2 and 5.2.5 – Automated decision-making based solely on automated processing.
[109] *Ibid.*, point 6.92.
[110] *Ibid.*, point 6.93.
[111] *Ibid.*, point 6.94.
[112] See Sections 5.2.5 – The right to obtain an explanation of the decision and AI explainability, and 5.3.4.

lead to discrimination. In fact, several data-mining projects pursued by the data miners solely aim at improving and ensuring data quality. Regarding social security infringements, we can only assume that any risk of inequality is discarded at the stage of the drafting of the data transfer protocol or at the stage of the obtention of the prior authorisation from the Information Security Committee.[113] However, due to the relative opacity in this regard, the existence of inequalities and discrimination cannot be excluded.

The principle of accessibility

The principle of accessibility of public services is enshrined in the Belgian Charter of the public service user.[114] This Charter provides that 'public services must be accessible in the broadest sense of the term, which goes beyond the problems of physical accessibility and proximity (...). It is also a question of the clarity of the texts. Administrative documents and legislation should not be drafted in such a way that the public has great difficulty in understanding them'.[115] In this regard, the application of the principle of accessibility of the public service precludes the introduction of AI that would make this public service excessively complex for users (Gérard, 2017). Therefore, the public administration wishing to rely on AI to fight social security infringements and tax fraud will have to ensure that the functioning of the algorithm is not too complex, and that the citizen can understand, at least at a high level of abstraction, how the algorithm works and why they are, for instance, considered a 'suspicious profile'. This links to the requirement of algorithmic transparency (Section 5.2.2 and 5.3.3).

5.3.3 Transparency of the public administration

The GDPR is not the only legal text that imposes transparency of the public administration for citizens. Indeed, citizens benefit from the fundamental right of administrative publicity, which is enshrined in Article 32 of the Constitution (Gérard, 2017). At the federal level, this constitutional provision is further specified in a Law of 11 April 1994,[116] which provides for two types of administrative publicity, namely active and passive publicity (Gérard, 2017).

In light of the obligation of active publicity, the federal public administration has to actively provide, independently of any request, a clear and objective information to the public about its actions, competences and functioning (Gérard, 2017).[117]

In light of the obligation of passive publicity, the federal public administration should, at the request of a citizen, allow them to consult an administrative document, to obtain explanations about it and to receive a copy of it (Gérard, 2017).[118] In order to broaden the scope of this publicity obligation

---

[113] See Chapter 4, Section 4.2.3

[114] Charte de l'utilisateur des services publics, 4 décembre 1992, *M.B.*, 22 janvier 1993.

[115] See Chapter II, Section A of the Charter. Author's own translation.

[116] Loi du 11 avril 1994 relative à la publicité de l'administration, *M.B.*, 30 juin 1994.

[117] Art. 2 of the Law of 11 April 1994.

[118] Arts. 4 and 5 of the Law of 11 April 1994.

as much as possible, an administrative document is defined as 'any information, whatever its form, held by a public administration' (Gérard, 2017).[119] For documents of a personal nature – e.g. administrative documents involving an assessment or value judgement relating to a named or easily identifiable natural person, or a description of a conduct the disclosure of which could manifestly cause harm to that person –,[120] the requester must justify an interest to access it.[121]

Regarding, more specifically, the explanations about the use of algorithms for decision-making pertaining to individuals, it is worth mentioning that a law proposal of 6 April 2021 provides that, in order to increase transparency, it should be compulsory for administrations to publish online the rules defining the main algorithmic treatments used in the performance of their tasks when these constitute all or part of the basis for individual decision.[122] Moreover, this proposal provides that, for any administrative document with an individual scope, the administration shall communicate to the person who is the subject of an individual decision taken in whole or in part on the basis of an algorithmic processing, at the latter's request, the characteristics of the algorithm in an intelligible form, provided that this communication does not infringe secrets protected by law.[123] This would cover the degree and type of contribution of the algorithmic processing to the decision-making; the data processed and their sources; the processing parameters and, where appropriate, their weighting applied to the individual's situation; and the operations carried out through the processing.[124] It remains to be seen whether this proposal will be adopted.

It should, however, be noted that Article 6.1.6° of this Law of 11 April 1994 provides that the federal public administration shall refuse a request for consultation, explanation or disclosure in the form of a copy of an administrative document if it is satisfied that the interest in disclosure does not outweigh the protection of a federal economic or financial interest (Degrave and Lachapelle, 2014). Such a decision will have to be formally motivated, as requested by the law of 29 July 1991[125] (Degrave and Lachapelle, 2014).

Public administrations wishing to rely on AI to fight social security infringements and tax fraud will thus have to be transparent about these missions, about the role played by the AI and by humans, and about the logic behind the decisions (active publicity) and will have to conduct a balance of interests in order to determine whether they need to provide a copy of these information to citizens that would request it (passive publicity) (Gérard, 2017). This will allow for a reduction in the opacity of the administration's actions in the eyes of the citizens (Committee of Ministers of the Council of Europe, 2021; Degrave, 2014).

---

[119] Art. 1.b).2° of the Law of 11 April 1994.
[120] Art. 1.b).3° of the Law of 11 April 1994.
[121] Art. 4 of the Law of 11 April 1994.
[122] Proposition de loi modifiant la loi relative à la publicité de l'administration du 11 avril 1994 afin d'introduire une plus grande transparence dans l'usage des algorithmes par les administrations, 6 avril 2021, *Doc. parl.*, Chambre, sess. ord., 2020-2021, no 55-1904/001, p. 6 and 8.
[123] *Ibid.*
[124] *Ibid.*, p. 8.
[125] Loi du 29 juillet 1991 relative à la motivation formelle des actes administratifs, *M.B.*, 12 septembre 1991.

5.3.4 Administrative decision-making by an AI system

Any use of AI for decision-making, irrespective of whether it is *solely* based on automated processing[126] (i.e. even if there is a human intervention), also generates challenges from an administrative law point of view. Indeed, a difference must be made between decisions where the AI simply applies a precise number of rules that it is bound to follow, and decisions where the AI has more leeway in taking its decision (Gérard, 2017). In the first case, resorting to AI might actually be extremely useful and efficient. For example, in terms of social security, an AI could be extremely efficient in crossing the citizens' data, in order to determine whether they are entitled to a certain subsidy. Here, the conditions to receive this subsidy are known in advance and easily verifiable, e.g. the citizen can receive the subsidy if conditions A, B and C are met, and the AI can easily check this and allow the subsidy based on those rules. In the second case, it might be more problematic to resort to AI decision-making if it has some leeway, as it will not always be possible to verify whether the AI's decision is compatible with the law, notably if the public administration is not able to check nor explain how the AI came to that decision.[127]

Additionally, the Belgian law of 29 July 1991 on the formal motivation of the administrative acts provides that all unilateral legal acts of individual scope emanating from an administrative authority, whose purpose is to produce legal effects in respect of one or more persons under its jurisdiction, have to be 'formerly motivated', which implies that the act must contain the legal and factual conditions that have led to the decision.[128] This formal motivation must be adequate,[129] which means that the person should be able to understand the reasoning that has led to the decision (Gérard, 2017). For decisions taken by an AI, this should be rather easy to do if it simply applies a precise number of rules that it is bound to follow, but it might be more problematic to explain the decision taken by the AI if it has some leeway in doing so (Gérard, 2017). Our developments above regarding the right to obtain an explanation about the AI's decision and about the explainability of an AI decision can be transposed here.[130] Finally, this formal motivation must satisfy the principle of accessibility outlined above, and must thus be written in such a way that the citizen can clearly understand the motives that have led to the decision (Gérard, 2017).[131]

5.4 Conclusions

As outlined throughout this chapter, public administrations are increasingly relying on artificial intelligence technologies and automated decision-making in order to provide public services and support their decision-making. Yet, while this can lead to significant benefits, notably in terms of efficiency, for these administrations, it must not be overlooked that the use of such technologies

[126] See Section 5.2.5 – Automated decision-making based solely on automated processing.
[127] See Section 5.2.5 – The right to obtain an explanation of the decision and AI explainability.
[128] Loi du 29 juillet 1991 relative à la motivation formelle des actes administratifs, *M.B.*, 12 septembre 1991, arts. 1, 2 and 3.
[129] Art. 3 of the Law of 29 July 1991.
[130] See Section 5.2.5 – The right to obtain an explanation of the decision and AI explainability.
[131] See Section 5.3.2 – The principle of accessibility.

can also generate significant impacts on the lives and fundamental rights of the citizens to which automated decision-making systems are applied. Therefore, administrations need to comply with the legal framework that aims at limiting the uses they can make of such technologies, in order to circumscribe these impacts. In this regard, the aim of this chapter was precisely to outline how this could be done in practice.

On that basis, several key takeaways pertaining to the use of artificial intelligence and of automated decision-making technologies by public administrations can be given. First, since data has to be processed fairly and in a transparent manner, public administration making use of individual decision-making based 'solely' on automated processes will have to inform the citizens about the existence of such processes, and will have to provide them, at least with meaningful information about the logic involved, as well as about the significance and the envisaged consequences of such processing.[132] In order to comply with this requirement, it is preferable for public administrations to make use of expert systems rather than neural networks, and to ensure that they hold the rights to the algorithm, rather than to grant those rights to the private entity that has been tasked, through the means of a public procurement, to develop the algorithm for them. Second, a public administration wishing to use AI applications should anticipate the potential data access and data erasure requests that it may receive from citizens, and it should thus make sure that the AI system is designed in a way that allows them to respond in a timely fashion to such requests.[133] Third, it is fundamental for a public administration willing to rely on AI techniques to respect the citizens' right not to be subjected to a decision based 'solely' on automated processing.[134] This right echoes the strong will of the human being not to be entirely subject to the machine, as the former does not accept the idea that a decision may be imposed on it solely on the basis of conclusions reached by that machine. In the same vein, it would not be acceptable for public services, based on AI or automated decision-making technologies, to become 'dehumanised' or no longer equally accessible for all.[135] Coming back to the right not to be subjected to a decision based 'solely' on automated processing, public administrations should refrain from including a fictitious or negligible human intervention in the automated decision process simply in order to avoid (potentially in bad faith) the application of this right.[136] They should also adopt suitable measures to safeguard the citizen's rights and freedoms and legitimate interests, which should, at least, provide the latter with the right to obtain human intervention on the part of the controller, to express his or her point of view and to contest the decision.[137] Last but not least, they should also be able to explain the decisions that have been taken by the algorithm.[138]

---

[132] See Sections 5.2.2 and 5.3.3.
[133] See Sections 5.2.3 and 5.2.4.
[134] See Section 5.2.5.
[135] See Sections 5.3.1 and 5.3.2.
[136] See Section 5.2.5 – Automated decision-making based solely on automated processing.
[137] See Section 5.2.5 – Exceptions to the right not to be subject to an automated individual decision-making, and appropriate safeguards.
[138] See Sections 5.2.5 – The right to obtain an explanation of the decision and AI explainability and 5.3.4.

An important common point between the personal data protection rules and the core principles of administrative law that have been presented in this chapter is that they both aim to give more control to the citizens, through increased transparency and understanding of the concrete uses of these technologies made by administrations. Furthermore, they aim to ensure that the human, rather than the machine, always remains at the centre. Indeed, it is fundamental for these public administration not to be blinded by the (efficiency) advantages that such technologies offer, and to always keep in mind that the primary goal of any public policy should be to improve the lives of their citizens.

In this regard, it is highly reassuring to see that the European Commission has proposed the adoption of an Artificial Intelligence Act (European Commission, 2021), which emphasises that any actor wishing to rely on AI techniques (including public administration using AI to support their policy- or decision-making) should question whether this entails a 'high risk' to people's rights and freedoms.[139] This risk-based approach is to be welcomed, as regulatory intervention on AI should be proportionate and should differentiate between categories of AI applications, focussing on those that generate the greatest risks for people. More concretely, the proposal first suggests to prohibit AI uses that create an unacceptable risk, notably because they would contravene EU values and/or would violate fundamental rights.[140] Second, the proposal lays down a risk methodology to define 'high-risk' AI systems that pose significant risks to the health and safety or fundamental rights of persons. The classification of an AI system as high-risk is based on the 'intended purpose' of the AI system, which implies that this classification 'does not only depend on the function performed by the AI system, but also on the specific purpose and modalities for which that system is used'.[141] These high-risk AI systems will have to undergo *ex ante* conformity assessment procedures,[142] and they will have to be registered in a public EU-wide database to increase public transparency and oversight and strengthen *ex post* supervision by competent authorities.[143] Furthermore, any AI system meeting this 'high-risk' threshold will have to comply with the legal requirements set out in the proposal, which pertain to the need: (1) to establish, implement, document and maintain a risk management system; (2) to produce technical documentation; (3) to ensure that high quality data is used; (4) to adopt an appropriate data governance allowing record-keeping – i.e. logs –, transparency and the provision of information to users; (5) to ensure human oversight; and (6) to ensure a certain level of accuracy, robustness, and cybersecurity.[144] Finally, the proposal suggests imposing transparency obligations for certain low-risk systems as well.[145]

[139] See Section 5.2.1.
[140] See Article 5 of the Proposal for an Artificial Intelligence Act.
[141] *Ibid.*, p. 13.
[142] *Ibid.*, p. 3. See also Articles 40 to 50, and Annexes VI and VII.
[143] *Ibid.*, p. 12-14. See also Articles 51 and 60, and Annex VIII.
[144] *Ibid.*, Articles 8 to 15 and Annex IV.
[145] *Ibid.*, p. 14. See Article 52 for more details.

References

Alston, P. 2019. Extreme poverty and human rights. Note by the Secretary-General of the United Nations. Available at: https://undocs.org/pdf?symbol=en/A/74/493.

Article 29 Working Party. 2018. Guidelines on Automated individual decision-making and Profiling for the purposes of Regulation 2016/679. Available at: https://ec.europa.eu/newsroom/article29/item-detail.cfm?item_id=612053.

Bibal, A., and B. Frenay. 2016. Interpretability of machine learning models and representations: an introduction. In M. Verleysen (ed.) *24th European Symposium on Artificial Neural Networks, Computational Intelligence and Machine Learning: ESANN 2016*. 77-82. CIACO, Bruges, Belgium.

Buolamwini, J., and T. Gebru. 2018. Gender shades: intersectional accuracy disparities in commercial gender classification. In S.A. Friedler, and C. Wilson (eds.) *Proceedings of the 1st Conference on Fairness, Accountability and Transparency, Proceedings of Machine Learning Research*. 1-15. New York University, New York, NY, USA.

Busuioc, M. 2020. Accountable artificial intelligence: holding algorithms to account. *Public Administration Review* 81: 825-836. https://doi.org/10.1111/puar.13293

Chantillon, M., R. Kruk, A. Simonofski, T. Tombal, J. Crompvoets, C. de Terwangne, N. Habra, M. Snoeck, and B. Vanderose. 2017. FLEXPUB Public e-Service Strategy – Work package 2 – Baseline Measurement. KU Leuven Public Governance Institute, Leuven, Belgium.

Committee of Ministers of the Council of Europe. 2018. Explanatory report of the Protocol amending the Convention for the Protection of Individuals with regard to Automatic Processing of Personal Data (ETS No. 108) – Modernised Convention 108, CM(2018)2-addfinal. Available at: https://ccdcoe.org/uploads/2019/09/CoE-180518-Explanatory-Report-to-the-Protocol-amending-the-Convention-for-the-Protection-of-Individuals-with-regard-to-Automatic-Processing-of-Personal-Data.pdf.

Committee of Ministers of the Council of Europe. 2020. Recommendation CM/Rec(2020)1 of the Committee of Ministers of the Council of Europe to member States on the human rights impacts of algorithmic systems. Available at: https://search.coe.int/cm/pages/result_details.aspx?objectid=09000016809e1154.

Committee of Ministers of the Council of Europe. 2021. Declaration by the Committee of Ministers of the Council of Europe on the risks of computer-assisted or artificial-intelligence-enabled decision making in the field of the social safety net, Decl(17/03/2021)2. Available at: https://search.coe.int/cm/Pages/result_details.aspx?ObjectID=0900001680a1cb98.

Committee on Civil Liberties, Justice and Home Affairs. 2013. Report on the proposal for a regulation of the European Parliament and of the Council on the protection of individuals with regard to the processing of personal data and on the free movement of such data (General Data Protection Regulation), A7-0402/2013 – 2012/0011(COD). Available at: https://www.europarl.europa.eu/sides/getDoc.do?pubRef=-//EP//TEXT+REPORT+A7-2013-0402+0+DOC+XML+V0//EN.

Consultative Committee of the Convention for the protection of individuals with regard to automatic processing of personal data. 2021. Guidelines on Facial Recognition, T-PD(2020)03rev4. Available at: https://rm.coe.int/guidelines-on-facial-recognition/1680a134f3.

Council. 2002. Council Framework Decision 2002/584/JHA of 13 June 2002 on the European arrest warrant and the surrender procedures between Member States. Available at: https://eur-lex.europa.eu/legal-content/EN/TXT/?uri=celex%3A32002F0584.

De Raedt, S. 2017. The impact of the GDPR for tax authorities. *Revue du Droit des Technologies de l'Information* 66-67: 129-143.

De Streel, A., A. Bibal, B. Frenay, and M. Lognoul, 2020. Explaining the black box – when law controls AI. CERRE Issue Paper. Available at: https://www.cerre.eu/publications/explaining-black-box-when-law-controls-ai.

De Terwangne, C. 2015. La réforme de la Convention 108 du Conseil de l'Europe pour la protection des personnes à l'égard du traitement automatisé des données à caractère personnel. In: C. Castets-Renard (ed.) *Quelle protection des données personnelles en Europe*? 81-120. Larcier, Brussels, Belgium.

Défenseurs des droits et CNIL. 2020. Algorithmes: prévenir l'automatisation des discriminations. Available at: https://www.defenseurdesdroits.fr/sites/default/files/atoms/files/synth-algos-num2-29.05.20.pdf.

Degrave, E. 2014. *L'E-Gouvernement et la protection de la vie privée. Légalité, transparence et contrôle*. Larcier, Brussels, Belgium.

Degrave, E, 2020. The use of secret algorithms to combat social fraud in belgium. *European Review of Digital Administration & Law* 1: 167-177.

Degrave, E., and A. Lachapelle. 2014. Le droit d'accès du contribuable à ses données à caractère personnel et la lutte contre la fraude fiscal. *Revue Générale du Contentieux Fiscal* 28: 322-335.

Edwards, L., and M. Veale. 2017. Slave to the algorithm? Why a 'right to an explanation' is probably not the remedy you are looking for. *Duke Law & Technology Review* 16: 77-79.

Eubanks, V. 2018. Automating inequalities: how high-tech tools profile, police, and punish the poor. *Law, Technology and Humans* 1: 162-164. https://doi.org/10.5204/lthj.v1i0.1386

European Commission. 2018. Communication from the Commission to the European Parliament, the Council, the European Economic and Social Committee and the Committee of the Regions. Artificial Intelligence for Europe. COM(2018) 237 final. Available at: https://eur-lex.europa.eu/legal-content/EN/TXT/?uri=COM%3A2018%3A237%3AFIN.

European Commission. 2020. White Paper on Artificial Intelligence – A European approach to excellence and trust, COM(2020) 65 final. Available at: https://ec.europa.eu/info/sites/default/files/commission-white-paper-artificial-intelligence-feb2020_en.pdf.

European Commission. 2021. Proposal for a Regulation of the European Parliament and of the Council laying down harmonised rules on artificial intelligence (Artificial Intelligence Act) and amending certain Union legislative acts, Brussels, COM(2021) 206 final. Available at: https://digital-strategy.ec.europa.eu/en/library/proposal-regulation-european-approach-artificial-intelligence.

European Data Protection Board. 2020. Guidelines 05/2020 on consent under Regulation 2016/679 (V.1.1). Available at: https://edpb.europa.eu/sites/edpb/files/files/file1/edpb_guidelines_202005_consent_en.pdf.

Fosch Villaronga, E., P. Kieseberg, and T. Li. 2018. Humans forget, machines remember: Artificial Intelligence and the Right to be Forgotten. *Computer Law and Security Review* 34: 304-313. https://doi.org/10.1016/j.clsr.2017.08.007

Gérard, L. 2017. Robotisation des services publics: l'intelligence artificielle peut-elle s'immiscer sans heurt dans nos administrations. In: H. Jacquemin, and A. De Streel, A. (eds.) *L'Intelligence Artificielle et le Droit*. 413-436. Larcier, Brussels, Belgium.

High-Level Expert Group on Artificial Intelligence. 2019. Ethics guidelines for trustworthy AI. Available at: https://ec.europa.eu/digital-single-market/en/news/ethics-guidelines-trustworthy-ai.

High-Level Expert Group on Artificial Intelligence. 2020. The assessment list for trustworthy artificial intelligence (ALTAI). Available at: https://ec.europa.eu/digital-single-market/en/news/assessment-list-trustworthy-artificial-intelligence-altai-self-assessment.

Information Commissioner's Office and the Alan Turing Institute. 2020. Explaining decisions made with AI (v. 1.0.11). Available at: https://ico.org.uk/for-organisations/guide-to-data-protection/key-data-protection-themes/explaining-decisions-made-with-artificial-intelligence/.

Misuraca, G., and C. van Noordt. 2020. Overview of the use and impact of AI in public services in the EU. Publications Office of the European Union, Luxembourg, Luxembourg, https://doi.org/10.2760/039619

Mittelstadt, B., C. Russell, and S. Wachter. 2019. Explaining explanations in AI. In: *Proceedings of the conference on fairness, accountability, and transparency (FAT)*. 279-288. Association for Computing Machinery, New York, NY, USA.

Nevejans, N. 2016. Study on the European Civil Law rules in robotics. Commissioned by the European Parliament's Legal Affairs Committee. Available at: https://www.europarl.europa.eu/RegData/etudes/STUD/2016/571379/IPOL_STU(2016)571379_EN.pdf.

Pasquale, F. 2015. *Black box society. The secret algorithms that control money and information*. Harvard University Press, London, UK.

Presidency of the Council of the European Union, 2015. Proposal for a Regulation of the European Parliament and of the Council on the protection of individuals with regard to the processing of personal data and on the free movement of such data (General Data Protection Regulation) – Preparation of a general approach, 9565/15. Available at: https://data.consilium.europa.eu/doc/document/ST-9565-2015-INIT/en/pdf.

Scarcella, L. 2019. Tax compliance and privacy rights in profiling and automated decision making. *Internet Policy Review* 8: 1-19.

Tombal, T., 2018. Les droits de la personne concernée dans le RGPD. In: C. De Terwangne, and K. Rosier (eds.) *Le Règlement Général sur la Protection des Données (RGPD/GDPR): analyse approfondie*. 407-557. Larcier, Brussels, Belgium.

Villani, C., M. Schoenauer, Y. Bonnet, C. Berthet, A-C. Cornut, F. Levin, and B. Rondepierre. 2018. Donner un sens à l'intelligence artificielle: Pour une stratégie nationale et européenne. Available at: https://hal.inria.fr/hal-01967551/document.

Wachter, S., B. Mittelstadt, and L. Floridi. 2017. Why a right to explanation of automated decision-making does not exist in the general data protection regulation. *International Data Privacy Law* 7: 76-99.

Zarsky, T., 2017. Incompatible: the GDPR in the age of big data. *Seton Hall Law Review* 47: 995-1020.

Chapter 6

Trustworthiness in an era of data analytics: what are governments dealing with and how is civil society responding?

B. Kleizen*, W. van Dooren and K. Verhoest

University of Antwerp, Faculty of Social Sciences, Department of Political Science, Research Group Politics & Public Governance (PPG); Centre of Excellence GOVTRUST, Sint Jacobstraat 2, 2000 Antwerp, Belgium; bjorn.kleizen@uantwerpen.be

Abstract

Implementing new technologies in government is almost inevitably risky. Even well-designed big data and/or AI projects run the risk of losing political or public confidence when encountering salient difficulties. Recent scandals have therefore led legal and social scientists to question the trustworthiness and security of data analytics projects in the public sector, in particular when these projects include sensitive data or result in outcomes such as profiling. However, the limited amount of empirical research into the way that governments attempt to generate trustworthiness and how the trustworthiness of data analytics projects are perceived in practice remains an Achilles heel of this literature. This chapter contributes to filling this gap by qualitatively examining a conceptualisation on trustworthiness dimensions and the imperfect translation of such strategies into externally perceived trustworthiness. More specifically, we conduct an exploratory analysis based on interview data with governments and civil society actors (such as NGO's). This exercise provides new insights into (1) how governments currently attempt to make their data analytics projects trustworthy and (2) how external actors perceive the trustworthiness of data analytics in government. It is concluded that governments place higher emphasis on trustworthiness and security than is frequently assumed, but that these efforts do not always align with external expectations. It is proposed that a more comprehensive focus, also incorporating stakeholder consultation and external communication through both sector-wide policies and project-level initiatives, could benefit perceived trustworthiness of data analytics projects.

Keywords: AI, trust in AI, trust in government, algorithm, data analytics

6.1 Introduction

As data analytics technologies such as AI and big data become increasingly ubiquitous in the public sector, so too does the importance of trust in these technologies and the ways in which they are governed. This is hardly a new or surprising insight: user trust and the ethical application

DOI 10.3920/978-90-8686-930-5_6, © Wageningen Academic Publishers 2022

of emerging technologies have become important topics of inquiry in data science, while citizen-government trust relationships are frequently featured in both social science and academic legal debates (Ahonen and Erkkilä, 2020; Ananny and Crawford, 2018; Busuioc, 2020). The topic has even reached mass media reporting, as is shown by major events such as the US Facebook-Cambridge Analytica scandal or the Dutch childcare benefits scandal, both of which featured major society-wide trust breaches that stemmed in part from the use of large-scale databases and AI (Busuioc, 2020; Isaak and Hanna, 2018; La Fors, 2020).[1] In Cambridge Analytica, the world was introduced to the prospect of AI influencing elections, while one component of the Dutch childcare benefits scandal featured an AI model that intentionally used nationality for fraud risk detection among childcare benefit applicants (Dutch Data Protection Authority, 2020; Dutch Parliamentary Investigation Commission, 2020). Given the salience that some governmental data analytics projects are thus gaining both within academia and in society, it is remarkable that only limited empirical research seems to exist on how governments are dealing with emerging societal trust issues regarding the use of data analytics and the degree to which these endeavours may be expected to be successful (Ahonen and Erkkilä, 2020; Sun and Medaglia, 2019).

This chapter presents an early step in addressing this gap in our knowledge by drawing on data from 17 interviews from Belgian and Dutch governmental organisations applying data analytics, developers cooperating with governments, intersectoral AI platforms and civil society organisations. We start by developing a theoretical framework on trust in governmental data analytics projects based on existing insights from social sciences, law and data science. This framework incorporates multiple project-level, organisational-level and inter-organisational features that can be incorporated into the design of data analytics project, but also emphasises that civil society is in most cases unlikely to directly perceive the trustworthiness features (Hardin, 1996, 2002: 152). We argue that at least two further elements are necessary: active engagement with civil society and the prevention of major trust breaches (on the latter see e.g. Kern *et al.*, 2015 Van der Cruijsen *et al.*, 2016). Although these factors may seem somewhat obvious at first glance, they are essential in understanding that trust is a dynamically evolving social phenomenon, influenced by not only the current actions of a public actor, but also by past actions and actions by other public actors (Bitektine, 2011). For instance, when one municipality is seen to be using AI in an illegal or unethical way (e.g. by engaging in discriminatory risk-profiling during fraud detection processes), societal trust in data analytics projects of other municipalities may decline as external actors generalise their views across governmental entities (Bitektine, 2011).

After establishing our theoretical framework, we continue by deductively drawing on interview data on the way that governmental data analytics projects attempt to achieve trustworthiness in practice. This is then compared with data from interviews with external actors scrutinising

[1] Cambridge Analytica revolved around an AI model using unethically obtained data to predict the profiles of potential voters, with the aim of improving targeted political advertisements. The Dutch childcare benefits scandal revolved around the (illegal) use of ethnicity to determine the risk of fraud among childcare benefits applicants. Although broader because it also encompassed non-AI elements, the childcare benefits scandal partially concerned an AI risk-classification model which incorporated ethnicity in its predictions on eligibility for childcare benefits.

government and what actions and design choices entities tend to expect from a trustworthy data analytics project in government. We show that ethics, technical and legal robustness, accountability, and trust are important considerations for the interviewed governments developing data analytics projects. This finding in itself holds relevance, as it shows – contrary to assumptions sometimes made in critical social sciences, law and mass media contributions (e.g. Freeman, 2016) – that many governments are actively engaging in the design of checks and balances that they hope will alleviate concerns regarding public sector use of data analytics. At the same time, we show that views on how to achieve trustworthy data analytics differs greatly between public sector actors. We then continue by presenting some insights from entities outside government, in particular civil society organisations (NGO's and a union). In line with our theoretical framework, interviews with NGO's suggest that governmental organisations' focus on internal trustworthiness measures is not sufficient to retain trust in the event of a major crisis, or to improve trust among already sceptical actors. One measure which did show some promise of improving trust is the active and transparent engagement of a governmental actor with civil society. We thus conclude with the finding that, while interviewed governments are actively aiming to foster trustworthiness, their actions may be insufficient to heighten external perceptions of trustworthiness.

The remainder of this chapter proceeds as follows. We begin the next section by presenting an overview of relevant trust dilemmas in governance by data analytics. Subsequently, we describe a number of dimensions that governments may address to enhance the trustworthiness of their data analytics programs. We then present a brief overview of the data used, before turning to our results. Finally, we round up with a number of concluding comments.

6.2 Theoretical framework

6.2.1 Trust problems in data analytics and government

As algorithms in general and AI more specifically begin to pervade society, so do the concerns over their potential misuse (Sun and Medaglia, 2019). With the private sector leading in the development of AI and big data, it is perhaps unsurprising that many frequently cited examples of issues pertain to major private-sector innovators, such as Facebook (e.g. Cambridge Analytica's use of Facebook profile data (Isaak and Hanna, 2018)) and Tesla (the reliability of image recognition to prevent accidents with self-driving cars (Stilgoe, 2018)). However, the public sector is less of a laggard than is often assumed. Governments around the globe are turning to data analytics to better analyse their – frequently substantial – data flows (Ananny and Crawford, 2018). In some cases, AI is not even required, with eligibility for certain permits, tax deductions or social benefits being possible through 'simple' matching algorithms. AI methods such as machine learning and deep learning become more useful where probabilistic predictions need to be made over large and complex datasets. Governments in the US and Western Europe (including Belgium and the Netherlands) are actively developing and using AI for problems ranging from the automatic interpretation of text (such as chat messages and applications from citizens or jurisprudence analysis) to the prediction of traffic flows, and from personalised recommendations to risk-detection in policing and administrative supervision.

These developments generate important new trust problems for society, a few of which will briefly be discussed here. Many non-AI and AI algorithms have in common that they require access to sufficiently large datasets, which in the public sector frequently incorporate sensitive personal data or pseudonymised data (McDermott, 2017). The use of such data is also changing, as algorithms concurrently analyse thousands or even millions of observations to generate predictions. By contrast, traditionally analysed databases (such as medical records examined by a doctor or criminal records used by an inspector) may incorporate high numbers of observations, but these observations are only analysed in small numbers and normally only with prior cause (e.g. the immediate relevance of the medical record of a patient that is currently under treatment). Depending on the goal of the data analytics project and the sensitiveness of the data involved, the large-scale use of such data may at least be uncomfortable for some members of society (Sun and Medaglia, 2019). This issue may be compounded where data is shared among multiple public entities or obtained from private entities, as more entities, developers and users gain access to more extensive (combinations of) data (Meuwese, 2020). Concerns that large-scale data processing may lead to privacy violations or discriminatory practices are no longer entirely theoretical either, given examples such as the Dutch tax service's intentional use of nationality data in its social benefits fraud detection algorithms (Dutch Data Protection Authority, 2020).

Where AI becomes involved, concerns regarding black boxes, biases and model drift are also introduced. The black box refers to the potential opaqueness of machine learning and deep learning algorithms regarding the factors that contribute most to a certain prediction. For instance, in a black box recidivism model, it may be unclear whether high predicted risk of committing further crimes is based on admissible criteria (e.g. prior record) or on potentially unlawful and/or discriminatory criteria (e.g. ethnicity) (Chander, 2017; Sandvig *et al.*, 2016). Both human biases in training the data and subsequent model drift caused by the model's self-learning features may however lead to unlawful and/or unethical criteria being used as a factor in making a certain prediction (Busuioc, 2020; Sandvig *et al.*, 2016). These biases need not even come from a variable in the dataset with direct information on ethnicity but could be implicitly introduced by incorporating data on a person's name or neighbourhood of residence (Sandvig *et al.*, 2016). Moreover, it must be mentioned that some controversial biases are introduced by way of policy choice rather than as part of the algorithm itself, as occurred for instance in the Dutch SyRi case, where a network of governments chose to focus fraud detection analytics on specific problem neighbourhoods (Meuwese, 2020). Where such biases eventually become public, they may give rise to major trust breaches among groups that the model was biased against, or even wider society. Moreover, where an algorithm is a true black box, it may be difficult even for developers to ascertain what factors are leading to certain predictors, which may make it difficult to take away suspicions among sceptical external actors, causing tension between such algorithms and the principle of transparency (Ananny and Crawford, 2018; Ahonen and Erkkilä, 2020). It should be noted, however, that 'the black box' presents a problem that data- and computer scientists are generally well aware of, and multiple techniques have been and are being developed to counter it (see Section 6.2.2) (Sandvig *et al.*, 2016).

Beyond these technical and data-related issues, we may also discern a number of organisational and governance challenges giving rise to potential trust problems. Several authors have noted the risk of automation bias occurring when algorithms support civil servants' decision-making, as civil servants may lack the data literacy, time or autonomy required to assess the limitations or design of a model (Busuioc, 2020; Giest and Grimmelikhuijsen, 2020; Peeters *et al.*, 2021). This may lead to an overreliance on AI predictions and a lack of internal signalling of issues with models (Busuioc, 2020). It should be noted, however, that no evidence was found of automation bias in a recent study (Alon-Barkat and Busuioc, 2021). An overreliance on black box AI, furthermore, may be an issue from the viewpoint of good governance principles, as it becomes increasingly difficult for civil servants to transparently motivate a decision towards citizens (Ahonen and Erkkilä, 2020). Existing accountability channels may also become ineffective, as actors in a position to provide oversight (e.g. senior management, political superiors and/or the judiciary) lack the required expertise to accurately assess potential issues with data analytics projects (Busuioc, 2020).

All these reasons may cause citizens and civil society organisations to reduce their level of trust in governmental data analytics processes. Lowered levels of trust, in turn, may impact the legitimacy of utilising data analytics and thus hinder governments from utilising advances in data analytics to their fullest benefit. Accordingly, governments have actively begun focussing on the trustworthiness, technical robustness and ethical features of their data analytics projects, with several major strategies being discussed in Section 6.2.3.

6.2.2 Trust and trustworthiness

Before continuing with the various strategies that governments may employ to enhance trustworthiness, it is useful to briefly elaborate on the notions of trust and trustworthiness. Trust has frequently been seen as a willingness to take risks, based on the expectation that the trustee will not behave opportunistically (Grimmelikhuijsen and Meijer, 2014; Hardin, 1996; Levin *et al.*, 2006). Trust is thus a relational state presupposing an interaction between trustor and trustee, which is not always the case when discussing trust in public sector actors. Trustworthiness, on the other hand, refers to the qualities that a trustee has or lacks in terms of their ability to deliver, their benevolence and their integrity. A trustee, such as a government, may actively attempt to foster such trustworthiness by emphasising activities and choices that display ability, benevolence and/or integrity (Hardin, 1996; Levin *et al.*, 2006).

Perceived trustworthiness then refers to the beliefs and expectations that a trustor has regarding the desirable qualities that a trustee possesses (Grimmelikhuijsen and Meijer, 2014; Levin *et al.*, 2006). This distinction between trustworthiness and perceived trustworthiness is essential here, as we emphasise the potential distinctions between governmental strategies to engender trustworthiness and civil society perceptions of such trustworthiness. Such discrepancies may arise due to a trustee lacking several desirable qualities (it is at least hypothetically possible that a government has malicious intent, for instance) or due to the characteristics of the trustor (e.g. different predispositions to trust (Grimmelikhuijsen and Meijer, 2014)). However, they may also emerge due to the bounded rationality of both trustor and trustee, with the former not always being capable of signalling trustworthiness perfectly, while the latter will usually not have the capacity to perfectly predict the trustworthiness of an actor (Bitektine, 2011; Hardin, 1996). Although such mismatches inherently arise in the social interactions between actors and are thus likely to never be resolved completely, matching strategies to enhance trustworthiness to the criteria on which the trustee evaluates trustworthiness may aid governments in enhancing the perceived trustworthiness in their data analytics projects.

6.2.3 Resolving trust problems by focussing on trustworthiness?

Given the many dilemmas associated with the rise of data analytics in the public sector, it is no wonder that much effort has gone into the way that data analytics projects can be designed to be trustworthy. We begin with a brief approach of measures that can be taken at the technical level, although this discussion will stay relatively general by discussing categories of measures. Subsequently, we discuss organisational aspects of designing trustworthy analytics. Finally, we discuss a recently emerging perspective in public administration, which more broadly examines the place that data analytics should take within public sectors. Although we discuss these as separate dimensions of trustworthy data analytics governance, it must be noted that dimensions are frequently complementary and – for specific measures – may sometimes overlap.

The technical dimension

Perhaps the most prevalent approach to attain trustworthiness for a data analytics project is to focus on technical features. Generally speaking, data scientists and computer scientists are well aware of the many technical problems associated with creating large databases of sensitive data, selecting data, training models and/or examining the robustness of results. As was mentioned earlier, one of the classic criticisms of AI models in particular is that they can function as black boxes, in which the relative weight of factors predicting a certain outcome is unclear even to developers (Stilgoe, 2018). The black box nature of many AI systems has led to a substantial thrust in research on methods to counter the black box problem (Emmert-Streib *et al.*, 2020; Gunning and Aha, 2019; Xu *et al.*, 2019). For instance, even deep-learning systems – which are noted for their complexity and opaqueness due to their use of multiple layers of data transformations – may nowadays be accompanied by explainer AI (so-called XAI algorithms (Emmert-Streib *et al.*, 2020; Gunning and Aha, 2019)), which are specifically designed to reveal the contributions of certain variables in the model. It should be noted, however, that XAI is a relatively recent area of inquiry, and that current models frequently face challenges in terms of e.g. accuracy/interpretability trade-offs or excessive computing power requirements (e.g. Gunning *et al.*, 2019). Another frequently used method is the generation of local explanations (i.e. explanations of a single or small group of predictions) to see whether these local predictions depend on factors that are deemed undesirable. Explaining which factors lead to certain predictions also aids in correcting emerging biases and model drift, suggesting ways in which datasets can be altered to prevent 'garbage-in-garbage-out' problems or suggesting a need to retrain the algorithm. For instance, should a natural language processing algorithm become less capable of analysing input from certain dialects, it may be possible to retrain the model with additional data from that dialect. Data scientists argue that XAI will be an essential tool to generate user trust in the future (Gunner and Aha, 2018), although it remains somewhat unclear how explainability should be used towards broader audiences (such as citizens or civil society organisations that are not immediate users).

Data security, data selection and anonymisation are other measures that can be taken to ensure the technical robustness of an algorithm. For instance, to reduce the impact on privacy, a public actor could choose to focus only on publicly available information (for instance, Google maps data instead of data gathered on-the-ground through photographs). Where more personal data is necessary, developers can train models on anonymised versions of a dataset. Even if de-anonymisation is necessary, as might be the case for risk-profiling algorithms in policing or supervision, this can be done only for observations of interest. It should be noted that preventing model drift and biases and ensuring technical robustness and data safety are also heavily emphasised by influential policy documents, the foremost of which is perhaps the EU High Level Expert Group ethics guidelines for trustworthy AI (High-level Expert Group on Aim, 2019).

The legal dimension

Another dimension to ensuring the trustworthiness of data analytics is ensuring it does not violate legal requirements and good governance principles. Although elaborated in greater detail in Chapter 5, it is useful to denote here briefly that ensuring compliance entails not only implementing requirements from the GDPR and future AI regulations, but also respect for various fundamental rights and requirements in administrative law (Donahoe and Metzger, 2019; Freeman, 2016; Meuwese, 2020). In practice, complying with legal norms frequently involves balancing different requirements, with a moderate breach of the right to privacy, for instance, being justifiable if it suitably serves for another goal (e.g. fraud detection) and does so in a proportionate manner (Meuwese, 2020). Similarly, a government aiming to engage in fraud detection may have to balance the effectiveness of its data analytics project with transparency requirements, as too much transparency could potentially aid violators in gaming the system (Chander, 2017). Malicious companies may, for instance, attempt to exhibit characteristics that make them unlikely to be seen as high risk by a risk-prediction AI, thus helping them evade detection (Chander, 2017). At the same time, although this is a frequently used explanation for limiting transparency in certain AI projects, one may question to what degree gaming the system will actually occur in settings such as social welfare fraud, and to what degree this can justify limiting the application of the transparency principle. It is therefore worth noting that a Dutch court was unconvinced by a similar reasoning in a case on geographically-focused tax and benefits fraud algorithms. In this case, the court noted that using highly sophisticated and new techniques with a substantial impact on citizens' legal positions places a heightened responsibility on governments to be transparent regarding the implementation of such techniques (Meuwese, 2020) – a relevant conclusion from a trustworthiness perspective.

The ethical dimension

Related to both the technical and legal dimensions is an area of inquiry which has seen substantial attention in recent years. Ethical AI is a broad term covering approaches and frameworks specifying that AI and associated data analytics projects must be employed according to norms of 'good' behaviour (Mittelstadt, 2019). Frequently, ethical AI incorporates principles and values such as the idea that data analytics projects may do no harm, should be fair, should promote well-being, should promote privacy and that humans retain their decision-making autonomy (Winfield, 2017). These principles are for instance visible in the EU High Level Expert Group's ethical AI guidelines, which aim to serve as a general framework for both private and public sector AI. Ethical AI as a field borrows from other professional codes, such as those employed in the medical sector (Mittelstadt, 2019), and have recently begun to see active implementation in governments (as will be seen in the results section). Despite its intentions, the ethical AI movement has not been without criticism. Principles are often relatively vague (Mittelstadt, 2019) and some may even be seen as somewhat arbitrary (e.g. is explainability always necessary, given the additional costs and burden that e.g. XAI may introduce?).

The organisational dimension

Although the aforementioned technical, ethical and legal measures seem to be the main focus of recent research into creating trustworthy data analytics initiatives, they provide a useful but arguably incomplete set of tools to address trust in governmental AI systems. An already somewhat more comprehensive view is gained when also taking into account organisational factors, in particular organisational values, culture and management processes (Brendel *et al.*, 2021). Organisations already possessing strong cultures focused on client-orientation, transparency and openness of decision-making may for instance have an easier time transitioning towards a transparent and explainable implementation of data analytics than entities traditionally seeing too much openness as a risk. Similarly, organisational cultures focussing more on values such as privacy and other fundamental rights could be better predisposed to proactively implementing GDPR requirements (such as comprehensive data protection impact assessments (DPIAs)) than counterparts focused more strongly on the value that algorithms may bring in terms of efficiency or effectiveness (Chua *et al.*, 2017). Given the legitimacy risks of inappropriately managing trust issues in the context of data analytics, fostering a culture that takes ethical, technical, and legal safeguards seriously is highly important (Brendel *et al.*, 2021). At the same time, the accompanying critical note should be that transitioning an entity towards such a culture may take years, with many aspects of cultures being extremely difficult to manage (Schein, 1990). This transition depends in part on the management structures in place at the entity. Agile management tools or similar management styles – if set up properly instead of being fitted onto existing structures as window dressing (see e.g. Molenveld *et al.*, 2020) – may be particularly suited to the open development with internal stakeholders. Such management styles allow the business side to provide feedback on missing elements and problems in the data or the algorithm and may foster internal trust in data analytics development. Moreover, codes of conduct and leadership styles may matter, providing a fixed frame of reference for employees (Brendel *et al.*, 2021). For more details on this, see Chapter 3.

The accountability dimension

Ever since the major incidents such as the COMPAS Scandal and the Dutch benefits affair, social science scholars have also begun to realise the ramifications that data analytics broadly and AI more specifically may have for public governance. Recently, focus has been placed on the blurring of accountability lines due to the rise of data analytics (Busuioc, 2020; Freeman, 2016). Essential components of accountability relationships, including transparency and the provision of information, debates on this information and justifications of the functioning of algorithms is often stymied due to technical complexities and information asymmetries (Ahonen and Erkkilä, 2020; Freeman, 2016). Accordingly, calls have been raised to avoid public sector-wide oversight gaps by introducing an entity with the capacity and competences to investigate data analytics projects (similar to existing data protection authorities), by ensuring that political representatives or their aides have sufficient proficiency in data and AI to evaluate information, and/or by introducing judges specialised in data, IT and data science matters. Others call for regulatory solutions, such as the EU's upcoming AI regulation or requirements for greater sector-wide transparency. It is certainly true that steps can be taken in this area, as the development and implementation of data analytics processes, including inherently value-based choices, remain fractured in most public sectors. Accordingly, while some entities may be proactively engaging with ethics and trustworthiness, other entities may remain less inclined to do so, for instance expecting citizens to 'game the system' if a government is too transparent regarding its algorithms. Such gaming involves knowing which data categories contribute to AI predictions and using this knowledge to window-dress your characteristics in such a way as not to raise suspicion. Fears of gaming may lead some public entities to engage in what has been called fuzzy transparency, in which general aspects of activities are divulged but with low levels of detail (Ananny and Crawford, 2018).

A final and frequently overlooked dimension? Communicating trustworthiness through murky lenses

It is argued here, however, that ensuring trustworthiness through technical robustness, legality, organisational procedures and accountability links may be insufficient – despite the substantial efforts that are sometimes invested in these dimensions. Although ensuring trustworthiness along these dimensions is certainly necessary, it is unlikely that civil society organisations, let alone other citizens, will directly evaluate how a data analytics project operationalises these dimensions (Kolkman, 2020). Instead, a small subset of easily interpretable project features will be filtered through a media lens, which will then be heuristically interpreted in light of existing public perceptions of data and AI, as well with regard to perceptions of the public entity involved (Bitektine, 2011). A similar argument has previously been made in the context of algorithmic transparency, where it was noted that even completely transparent data science projects often lack critical audiences with either the ability or the interest to closely scrutinise algorithmic decision-making (Kemper and Kolkman, 2019; see also Stilgoe, 2018 and Burrell, 2016). Even data scientists themselves will frequently have difficulty interpreting the full scope of a peer's model, regardless of the original developer's transparency (Kemper and Kolkman, 2019).

Perceived trustworthiness is thus best seen as a complex and dynamic social construct, influenced by a wide variety of social and cultural factors beyond the immediate control of public sector organisations. Bitektine (2011) argues that in such situations of incomplete knowledge regarding the organisation, project and algorithms involved, evaluators will fall back on status judgments, in which attributes of an entity are predicted on the basis of its membership of a broader group of similar organisations. The point is that specific policies or actions designed to enhance trustworthiness, especially when these policies are technically complex, may fail to generate perceived trustworthiness among non-expert audiences (see e.g. Kolkman, 2020) – in particular when these audiences are somewhat distanced from a policy, such as civil society actors. Moreover, when some public organisations have engaged in salient unethical data analytics practices, the social judgment process implies that governments as a whole may be evaluated based on these practices. To complicate matters further, cultural associations of algorithms may also affect social judgments. The analogy with nuclear energy is striking here, with the latter having lost significant legitimacy in part due to cultural fears and associations with disasters and danger (Koerner, 2014). Similarly, terms such as big data and AI sometimes seem to invoke cultural associations with a police state in which all actions are monitored, or with sentient 'general AI's' capable of sentience and intelligent action. Given such social interpretation processes, there may be a substantial disconnect between the policies designed to secure data analytics processes and the trustworthiness that external actors attribute to such processes externally.

Translating a project's trustworthiness into perceived trustworthiness among civil society and citizens is therefore in part also a communication challenge. A combination of organisational transparency, project-level motivation procedures and algorithmic explainability can help here but are unlikely to resolve the issue on their own. Perhaps greater promise lies in active communication by signalling moral choices, and trustworthiness measures, such as ethical principles already being implemented or the applications of data analytics that are precluded by policy. At the same time, it must be realised that some segments of society may remain averse to even the most well-designed data analytics processes.

6.3 Data and methodology

To examine how public organisations approach trustworthiness of data analytics projects in practice, we draw on 17 semi-structured interviews with Dutch and Belgian public organisations, developers from other sectors working with governments, as well as civil society organisations (mainly NGO's, but also one union that involved itself with scrutinising AI in government). Interviews were held between November 2020 and May 2021 and focused on questions regarding the workings of a data analytics project (where relevant) before examining different dimensions of trustworthiness. All interviews were anonymised on the organisational level, as the subject matter was sensitive to some entities involved.

First, of the 11 public sector interviews performed, nine were held with six governmental organisations applying data analytics in Belgium and the Netherlands and two were held with external developers co-producing AI tools with a public sector organisation. One public organisation had a wider portfolio of data analytics projects with more departments involved than its counterparts, allowing us to perform four interviews in this organisation (this is also the reason that our sample contains a lower number of organisations than interviews). The organisations' policy domains range from tax and social security to policing, tourism, waste collection and transport, while applications of data analytics include data sharing platforms (non-AI), natural language processing of data (both on existing databases and new requests by citizens), the prediction of crowd movements and risk-score based fraud detection processes. Within the interviewed entities, interviews were usually held with respondents fulfilling either a data scientist or product owner role, although the sample also incorporates two data protection officers (DPO) and two strategic advisers in the areas of data and AI. Most entities in the sample either applied data analytics tools such as AI for several years or were in an advanced development state (the proof of concept phase). The reasoning underlying our sampling strategy of approach with a heterogeneous set of interviewees is that different organisations are likely to vary in terms of both policy goals and measures used to secure their data analytics projects. Additionally, most entities working on advanced data analytics tools do so in moderately sized project groups, making it practically difficult to hold all interviews in one or two entities.

Second, six interviews were held with civil society actors, five of these with Dutch and Belgian civil society organisations. The final interview was held with a Dutch union that had aligned itself with NGOs in scrutinising governmental use of data analytics. These interviews were held to obtain an external view on the perceived trustworthiness of data analytics projects in government, from organisations that fulfil a societal check and balancing role. This allows us to evaluate what civil society organisations expect from governmental data analytics projects in terms of trustworthiness and where potential caveats may be remaining in governmental projects and policies.

6.4 Results

In discussing the results, we first present a subsection on results from interviews with governmental organisations, illustrating how governmental organisations or the developers they cooperate with attempt to secure the trustworthiness of their data analytics projects in practice. In Section 6.4.2, this is contrasted with the viewpoints of external actors. This then forms a basis for an exploration on the matches and mismatches between trustworthiness as implemented in governments and the actions that external actors may expect.

6.4.1 Trustworthiness measures by governments and developers

All four theorised dimensions of measures that can be used to enhance the trustworthiness of data analytics projects were visible in the 10 governmental interviews. Generally speaking, and contrary to some of the extant literature that suggests that governmental organisations or their developers are uncritical of their own projects, all respondents seemed highly aware that their analysis techniques had to be deployed responsibly and that downsides of certain techniques – such as black box issues, model drift and bias – should be countered proactively. Similarly, all respondents realised that using sensitive information such as personal data has its risks, requiring procedures such as anonymisation to be implemented securely. Notable is, however, that emphases on the way that trustworthiness can be achieved diverged strongly between different interviews and organisations. Whereas some respondents displayed narratives that relied heavily on the technical features of models, others focused on achieving transparency towards the outside world, still others emphasised legal aspects and a final group focused on ethics and public value choices.

The technical dimension of achieving trustworthiness was often subsumed in a broader data science paradigm. Somewhat unsurprisingly, respondents answering on the basis of this perspective were often linked to the development process in a data analytics process themselves. Such respondents frequently focus on the ways in which models can be made explainable (e.g. by creating local explanations and explainer algorithms for developers and users), and on how bias may be prevented, preventing garbage-in-garbage-out issues and gathering high quality and suitable data to train models appropriately. In these narratives, trustworthiness is perceived as a function of model quality and the added value of a model in terms of efficiency, effectiveness, and objectiveness. Whether a model should be used or not is therefore evaluated based on the accuracy of its predictions and potential blind spots in terms of missing data. In line with this view, respondents emphasising the technical dimension note that public service delivery may be enhanced through increases in efficiency and effectiveness, which may in itself lead to greater trustworthiness. For that, however, it is necessary to achieve user trust in the data analytics system.

It is also in such interviews that a reversal of the bias argument is sometimes made. Where AI-based governance is frequently criticised on the basis that AI models may become biased towards vulnerable groups, these respondents instead see AI as a potential contribution to the objectiveness of public administration. Some respondents argue that by infusing procedures previously based on professional experience and gut feelings (e.g. risk selection in inspection processes) with evidence from an AI trained on thousands of observations (even though they realise that AI models are also based on developer choices), the objectiveness of governmental procedures may be improved. Through greater objectiveness in citizen-state interactions, it might then also be possible to increase the trustworthiness of governance arrangements. Respondents relying on this narrative are generally supportive of transparency and consultation with the client-side, usually from the viewpoint that this may improve modelling choices during the development process, as well as civil servant trust and uptake of the AI model during the implementation phase. At the same time, respondents developing risk detection algorithms also argue that unfettered transparency may be undesirable, as this could allow regulatees to game the system.

The legal perspective was also visible in a number of interviews. Interviews in this dimension frequently revolve more around the data-gathering and data-sharing processes, with the narrative focussing strongly on GDPR requirements and the values underlying these. Respondents with a DPO role in particular emphasise the importance of projects going through stringent evaluation based on GDPR principles such as meaningful human intervention. Moreover, they argue that data scientists should not be afraid of compliance procedures such as data protection impact assessments (DPIAs), as their intent is not to stymie development but to ensure that development occurs safely. All narratives incorporating a discussion of legal aspects agree that ensuring legality is an important way of ensuring that data analytics projects remain trustworthy, although one data scientist also notes that further increasing the stringency of data protection norms may make the development and deployment of AI disproportionately cumbersome. Another data science department working on valuation of goods furthermore makes an interesting link between the technical and legal dimensions: by offering frontline civil servants not just the requested value prediction, but also several 'neighbouring' predictions (i.e. other goods in the same class with a similar predicted value) which may allow civil servants to better motivate their decisions, and increase adherence to the motivation principle. This statement is noteworthy, as it illustrates how enhancing technical features can also frequently make achieving legal compliance and – in turn – trustworthiness from a legal perspective – easier instead of more difficult.

In the area of ethical and value-based narratives, we find that interviewees focus strongly on what governments should do from a normative perspective, although multiple threads may be distinguished here. Several interviews focus on the methods that government should and should not employ, warning against the overuse of data-gathering techniques such as facial recognition or the use of data unnecessary to achieve a certain goal. Another commonly used narrative focuses more on the development process, arguing that governmental organisations should be imbued with ethics procedures (e.g. an ethical board or an ethics-by-design philosophy) and a culture that emphasises openness and active debate on the ethical choices underlying a project. One respondent in this vein also mentions the importance of developing ethical procurement standards to prevent tensions between proprietary development in the private sector and public sector good governance principles (such as the motivation principle, due diligence, and transparency). In perhaps the most far-reaching ethical narrative, one interviewed organisation focused strongly on no-harm principles, subscribing to the policy that their AI projects should never come to the harm of citizens, for instance by precluding the application of risk-profiling AI that directly leads to sanctions. A similar narrative focused on avoiding hyper-efficiency in enforcement agencies such as the police, as hyper-efficient enforcement of (especially minor) crimes will no longer be seen as proportionate by society. A common thread in most of the narratives focussing on ethics and values, however, is that more attention should be devoted to balancing the values that data analytics governance aims to strive after, and that governments should take trustworthiness seriously to prevent major trust breaches. This may incorporate a dialogue on where AI and data analytics can contribute to the organisation and where it may be undesirable.

The organisational dimension is also mentioned frequently and displays some overlap with the ethical and value-based dimensions when respondents discuss ethical procedures. At the same time, the organisational dimension is broader, as it also includes other aspects of organisational cultures and procedures. Regarding cultures, several respondents argue that values should be internalised that emphasise openness, communication with both domain experts and experts in areas such as GDPR requirements and openness to external scrutiny. Other recurrently mentioned values are data minimisation and proportionality, with several respondents hoping that at some stage, data science departments will proactively begin wondering whether some categories of data are truly necessary to incorporate in a model. In this vein, one respondent mentions that the GDPR may have had beneficial organisational-level effects, with data scientists gradually displaying a greater degree of internalisation of its underlying values. Conversely, a respondent from a different entity warns of the dangers of groupthink when such values are not internalised. In such cases, civil servants and sometimes even politicians above them no longer see the potential issues that emerge during the implementation of an algorithm. When this occurs, the promise of AI and other analytics tools may no longer be sufficiently nuanced by issues and value trade-offs, potentially leading to major policy crises. The respondent argues that crises emerging from data analytics projects are therefore frequently an organisational issue, rather than a technical issue with e.g. an algorithm itself. Finally, several respondents mention the relevance of procedures with external actors, including the establishment of appropriate data transfer protocols and – again displaying some overlap with the ethical dimension – ethical procurement rules.

Least mentioned, but still present in our interviews with governmental actors and their private developers, is the accountability perspective. Two respondents noted the importance of political oversight on certain decisions, in particular where the policy choices become strategic and political in nature. One of these respondents for instance mentions an example of a profiling algorithm designating individuals as groups in need of additional support. This was a departure from previous working methods: instead of policymakers designating certain groups and asking for approval from the political level, the model could now predict such choices for the organisation, without there being a clear political rationale for such selections. As this blurred the accountability relationship between politics and administration, choices made by the algorithm also lacked a degree of legitimacy which could be achieved by obtaining political approval for policy choices. Relatedly, a second respondent notes that if an algorithm begins taking automatic decisions, this algorithm should have a legal basis that goes through regular democratic legislative procedures to gain the necessary legitimacy. A third respondent indicated that they receive parliamentary questions on the role of AI within the entity, noting that this required consistent communication that AI was a supplementary tool in the organisation's services.

Beyond such vertical accountability, a number of interesting comments were made regarding internal accountability mechanisms. One organisation was organising the creation of an ethics board of data analytics processes, while another was working on obligatory ethics work processes. Both processes, while also aligning with the ethical dimension, were notable from an accountability viewpoint as they incorporated independent experts in the process of reviewing AI projects. Finally, a number of mentions were made regarding accountability directly towards citizens, with one respondent noting that citizens should not be afraid to utilise their GDPR rights in asking where certain data is being used. A common thread in all narratives incorporating accountability therefore seems to be the belief that public sector entities should be ready to face scrutiny over their data analytics processes.

In summary, interviewed developers and governmental organisations in both Belgium and the Netherlands place substantial interest on trustworthiness, although the dimensions emphasised sometimes differ greatly between respondents. Perhaps unsurprisingly, data scientists seem to favour the technical dimension in generating trustworthiness, while respondents with a DPO role focus strongly on legality. More surprising was perhaps that openness and organisational culture was emphasised as essential in nearly all interviews, with this emerging as a cross-disciplinary point of consensus. Furthermore, while the legal and technical dimensions vary strongly based on respondent discipline, emphases on the ethical, value-based and accountability aspects seem to differ strongly across organisations, potentially suggesting that prior organisational culture is a relevant predictor of the degree to which such values are implemented in data analytics projects.

6.4.2 An external perspective: perceived trustworthiness of governmental data analytics processes

In the previous subsection, we discussed how governments and entities developing data analytics tools for governments use strategies based on five analytically separable dimensions, with specific respondents displaying a preference for certain strategies. In this section, we present an external perspective, which illustrates how civil society organisations perceive the trustworthiness of governmental data analytics projects. The aim here is to make clear how narratives external to government may differ from internal narratives, pointing towards a potential mismatch between strategies civil servants frequently deem important and measures expected by civil society.

Perhaps the most emphasised element by civil society organisations is the threat of the unchecked government expansion of data-gathering and AI tools. A commonly mentioned theme is the loss of citizen control over what (sometimes sensitive) data governments have of them, how much data of them is available, how this data is used, and which entities have access to this data. Related to this, civil society organisations frequently mention the prospect of function creep, i.e. that once technology or data is available for one purpose, it is likely that it will gradually gain applications beyond its original intended use. An example mentioned independently by both a Dutch and a Belgian NGO concerns the digital electrical meters being installed in many Western European households. Although at first glance an innocuous development, these digital meters may in the future be used to analyse anomalies in household energy consumption (which in turn may be indicative of offenses such as producing narcotics). In line with such examples, the general consensus among all the interviewed civil society organisations seems to be that the so-called goal binding principles underlying the GDPR and the principle of proportionality are frequently interpreted too loosely. According to this viewpoint, reasons for accessing data are often defined in a relatively vague manner, allowing public entities to access data gathered for different purposes too easily. Instead, civil society organisations seem to favour that the use of gathered data is clearly bound to goals specified beforehand. Moreover, civil society organisations argue that governmental entities may not always ask themselves whether similar results could be gained with less extensive data-gathering and analysis tools, and whether big data and/or AI solutions are effective at all. It is important to note that not all civil society organisations interviewed assume malicious intent on the part of governments when engaging in such function creep, with some suggesting that overbearing AI and data systems likely result from a subconscious path dependence process. This process results from the increasing availability of data, which in turn gradually encourages the use of AI systems on this data.

Another frequently recurring theme is the lack of scrutiny regarding which data flows are used and which AI solutions are applied, the argument being that neither average citizens nor professionals such as managers, journalists, politicians and judges possess the necessary data literacy to appropriately scrutinise such developments. As both government and private developers possess a greater amount of trained computer scientists and data scientists, this results in information asymmetries that may render political action, media scrutiny or (civil society) litigation against governments applying data analytics less effective. What's more, there may be tension between big data and AI-based decision-making and administrative appeal procedures, as some decisions become difficult to explain due to black box problems.

Traces of trust breach phenomena can be gleaned from two Dutch NGO interviews. In one, the civil society entity involved noted that the Dutch government had launched new legislation allowing government organisations to link multiple data sources in the same month that the Dutch childcare benefits scandal (which itself contained elements of both too strict enforcement cultures and an AI risk profiling tool that was partially based on nationality). To provide some context, The Law Data Processing for Partnerships was criticised by the Council of State for being insufficiently precise and repeated mistakes of prior legislation that contained insufficient guarantees to safeguard the right to privacy. The fact that the Dutch government was attempting to pass this new legislation while at the same time apologising for too strict enforcement (in part based on over-reliance on data and directly discriminatory algorithms) lead to one civil society organisation characterising the government's apologies as 'crocodile tears'. The assumption was that government will continue with its attempts to introduce further data analytics systems without sufficient guarantees to respect fundamental rights regardless of the outcome of the benefits scandal. Interestingly, such breaches of trust were not observed in the Belgian data, with the Belgian government not having experienced a major data analytics-related crisis such as the benefits scandal.

As mentioned, most civil society organisations interviewed were not inherently opposed to the use of data analytics in government. A varied palette of civil society organisations can be seen, with some not doubting the intentions of most governmental data analytics programs, and others experiencing a low trust in government, e.g. due to trust breaches flowing from Dutch crises. The latter category of civil society organisations therefore seemed doubtful regarding the motivations behind governmental action, which seems to lead to a less open general attitude towards the usage of data analytics in governments. The interviewed civil society organisations did not emphasise specific technical and ethical AI solutions as strategies to ensure trustworthiness, and instead focused more on avoiding negative outcomes for citizens, maintaining proportionality in digital governance, preventing function creep and addressing accountability gaps. Accordingly, when governments focus on technical solutions to AI problems (e.g. addressing biases in data, using XAI to explain regular AI outcomes) they may be leaving out several important areas of attention to obtain trustworthiness. Although multiple governments also mentioned legal compliance and ethical measures to attain trustworthiness, the relative absence of accountability and proportionality-based narratives is noteworthy and may point to a mismatch between current governance strategies and stakeholder expectations.

6.5 Discussion and conclusions

The use of datasets incorporating large segments of society, the possibility of function creep and the unfamiliarity of emerging technologies such as AI combine to create new trust issues in the relationship between governments and citizens (Busuioc, 2020). After several scandals, both governments and academic fields such as data and computer science, public administration and law are actively working to mitigate such issues and to facilitate trustworthy AI projects (Busuioc, 2020; Gunning and Aha, 2019). However, the results described in the previous section suggest that these efforts may not yet be sufficiently aligned with what civil society actors expect of governments. Although all public sector respondents sought to attain trustworthiness, they seemed do so through a varied palette of strategies. This article distinguished between technical strategies, legal strategies, organisational strategies, ethical and AI-based strategies, and accountability strategies, with each respondent emphasising different combinations of these strategies as important components of their projects. The interviews with civil society organisations suggested that these strategies do not always succeed however, in making governmental use of data analytics trustworthy. Particularly noteworthy was that civil society organisations seemed to emphasise legal, value-based and accountability strategies, noting the importance of compliance to the GDPR and fundamental rights, the importance of governments not interfering too much in citizens' private lives and the lack of effective scrutiny mechanisms. Although there is strong overlap between the emphases placed on legality by both public sector actors and civil society, there currently seems to be less overlap on value-based and accountability-based emphases.

With regard to accountability, we do see several mentions of the need for transparency and effective oversight among public organisations. One entity in particular was experimenting in this area, establishing an internal ethics board composed in part of external and independent experts. Nevertheless, the relatively low amount of attention devoted to accountability processes among other entities may point to a relevant area in which governments can further enhance the trustworthiness of their projects. One avenue to strengthening trustworthiness in this dimension could be to introduce greater openness to the involvement of civil society organisations and external experts, a measure which seemed to enhance trust among some of the interviewed civil society organisations. Concerning value-based policies, the issue is not so much that governments do not attempt to identify the values that their data analytics tools should strive for, but that these values may differ strongly from those preferred by civil society actors. Governments are frequently focused on discussions of objectivising public services, preventing harm to citizens, maintaining equality or reducing throughput time by appropriately designing AI (as e.g. recommended by the EU's ethical guidelines (High-Level Expert Group on AI, 2019)). At the same time, civil society organisations frequently advocate taking a step back and question the need for big data and/or AI solutions altogether, frequently viewing such solutions as disproportionate. This perception of disproportionality stems from the idea that governments should not be able to combine large, varied and sometimes sensitive data flows as this breaches the right to privacy too severely and might lead to function creep. As such, this viewpoint also illustrates the limitations of various solutions proposed for societal trust problems in the data science and computer science literatures. It suggests, for instance, that technical methods such as XAI algorithms or blockchain systems registering access to data in tamper-proof ways, however useful from a data science perspective, are unlikely to succeed in generating high trust among citizens and/or civil society, as the underlying normative criticism of governments relying on excessive amounts and invasive forms of data is left unaddressed.

The conclusion that technical, ethical, and organisational strategies are less emphasised in the narratives of external actors also relates to a core point in the theoretical section, namely that specific trustworthiness measures taken internally are unlikely to be directly perceived by external actors (Bitektine, 2011; Hardin, 1996, 2002). What is more, while data science projects in governments are currently splintered in terms of activities and communication across multiple organisations, civil society organisations tend to emphasise general trends in government when assessing trustworthiness. In the absence of specific knowledge regarding every governmental organisation in a country, civil society respondents heuristically extrapolate general expectations on data science in government from characteristics that they perceive as relevant for classes of governmental organisations (Bitektine, 2011), such as municipalities, security agencies or federal-level governments. We argue therefore that locally-taken internal technical, organisational and ethical measures to ensure trustworthiness are necessary but insufficient, as they may be unlikely to achieve wide-spread influence on external perceptions of trustworthiness. Project- or organisational-level transparency has been similarly viewed as a necessary precondition for societal trust in algorithmic decision-making but has its own pitfalls when being operationalised too passively or in an imprecise and 'fuzzy' way (Chander, 2017; Ananny and Crawford, 2018). Again, greater proactive communication towards civil society organisations and journalists may offer one way of translating internal trustworthiness strategies to externally perceived trustworthiness. Moreover, although central frameworks can themselves be stifling to development in specific governmental organisations, a base-level set of best practices for groups of public organisations may help in maintaining perceived trustworthiness among civil society entities. This could take the form of a digital good governance framework or a framework law, although care must be taken to ensure its government-wide application on the one hand, and sufficient space for governments to experiment, develop and acquire data analytics techniques on the other.

Furthermore, our results support extant literature in the idea that trust and perceived trustworthiness evolve dynamically, in part based on major events (e.g. Kern *et al.*, 2015). It was observed that violations of civil society organisation values during a data analytics crisis may lead to severe trust breaches (as was seen following the Dutch child benefits scandal, during which the incumbent government violated values by submitting new data-sharing legislation to parliament with limited attention to citizen rights). Following this trust breach, several interviewed civil society organisations were not merely critical of policy choices, but also doubted the motives underlying such choices, signalling lowered perceived trustworthiness regarding the trustee's benevolence and integrity. Such doubts regarding integrity and benevolence were less visible in the Belgian context, where governmental data analytics crises have not occurred at the level of their Dutch counterparts. At the same time, unlike the actual harm caused by some trust breaches, it must be noted that the lowered degree of perceived governmental trustworthiness following a trust breach is not necessarily dysfunctional. Such drops in perceived trustworthiness may stimulate civil society organisations to become more critical of potentially harmful practices and may stimulate governments to align their data analytics policies more closely with external expectations (Oomsels *et al.*, 2019).

The research presented here can also be seen as a call for greater attention towards the societal aspects of emerging data analytics technologies. Further work can explore whether our initial findings on mismatches between civil society expectations and governmental trustworthiness strategies extends to other states and cultures. Moreover, the research approach adopted here, while very relevant to exploring the interpretative lens of societal actors that actively engages with AI in government issues, is less suited to exploring the degree to which citizens trust AI-projects in government. Lay citizens are currently a relatively neglected audience in AI research, with most extant research focussing on direct users of algorithms instead of citizens whose data is in most cases analysed without them being aware of these analyses or their outcomes. Future survey-based studies may therefore be relevant to explore such gaps in the extant literature.

Finally, there are tentative indications for the intuitive notion that goals may matter regarding data analytics processes. Civil society interviews suggested that, although disproportionate data-gathering and analysis may take multiple forms, risk-profiling applications may have the most severe consequences for citizens. Similarly, even several governmental organisation and co-operating developer respondents noted their mistrust of the values behind data applications that may cause harm or that supervise citizens hyper-efficiently. While most respondents agreed the technology itself was not undesirable, several values and goals underlying specific projects can be. In this sense, the proposed approach in the new EU AI Regulations to designate particularly invasive applications of AI as 'high risk' may be a step forward.

References

Ahonen, P., and T. Erkkilä. 2020. Transparency in algorithmic decision-making: Ideational tensions and conceptual shifts in Finland. *Information Polity* 25 (4): 419-432. https://doi.org/10.3233/IP-200259

Alon-Barkat, S., and M. Busuioc. 2021. Decision-makers processing of AI algorithmic advice: automation bias versus selective adherence. *arXiv*: 2103.02381

Ananny, M., and K. Crawford. 2018. Seeing without knowing: limitations of the transparency ideal and its application to algorithmic accountability. *New Media and Society* 20(3): 973-989. https://doi.org/10.1177%2F1461444816676645

Bitektine, A. 2011. Toward a theory of social judgments of organizations: the case of legitimacy, reputation, and status. *Academy of Management Review* 36(1): 151-179. https://doi.org/10.5465/amr.2009.0382

Brendel, A.B., M. Mirbabaie, T.B. Lembcke, and L. Hofeditz. 2021. Ethical management of artificial intelligence. *Sustainability* 13(4): 1974. https://doi.org/10.3390/su13041974

Burrell, J. 2016. How the machine 'thinks': understanding opacity in machine learning algorithms. *Big Data & Society* 3(1): 2053951715622512.

Busuioc, M. 2020. Accountable artificial intelligence: holding algorithms to account. *Public Administration Review* 81: 825-836. https://doi.org/10.1111/puar.13293

Chander, A. 2017. The racist algorithm? *Michigan Law Review* 115(6): 1023-1045.

Chua, H.N., A. Herbland, S.F. Wong, and Y. Chang. 2017. Compliance to personal data protection principles: a study of how organizations frame privacy policy notices. *Telematics and Informatics* 34(4): 157-170. https://doi.org/10.1016/j.tele.2017.01.008

Donahoe, E. and M.M. Metzger. 2019. Artificial intelligence and human rights. *Journal of Democracy* 30(2): 115-126. https://doi.org/10.1353/jod.2019.0029

Dutch Data Protection Authority (Autoriteit Persoonsgegevens) 2020. 'Belastingdienst/Toeslagen: De verwerking van de nationaliteit van aanvragers kinderopvangtoeslag'. Available at: https://autoriteitpersoonsgegevens.nl/sites/default/files/atoms/files/onderzoek_belastingdienst_kinderopvangtoeslag.pdf.

Dutch Parliamentary Investigation Commission 2020. 'Ongekend Onrecht'. Available at: https://www.tweedekamer.nl/kamerleden_en_commissies/commissies/pok.

Giest, S., and S. Grimmelikhuijsen. 2020. Introduction to special issue algorithmic transparency in government: towards a multi-level perspective. *Information Polity* 25 (4): 409-417. https://doi.org/10.3233/IP-200010

Gunning, D., and D. Aha. 2019. DARPA's explainable artificial intelligence (XAI) program. *AI Magazine* 40(2): 44-58. https://doi.org/10.1609/aimag.v40i2.2850

Gunning, D., M. Stefik, J. Choi, T. Miller, S. Stumpf, and G.Z. Yang. 2019. XAI – explainable artificial intelligence. *Science Robotics* 4(37): eaay7120. https://doi.org/10.1126/scirobotics.aay7120

Grimmelikhuijsen, S.G., and A.J. Meijer. 2014. Effects of transparency on the perceived trustworthiness of a government organization: Evidence from an online experiment. *Journal of Public Administration Research and Theory* 24(1): 137-157. https://doi.org/10.1093/jopart/mus048

Emmert-Streib, F., O. Yli-Harja, and M. Dehmer. 2020. Explainable artificial intelligence and machine learning: a reality rooted perspective. *Wiley Interdisciplinary Reviews: Data Mining and Knowledge Discovery* 10(6): e1368. https://doi.org/10.1002/widm.1368

Freeman, K. 2016. Algorithmic injustice: how the Wisconsin Supreme Court failed to protect due process rights in State v. Loomis. *North Carolina Journal of Law and Technology* 18(5): 75.

Hardin, R. 1996. Trustworthiness. *Ethics* 107(1): 26-42.

Hardin, R. 2002. *Trust and trustworthiness.* Russell Sage Foundation, New York, NY, USA.

High-level Expert Group on AI. 2019. Ethics guidelines for trustworthy AI. Available at: https://digital-strategy.ec.europa.eu/en/library/ethics-guidelines-trustworthy-ai.

Isaak, J., and M.J. Hanna. 2018. User data privacy: Facebook, Cambridge Analytica, and privacy protection. *Computer* 51(8): 56-59. https://doi.org/10.1109/MC.2018.3191268

Kemper, J., and D. Kolkman. 2019. Transparent to whom? No algorithmic accountability without a critical audience. *Information, Communication and Society* 22(14): 2081-2096. https://doi.org/10.1080/1369118X.2018.1477967

Kern, A., S. Marien, and M. Hooghe. 2015. Economic crisis and levels of political participation in Europe (2002-2010): the role of resources and grievances. *West European Politics* 38(3): 465-490. https://doi.org/10.1080/01402382.2014.993152

Koerner, C.L. 2014. Media, fear, and nuclear energy: a case study. *The Social Science Journal* 51(2): 240-249. https://doi.org/10.1016/j.soscij.2013.07.011

Kolkman, D. 2020. The (in)credibility of algorithmic models to non-experts. *Information Communication and Society* 25: 93-109. https://doi.org/10.1080/1369118X.2020.1761860

La Fors, K. 2020. Legal remedies for a forgiving society: children's rights, data protection rights and the value of forgiveness in AI-mediated risk profiling of children by Dutch authorities. *Computer Law and Security Review* 38: 105430. https://doi.org/10.1016/j.clsr.2020.105430

Levin, D.Z., E.M. Whitener, and R. Cross. 2006. Perceived trustworthiness of knowledge sources: the moderating impact of relationship length. *Journal of Applied Psychology* 91(5): 1163. https://psycnet.apa.org/doi/10.1037/0021-9010.91.5.1163

McDermott, Y. 2017. Conceptualising the right to data protection in an era of Big Data. *Big Data and Society* 4(1). https://doi.org/10.1177%2F2053951716686994

Meuwese, A. 2020. Regulating algorithmic decision-making one case at the time: a note on the Dutch 'SyRI' judgment. *European Review of Digital Administration and Law* 1(1): 209-212.

Mittelstadt, B. 2019. Principles alone cannot guarantee ethical AI. *Nature Machine Intelligence* 1(11): 501-507.

Molenveld, A., K. Verhoest, J. Voets, and T. Steen. 2020. Images of coordination: how implementing organizations perceive coordination arrangements. *Public Administration Review* 80(1): 9-22. https://doi.org/10.1111/puar.13136

Oomsels, P., M. Callens, J. Vanschoenwinkel, and G. Bouckaert. 2019. Functions and dysfunctions of interorganizational trust and distrust in the public sector. *Administration and Society* 51(4): 516-544. https://doi.org/10.1177%2F0095399716667973

Peeters, M.M., J. van Diggelen, K. Van Den Bosch, A. Bronkhorst, M.A. Neerincx, J.M. Schraagen, and S. Raaijmakers. 2021. Hybrid collective intelligence in a human-AI society. *AI and Society* 36(1): 217-238. https://doi.org/10.1007/s00146-020-01005-y

Sandvig, C., K. Hamilton, K. Karahalios, and C. Langbort. 2016. Automation, algorithms, and politics| when the algorithm itself is a racist: Diagnosing ethical harm in the basic components of software. *International Journal of Communication* 10: 19.

Schein, E.H. 1990. Organizational culture. *American Psychologist* 45(2): 109-119. https://doi.org/10.1037/0003-066X.45.2.109

Stilgoe, J. 2018. Machine learning, social learning and the governance of self-driving cars. *Social Studies of Science* 48(1): 25-56. https://doi.org/10.1177%2F0306312717741687

Sun, T.Q., and R. Medaglia. 2019. Mapping the challenges of Artificial Intelligence in the public sector: evidence from public healthcare. *Government Information Quarterly* 36(2): 368-383. https://doi.org/10.1016/j.giq.2018.09.008

Van der Cruijsen, C., J. de Haan, and D.J. Jansen. 2016. Trust and financial crisis experiences. *Social Indicators Research* 127(2): 577-600.

Winfield, A. 2019. Ethical standards in robotics and AI. *Nature Electronics* 2(2): 46-48.

Xu, F., H. Uszkoreit, Y. Du, W. Fan, D. Zhao, and J. Zhu. 2019. Explainable AI: a brief survey on history, research areas, approaches and challenges. In *CCF international conference on natural language processing and Chinese computing*. 563-574. Springer, Cham, Switzerland.

Chapter 7

Supporting public sector innovation through a living lab approach: the use of new technologies in administrations

M. Petit Jean[1,2*], L. Bechoux[1], C. Fallon[1] and M. Sabbe[1]

[1]Spiral research centre, University of Liege, Quartier Agora, Place des Orateurs 3, Bât. B-31, Bte 8, 4000 Liège, Belgium; [2]High Strategic Council – Walloon Region, Place Joséphine Charlotte 2, 5100 Jambes, Belgium; maxime.petitjean@spw.wallonie.be

Abstract

Implementing data technologies such as big data, artificial intelligence and blockchain in the public sector is no easy task. It requires the collaboration of multiple actors, displaying various viewpoints and often divergent interests and unequal knowledge vis-à-vis technologies. Considering the implementation of new digital technologies as a public sector innovation, this chapter proposes a participatory methodology to support such process. The proposed methodology relies on a living lab approach supported by scenario thinking. A living lab is an innovation space in which stakeholders are engaged in a long-term bottom-up collaboration. It aims at fostering open innovation through knowledge exchange, co-creation techniques and participatory methods, bridging the gap between stakeholders that do not usually meet. In parallel, scenario thinking provides room for debate and convergence of multiple viewpoints of the future of an issue, by challenging the business-as-usual thinking. As the future is uncertain, it helps in considering a broader range of possible and plausible futures and to think creatively about them. Therefore, scenarios are a heuristic tool for one to think differently about the future(s), considering the future as an open yet not empty space. This chapter relies on data from a living lab process centred around the development of a governance model regarding the integration of big data, artificial intelligence and blockchain in Belgian federal public policies. This process was mostly based on scenario workshops that brought together civil servants and major stakeholders, such as technology providers, business federations, workers federations, sectoral federations, academics and NGOs. This chapter concludes with some recommendations for further use of participatory innovation in the public sector.

Keywords: living lab, scenario thinking, participatory methods, public sector innovation, new technologies

DOI 10.3920/978-90-8686-930-5_7, © Wageningen Academic Publishers 2022

7.1 Introduction

Implementing data technologies such as big data (BD), artificial intelligence (AI) and blockchain (BCT) in the public sector is no easy task. It requires the collaboration of multiple actors, displaying various viewpoints and often divergent interests and unequal knowledge vis-à-vis technologies.

Considering the implementation of data technologies as a public sector innovation (PSI), the theoretical starting point of this chapter is the growing evidence showing that PSI is better supported by collaborative and participative processes.

Building on the experience of the DIGI4FED research project, the aim of this chapter is to introduce the living lab approach as a participatory and deliberative methodology to support PSI while reflectively assessing the conduct of its implementation. A living lab can be seen as an innovation space in which stakeholders are engaged in a long-term bottom-up collaboration. It aims at fostering open innovation through knowledge exchange, co-creation techniques and participatory methods, bridging the gap between stakeholders that do not usually meet.

This chapter is divided into several sections. The next section (Section 7.2) offers a literature review of public sector innovation, in link with new digital technologies. Section 7.3 introduces the living lab approach to fostering public sector innovation and its use in developing public sector innovation. The following sections (Section 7.4-7.6) present the methodology to implement living labs, based on our DIGI4FED experience, including the use of scenario thinking to support such process. The final section (Section 7.7) assesses the proposed approach and provides theoretical and practical recommendations for further implementation.

7.2 Public sector innovation and new technologies

Since the early 2010s, there has been a renewed interest in public sector innovation. The OECD has, for example, published several reports and organised multiple workshops on the topic (OECD, 2011, 2014, 2015), defining an 'innovation agenda' for the public sector and calling for the 'innovation imperative'. Innovation has therefore become an instrument to make the public sector more open, more collaborative, more participative, and more productive. From 2017 onwards, the OECD Observatory for Public Sector Innovation (OPSI) has annually published a report of public sector innovation trends with the United Arab Emirates (UAE) Mohammed Bin Rashid Centre for Government Innovation. Similarly, the European Commission has funded several studies on the topic, considering innovation as a means to make the public sector more effective, more efficient and more responsive (Bason *et al.*, 2013; Hollanders *et al.*, 2013).

As any policy instrument, innovation is a construct and reflects the choices of those that develop and mobilise it (Lascoumes and Le Galès, 2007). Therefore, public sector innovation (PSI) is a fluid concept that is dependent on the values, interests and objectives of the actors involved in the innovation process.

The literature on PSI often refers to the writings of Schumpeter (1942). He sees innovation as a process of creative destruction in which new combinations of existing resources are formed. In his view, innovation is intrinsically linked to the notion of entrepreneurship. Another commonly cited definition is from Rogers (2003), for whom innovation is an idea, practice or object perceived as new by an individual or other entity adopting it. In this sense, it is the perception of the new nature of the idea that matters, as opposed to its objective novelty. Innovation is always contextualised. This is also reflected in the vision of Koch and Hauknes (2005) who propose to consider PSI as a tool for analysing social activities and interactions rather than as a marker of an objective reality. Finally, Bekkers *et al.* (2011a: 6) point out that:

> The need for public innovation can be defined as the search for new ideas and concepts, technologies, techniques and methods, forms, systems, and procedures to create meaningful interactions between the government and society in order to deal with a number of societal challenges.

Innovation is also based on a rather positive normative *a priori* and the added value is often left unquestioned (Fuglsang and Pedersen, 2011; Koch and Hauknes, 2005; Pollitt, 2011; Sveiby *et al.*, 2012). Therefore, taking a reflective stance on PSI, we can highlight four distinguishable features of PSI compared to other policy instruments:

1. PSI includes the notion of novelty; it produces a new idea, practice, or object (Bekkers *et al.*, 2011b; De Vries *et al.*, 2016; Hartley, 2005; Moore and Hartley, 2008). Some authors therefore distinguish between an objective innovation, in the sense that it is the very first time it appears, and a subjective innovation, appearing for the first time in a given context (Brown and Osborne, 2012).
2. PSI must be implemented in a specific context within which it is new. More than just ideas or inventions, innovations are new ideas and practices that are contextualised (De Vries *et al.*, 2016).
3. PSI involves a discontinuity with the past (and is therefore different from continuous improvement). If innovation is seen as an organisational change that can lead to significant changes in the organisation's environment, it also leads to a paradigmatic shift, in contrast to a classical organisational development. This shift changes the nature of the product, service or process concerned by positioning it as a disruption from previous practices (Brown and Osborne, 2012). However, it is possible to distinguish between incremental and radical innovations (Bekkers *et al.*, 2011b; Fuglsang and Pedersen, 2011).
4. PSI is both a product and a process. Thus, innovation is both the product of a process but also a process of transformation as such (Alter, 1993; Brown and Osborne, 2012). As a process, Schuurman *et al.* (2016) distinguish between internal and open innovation. While internal innovation processes only include members of an organisation, open innovation processes involve actors external to an organisation and their knowledge.

Novelty, context, discontinuity, and the productive and processual nature of innovation are therefore distinctive features of public sector innovation. Considering BD, AI and BCT as defined in Chapter 2 and these characteristics, implementing these technologies can clearly be seen as public sector innovation.

Moreover, the issue is also touched upon by the OECD Observatory for Public Sector Innovation that published a working paper in 2019 specifically oriented towards AI in the public sector (Berryhill *et al.*, 2019). It aimed at improving government officials' understanding of AI and its considerations within the public sector.

In the current context, public sector organisations are constantly engaged in innovation processes. In that regard, Hartley *et al.* (2013: 825) state the following:

> There is growing evidence that collaboration can spur public innovation [...] Theories of collaborative innovation in the public sector derive both from theories of network governance, which emphasise the role of collaborative networks in finding innovative solutions to complex problems [...], and from theories of learning that conceptualise step change as occurring through interorganisational interaction and collaborative processes [...]. Theories of collaborative innovation also echo insights from management theories about private sector innovation, where it focuses on 'social innovation' [...], 'co-creation' [...], and 'open innovation' [...].

7.3 Spurring public sector innovation through living labs

As mentioned in the previous section, open innovation, social innovation, and co-creation are pointed out as key principles of public sector open innovation processes. Involving actors external to the public sector becomes therefore a key element of such processes.

However, Schuurman *et al.* (2016: 7) specifically point out the 'lack of adequate management models'. Additionally, Gascó (2017: 90) states that the way 'open innovation can become a true and effective tool for governments is still an underexplored topic' as the literature is especially looking at ways to implement a successful private sector practice. in the public sector.

According to Gascó (2017), the role of agents and intermediaries is crucial to foster innovation. In the private sector, innovation networks of actors aiming at common goals have become the norm. However, such practices are difficult to directly implement in the public sector as actors are not competing for market shares: for example, the goals of policymakers, citizens, enterprises and NGOs are not necessarily aligned. This difficulty of alignment reinforces the need for intermediaries in the public sector. Examples of intermediaries supporting innovation networks can be specific actors (innovation consultants, civic innovators), permanent installations (FabLab, incubators), living labs and ad hoc events (workshops, conferences, networking events, hackathon) (Bakici *et al.*, 2013; Brandsen *et al.*, 2018; Gascó, 2017; Yuan and Gasco-Hernandez, 2021).

Specifically, in this chapter, we focus on one of these intermediaries: living labs. As noted by Bergvall-Kareborn and Stahlbrost (2009: 357), 'living lab has become an umbrella concept for a diverse set of innovation milieus'. We conceptualise the living lab as an 'innovation space' in which stakeholders are engaged in a long-term bottom-up collaboration. Such innovation space can take the form of one or several workshops, of an open space, or of any space that allows for dialogue between stakeholders (Alavi *et al.*, 2020). Living labs can be defined as:

> [...] constituting a setting for collaborative innovation by offering a collaborative platform for research, development, and experimentation with product and service innovations in real-life contexts, based on specific methodologies and tools, and implemented through concrete innovation projects and community-building activities (Schaffers and Turkama, 2012: 25).

Dell'Era and Landoni (2014: 139) provide a shorter definition, as 'a design research methodology aimed at co-creating innovation through the involvement of aware users in a real-life setting'. Considering these definitions, living labs are relevant intermediaries to foster innovation regarding new digital technologies. They allow collaborative innovation in the public sector, bringing stakeholders that usually do not meet in such a setting. This chapter focuses on various stakeholders involved in tax fraud policy and in social security infringement policy.

In methodological terms, living labs aim at fostering open innovation through knowledge exchange, co-creation techniques and participatory methods, bridging the gap between stakeholders that do not usually meet. They thus act as an intermediary for innovation, connecting actors and supporting the knowledge exchange. In that regard, the process matters often as much as the end-result (Gascó, 2017).

Nevertheless, such an approach cannot be decreed. Co-creation in open innovation requires an open mind-set towards sharing and collaboration, which can be supported by techniques such as 'context-mapping', which involves users intensively in creating an understanding of the contexts of service use (Sleeswijk Visser *et al.*, 2005), and 'generative' techniques, which can reveal tacit knowledge and expose latent needs (Dell'Era and Landoni, 2014). Such techniques include among others, scenarios thinking.

The real-life context of the innovation process contributes to a better understanding of the tacit and domain-based knowledge needed to assess needs and build feasible, appropriable solutions. The participatory techniques must respect some principles to ensure the sensitisation of participants, but they are to be adapted by the research team to the specific context and issues at stake.

In terms of living lab, several examples and processes are cited in the literature. Ruijer and Meijer (2019) describe, for instance, a living lab approach for open government data, based on several workshops and experiments with civil servants, students and researchers. Similarly, Bergvall-Kareborn and Stahlbrost (2009) offer insights on a living lab whose goal is to develop innovative

IT services and products with real users and their needs. Contrary to the previous example, this living lab is a continuous endeavour rather than a punctual project, and it has evolved with time to building a large community of users that test services and products.

A third example is provided by Dekker *et al.* (2021) who researched quite a different form of living lab, aimed at innovating alternatives in housing for asylum seekers to the classical large-scale facility housing. In a specific housing complex, a reduced number of asylum seekers lived together with locals and had access to learning activities. The living lab context was therefore the housing complex in which residents were innovating through living their life in the facilities. A final example is the concept of urban living labs, bringing together citizens, local authorities, businesses, and research organisations to co-produce innovative urban services. It is mostly based on meetings and workshops in different urban locations, sometimes using the city as a living environment (Nesti, 2017).

Nevertheless, none of these works provide a detailed approach to setting up a living lab. Hence, building upon these insights, we propose methodological steps to set up a living lab in the next section. We insist that what we provide here are guidelines rather than a recipe book, and recommend that the reader keep a critical eye on the creation and implementation of his/her own living lab, to keep it aligned with its objectives.

7.4 Implementing a living lab approach around new technologies in the public sector

7.4.1 The general methodological approach

Seeing the living lab as an innovation space that is user-centric in real-life settings, multiple methodological questions must be answered: Who to involve? When? What for? And how? The answers to these questions come from multiple sources including literature reviews and interviews, as shown below.

Our approach to living labs is designed in three phases: exploring, co-creating, and testing and evaluating (Zwetkoff *et al.*, 2018).

1. The first phase is about exploring problems and opportunities to be addressed regarding the issue at stake (in our case, integrating BD, AI and BCT in Belgian federal public organisations), as well as identifying emerging ideas and breakthrough scenarios.
2. The second phase is to support co-creating a product or a service, through deliberative processes including several diversified stakeholders. In the DIGI4FED project, we took the development of an open governance model for the use of new digital technologies in the fight against fraud as the policy innovation objective in the co-creation phase
3. The third phase (testing) allows stakeholders to test the product or the service. In a real-life context, in terms of performance and potential adoption, to identify incentives, possible risks and the discretionary space of professionals and users which can be specified (Brandsen *et al.*, 2018). Again, in the DIGI4FED project, the aim is to test key dimensions related to an open governance model developed thanks to the co-creation phase as well as to other research

outputs. In this project, testing involves the selection of adapted evaluation criteria to test the proposed governance model with the help of the Delphi survey method. Such criteria include the consistency, the feasibility, as well as the overall efficacy and suitability of the model for instance. These were selected after an assessment of the common criteria used in public policy evaluation, planning, and design science research literature.

To set up a living lab, several elements must be taken into account: users, an application environment, Information and Communication Technologies (ICT), methods, and partners bringing expertise (Ruijer and Meijer, 2019). These five elements are described in the following sections, grouping though users and partners within a single 'Participants' category, combined with ICT technologies.

7.4.2 The application environment

The application environment includes all organisations and stakeholders that could be impacted by the innovation. In the DIGI4FED project, it was centred on the Belgian federal administration. This includes all public organisations at the federal level. Such focus does not, however, exclude other politico-administrative levels (that is, for Belgium, European, regional, community, provincial and municipal levels), although the main focus is not on them.

In addition, if the topic at stake is more specific, it could be interesting to specify and/or restrain the organisations and stakeholders to be considered as relevant. For example, the DIGI4FED project mainly considered two policy domains: the fight against tax fraud, and the fight against social security infringements. These two policy domains were chosen as their stakeholders have shown sustained interest in the use of data for public policy. These two policy domains display quite distinctive configurations in terms of stakeholders.

On the one hand, tax policy concerns both direct and indirect taxation at the federal level. Direct taxation notably includes personal and corporate income tax, while indirect taxation comprises, among others, value-added tax and import and excise duties. Tax policy is a rather centralised policy domain, organised around the Belgian federal tax administration, that is in charge of both policy design and is a part of policy implementation. Other actors, such as the financial intelligence processing unit, several services of the police forces, the national bank and the financial services and markets authority, can provide information regarding potential infractions to the central administration. In that regard, the living lab will focus on topics that relate to the fight against tax fraud and tax evasion; the former is always illegal while the latter can be legal or illegal (based on several specific conditions). The participants (see below) were identified and chosen within this realm.

On the other hand, social security policy aims at providing several social services to citizens such as benefits (e.g. unemployment, retirement, children, maternity leave) and healthcare reimbursement. This policy domain is rather fragmented. The social security central administration is in charge of part of the policy design, while most of the implementation, including the fight against social security infringements, is led by a multitude of public bodies, each with their own specific area, such as employment, healthcare insurance, social security for contractual workers, social security for self-employed workers and control of social legislation. There are other areas, but these five areas are also those in which there is a specific work to combat what is broadly defined as social security infringements. In addition to these several administrative actors, there are a series of other institutions involved, as the federal IT support service and the federal platform for social security data exchange, as well as private entities such as social partners, social secretariats, and mutual health funds.

Finally, other more horizontal political and administrative actors are also present in these two policy domains, such as the federal government, the federal parliament, the data privacy authority, and the Court of Audit.

7.4.3 Participants and ICT technologies

Participants and ICT technologies are the key interacting elements of an innovation process within a living lab.

The definition of the participants is one of the crucial elements in implementing a living lab. In our research, we worked with a closed collaborative network, rather than with an open one. This means participants will be pre-selected based on specific criteria in order to generate more focused inputs from them as users. The closed approach allows to choose the size of the group and the profile of participants, but is slightly more complex to implement than an open approach, especially in identifying, contacting and attracting potential participants (Dell'Era and Landoni, 2014).

In the DIGI4FED project, participants were members of an organisation that is (more or less closely) linked to one of the two topics that were specifically researched in the application environment (either tax policy or social security policy). In that regard, they all had an interest in participating as well as a particular expertise to bring to the debate. Similarly, the ICT technologies that are considered are, as stated before, BD, AI and BCT. Their development will have an impact on any identified participant to the living lab.

7.4.4 Methods

Living labs methodologies use user-centred co-creation methods, that are mostly participatory and qualitative. These methods also aim at engaging stakeholders in a real-life setting.

In DIGI4FED, the living lab approach we designed combined several qualitative and participatory methods, used in the three previously mentioned phases (exploration, co-creation and testing). The phases of exploration and co-creation strongly rely on scenarios thinking to position participants in real-life settings. In the exploration phase, the main methods used were literature review and semi-structured interviews, while the co-creation phase relied on four scenario workshops. The testing phase was supported by a Delphi survey. The whole process can be summarised as shown in Figure 7.1.

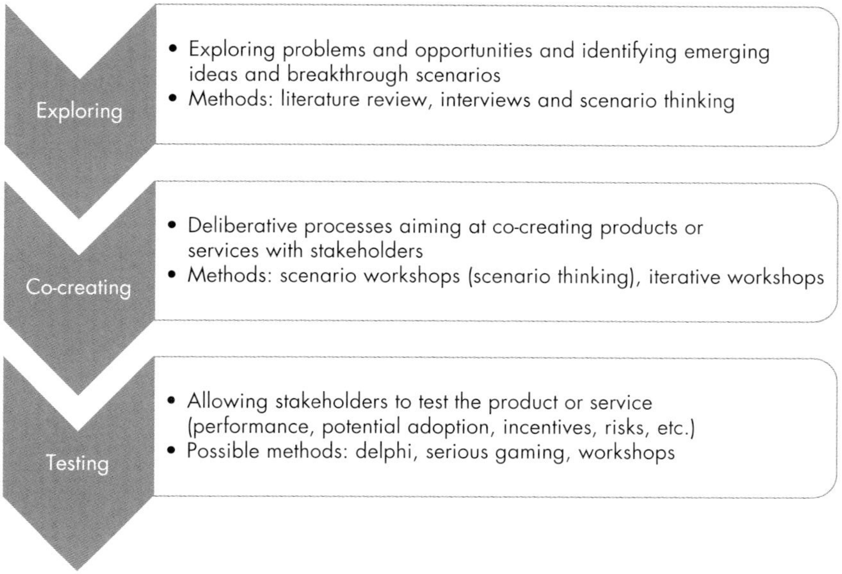

Figure 7.1. The living lab process.

In a nutshell, our testing phase was supported by a Delphi survey submitted to all the key stakeholders that were involved in the previous research steps of the DIGI4FED project (including the living lab workshops). The Delphi method is a social research technique that relies on a multi-round survey process to obtain positions on a topic of interest from a panel of experts (Brady, 2015; Landeta, 2006; Linstone and Turoff, 1975). By eliciting such positions, the goal is to determine the level of consensus or dissensus among experts regarding a topic of interest. In view of testing our model, we took inspiration from various approaches (such as design science research and public policy evaluation) to develop a list of ten evaluation criteria. To assess the general validity of the model, we used some wide-ranging criteria, such as the model's overall simplicity, consistency, and suitability. Some more refined criteria were also used to assess specific aspects of the model, such as the efficacy and consistency of specific design solutions in day-to-day practice.

To exemplify how participatory innovation can support public sector innovation, we will focus on the exploration and co-creation phases of the living lab process. The following section describes scenario thinking while the subsequent sections provide insights on the exploration and co-creation phases.

7.4.5 Scenario thinking as a support for the exploration and co-creation phases

As mentioned in the previous section, the exploration and co-creation phases rely on scenarios thinking. As Cairns and Wright (2018: 1-2) point out:

> Scenario thinking offers a way for individuals and groups to face up to the threats and opportunities of the future and to their potential impact upon the organisation or community. As a decision-maker in this sort of situation, you may not fully understand the complexities and ambiguities that the future may hold.

Scenario thinking provides room for debate and convergence of multiple viewpoints of the future of an issue, by challenging the business-as-usual thinking (Cairns and Wright, 2018). As the future is uncertain, it helps to consider a broader range of possible and plausible futures and to think creatively about them (Bishop *et al.*, 2007). Specifically, scenarios are a heuristic tool for one to think differently about the future(s), considering the future as an open but not empty space (Adam and Groves, 2007). They often take the form of narratives or future images that are consistent and plausibly possible (Spaniol and Rowland, 2019).

Such scenarios (and, more broadly, foresight methods) can fulfil multiple objectives: they can support decision-making, directly or conceptually, by building anticipatory knowledge about futures (Fobé and Brans, 2012). They can also provide information to decision-makers and help translate it within a particular policy (Havas *et al.*, 2010) while they also help to position some issues higher on the political agenda (Van der Steen and Van Twist, 2012). In addition, such a scenario approach allows a broad participation of stakeholders in the formulation of a public policy, potentially facilitating policy implementation (Havas *et al.*, 2010; Van der Steen and Van Twist, 2012). It thus helps build a network of stakeholders (Koschatzky, 2005), fosters organisational learning (Bootz, 2010), and facilitates the identification of barriers and obstacles to a particular vision (European Commission, 2009).

As Cairns and Wright (2018: 50-51) point out:

> Scenario stories in isolation serve no purpose per se. While some approaches to scenarios work place an emphasis on the scenario stories themselves as narratives of some 'real world' futures, we see them as providing primarily a better understanding of, and a broader range of perspectives on the present.

In sum, both the scenario process and its outcome are key features of the living lab approach we designed, in that they allow the focus to be on the experience of participants in (future) real-life settings.

7.5 Exploring big data, artificial intelligence and blockchain futures in public organisations

The exploration phase is the starting point of a living lab process; it aims at engaging stakeholders in the living lab and to providing insights to feed into the co-creation phase. In DIGI4FED, this phase was also used to build a set of four future scenarios.

7.5.1 Engaging with stakeholders

First, stakeholder engagement is vital for any living lab processes and must start as early as possible. Early engagement favours the opening of new research avenues and of emerging and creative ideas while it also incentivises these stakeholders to take part in later activities organised in the co-creation phase. More than a research activity per se, engaging with stakeholders is a mind-set, geared towards collaboration with multiple potential stakeholders.

In this regard, early engagement with stakeholders supports the development of scenarios by taking on board insights from various public and private stakeholders. They can be combined to try 'to think out of the box' for the mid-term and longer-term future. In DIGI4FED, we mostly relied on the follow-up committee meetings to discuss and test specific ideas and on semi-structured interviews to engage with users and stakeholders. This approach placed them early in the process and helped us to better define and refine the extent and the range of our research process.

This 'engagement' mind-set also oriented the scenario-building process of this exploration phase (see below).

7.5.2 Building a first set of scenarios

Several steps can be taken to build a set of future scenarios: clarifying the role of scenarios in the living lab process, eliciting information in order to build scenarios, and the building of the scenarios per se.

7.5.3 Clarifying the role of scenarios

Building scenarios is no easy task. As previously said, it is not possible to predict the future and scenarios must be aimed at broadening the range of thinking among stakeholders and living lab participants, thereby fostering innovation.

In that regard, scenarios are clearly built towards a specific goal, that is supporting the process of identifying conditions, problems, and opportunities to be addressed in building the open governance model, through the living lab process. In other words, scenarios must be built as cognitive tools that spur debate and exchanges within the living lab process. As said before, scenarios are not an end but a means to reach this objective of opening the debate.

In the DIGI4FED context, the choice has been made to develop scenarios in the exploration phase, rather than in the co-creation phase, for practical reasons. As we needed to bring 'time-poor, senior stakeholders' in the living lab process to ensure its relevance, designing scenarios *ex ante* favours the organisation of short and intense workshops (Cairns and Wright, 2018).

7.5.4 Eliciting information to build scenarios

There are many methods and techniques to build scenarios (Bishop *et al.*, 2007). For example, their starting point can be an individual judgement, dominant trends, particular logics, uncertainties, specific end-states, potential future events or system models. Depending on these different starting points, both the process and the products of scenario development can vary, for instance going from a trend-based (or baseline) single scenario to multiple scenarios or to alternative end-states. In this regard, techniques like visualisation, probability trees, backcasting, cross-impact analysis and modelling are often cited.

In DIGI4FED, we built scenarios on the intuitive logic tradition, also called the matrix approach, that leads to the definition of four scenarios based on the combination of two very impactful and two very uncertain external macro factors (Bishop *et al.*, 2007; Bradfield *et al.*, 2005; Cairns and Wright, 2018).

In order to identify these external macro factors, we relied on several methods: literature review, interviews, and brainstorming within the DIGI4FED research group. We also relied on research notes written by other members of the DIGI4FED team on the subject matter.

First, the literature review, which included both academic and grey literature,[1] had two objectives: it aimed at identifying, broadly, the driving forces of change regarding technology, public (digital) management and public policy, and at detecting possible use cases to feed into scenario narratives. The literature review also helped to prepare the semi-structured interviews (see below). It led to identifying particular driving forces such as skills and competences, organisational culture, politico-administrative support, socio-technical infrastructures (both software and hardware) and the legislation concerning, for instance, privacy rights and collaboration framework. It also highlighted the necessity to consider multiple levels of change (individual, organisational, policy and international), and it finally helped in identifying relevant use cases from other countries and contexts.

The (grey) literature review also supported the mapping of the key actors within the two policy domains, to build a rigorous multidisciplinary view regarding each policy, notably in terms of political and administrative policymaking, knowledge production and use, as well as legal issues and organisational, cultural and institutional matters. Such contextual knowledge is essential to be able to genuinely engage with stakeholders when meeting them.

[1] Grey literature includes in this case government/public institutions communication documents, such as political orientation notes and parliamentary debates, as well as documentation written by major technology providers, such as SAS, IBM and Microsoft.

Second, semi-structured interviews were conducted among stakeholders. These interviews aimed, among other project-related objectives, at identifying drivers of change regarding technology, public (digital) management and public policy, and at eliciting future visions leading up to 2030. Insight was gained on other topics, such as data management processes, trust-based issues, ethical and legal dimensions (e.g. privacy and control), transparency and public opinion.

Four categories of actors were surveyed: public stakeholders directly involved in the use of data, public stakeholders that are impacted by or have an impact on the use of data in the public sector, private stakeholders and academics. Stakeholders were identified, based on the mapping, in the two policy domains (i.e. the fight against tax fraud and the fight against social security infringement).

In the first category, public stakeholders directly involved in the use of data are the central administration and agencies that collect and/or use data in policy implementation either to fight tax fraud or social security infringements.

The second category is of public organisations that are impacted by such use of data. They can either control the use of data (such as data privacy authorities), support their exchange (such as the crossroad bank for social security which coordinates and controls a decentralised system of data exchange between social security public actors) or be compelled to transfer personal data when specific conditions are met. The latter case notably concerns the financial intelligence processing unit that collects financial data from various sources and can inform the tax inspection service within the central administration.

Private stakeholders, the third category, are actors that are affected in one way or another by the policies that are implemented by actors from the first categories. In a political system like Belgium in which social concertation is central, social partners, representing workers and employers, must be taken into account. Other transversal actors (in terms of policy domain) are technology providers (either big companies or small tech start-ups) and non-governmental organisations active in domains like human rights (including privacy). In addition, each policy domain has its own private stakeholders. For example, in terms of tax policy, stakeholders included social secretariats, business federations in highly monitored sectors (such as construction, food and drinks, and cleaning), information providers (such as banks, accountants and lawyers) and, obviously, potential fraudsters.

Finally, the fourth category is made up of academics who were chosen specifically to elicit their views of the future of BD, AI and BCT, both as technologies and in their projected use in the public sector by 2030.

Among these interviews, future expectations and views about the future of new technologies in government by 2030 were specifically used in identifying the driving forces.

By the end of the exploration phase, almost all the key stakeholders had been involved in talks or interviews, easing their subsequent invitation to the living lab workshops in the co-creation phase.

7.5.5 Building scenarios

As previously mentioned, the process of building scenarios relied on the matrix approach proposed by Cairns and Wright (2018), although slighted adapted to our purpose, as described below.

Theoretically, this approach ...

> ... is focused on development of multiple scenarios that explore the 'limits of possibility' for the future, rather than on the development of singular, 'normative' scenarios of some ideal future to be aspired to.

It is composed of 8 stages: (1) setting the agenda, (2) determining the driving forces, (3) clustering the driving forces, (4) defining the cluster outcomes, (5) determining the key scenarios factors, (6) framing the scenarios, (7) scoping the scenarios and (8) developing the scenarios (Cairns and Wright, 2018). Such a structured process helped to document and provide methodological transparency regarding the way the scenarios were developed. However, researchers must implement such an approach with flexibility, creativity and an open mind to new ideas which could be added at any stage of the process.

In DIGI4FED, we conducted the scenario-building process in the exploration phase within the research team without involving many external actors, as many of these were evolving in a time-limited context. Some phases were therefore reduced or merged with others. It is also important to note that due to the Covid-19 rules, all meetings were conducted online, sometimes using specific facilitation tools such as Mural.

As a matter of fact, setting the agenda (phase 1) was fairly easy: the focal issue was to build scenarios about the future integration of BD, AI and BCT within federal public administrations by 2030. The choice of 2030 was a pragmatic choice, guided by two elements: on the one hand, the time horizon needed to be far enough in advance in order to provide room for multiple visions of the future without being locked into technical and legal considerations; on the other, 2030 is also very close, and allows participants to easily project themselves to less than ten years from now.

The second phase was focused on determining the driving forces. The collected data described in the previous section were systematically analysed in order to identify the insights to be transformed into possible driving forces. Their identification relied on a PESTEL approach, to highlight the political, economic, social, technological, environmental, and legal dimensions. Driving forces can be defined as 'underlying and impacting factors that set the pattern of events and determine outcomes in the business environment and timescale being considered – the forces that make things happen' (Van der Heijden *et al.*, 2006: 202). These driving forces must thus be formulated so as to highlight an uncertain impact without providing information on the orientation of such an impact.

A first proposition of possible driving forces was discussed among the research team, adding new driving forces and reformulating existing ones. This process led to the identification of 74 driving forces. To ensure traceability, all driving forces were systematically referenced in a table, linking them with the source they come from. Examples of such driving forces were:

- Degree of use of new technologies within public administration and degree of international cooperation in policy domains (political driving forces).
- Degree of data availability to develop particular AI and BCT applications, and degree of transparency regarding data, processes and objectives (technological driving forces).
- Extent of R&D funding related to AI and BCT and degree of involvement of the private sector in technology development in public administration (economic driving forces).

All the driving forces were classified according to one of the PESTEL dimensions, in an online interactive manner, that led to such a scheme. The interest of the scheme is not to provide all the driving forces to the reader but rather to illustrate the research process using online facilitation and visualisation tools. Table 7.1 presents the PESTEL dimensions and their associated driving forces.

The driving forces were then clustered (phase 3), based on a preliminary proposition that was discussed within the research group. Such clustering process was based on a chronology analysis, considering which driving forces would lead to an impact that would lead to a change in other driving forces.

In total 8 clusters were defined:

1. political support in using of AI and BCT in public administration;
2. availability of skilled IT profiles for public organisations;
3. trust in public sector using new ICT technologies;
4. technological progress in AI and BCT applications for public sector;
5. Belgian federalisation process;
6. environmental regulation regarding new digital technologies;
7. development, compliance and enforcement of a regulatory framework;
8. solidarity within society.

Similar to the identification process, the Figure 7.2 and 7.3 provide some examples which illustrate the results of the clustering process. The full clustering process results are available as Electronic Supplementary Material.

To provide the reader with an example, we consider the cluster 4 regarding the availability of skilled ICT profiles for public organisation. The establishment of the cluster was inductive, based on a chronology analysis (ordering driving forces based on their influence on subsequent driving forces). We started this process looking at the driving force 2 as we thought it could be a key element of any process. We identified that driving forces 3, 8 and 19 preceded and led to the degree of political financial support towards new technology. Concretely, we hypothesised that a high degree of budgetary constraints, a high degree of risk aversion towards new technologies among politicians, and a low degree of political will to implement new technologies could lead to a low

Table 7.1. Driving forces classification.

PESTEL dimensions	**Driving forces**
Political driving forces	1. Degree of international cooperation about AI / BCT development; 2. Degree of political support towards new technologies (money); 3. Extent of the Belgian federalisation process; 4. Degree of collaboration between federal / federated entities; 5. Extent of proactive attitude in policy orientations; 6. Degree of use of new technologies (BD, AI, BCT) within PA; 7. Degree of international cooperation in policy domains; 8. Degree of political will to implement new technologies (discourse); 9. Extent of procedures regarding personal judgement in data processes; 10. Extent of transparency in policy orientations; 11. Degree of political support to open data policies; 12. Changing policy orientations in the two policy-domains; 13. Extent of access to public services by citizens / enterprises; 14. Extent of surveillance in policy orientations; 15. Outcome of a national data strategy; 16. Extent of involvement of the state in society; 17. Extent of horizontality between public organisations; 18. Outcome of a particular technology event (scandal / breakthrough); 19. Extent of risk aversion regarding new technologies among politicians; 20. Changing general policy in terms of public management; 21. Extent of international policy diffusion; 22. Degree of populism in politics; 23. Outcome of a particular non-technological event (e.g. terrorism); 24. Extent of harmonisation at the EU level of the internal market; 25. Extent of digital literacy in policy orientations
Environmental driving forces	1. Extent of the interest in environmental consequences of new technologies; 2. Degree of resilience vis-à-vis environmental / health crises; 3. Importance of environmental consequences of new technologies; 4. Extent of new EU energy regulation
Socio-cultural driving forces	1. Degree of public sector legitimacy (trust in public sector) among citizen / stakeholders; 2. Degree of interest in digitalisation among stakeholders; 3. Degree of technological literacy / interest in society; 4. Degree of citizen / stakeholders trust in new technologies; 5. Degree of acceptance of freedom restrictions; 6. Extent of importance of leisure time; 7. Extent of importance of work in life; 8. Outcome of digital divide; 9. Degree of data-led applications acceptance in society; 10. Degree of solidarity within society; 11. Degree of legitimacy of public management ideas in society; 12. Distribution of IT / data science skills between public and private sectors; 13. Extent of privacy consciousness

Table 7.1. Continued.

Technological driving forces	1. Degree of security of technologies vis-à-vis hacking and leaks; 2. Degree of data availability to develop particular AI and BCT applications; 3. Degree of transparency regarding data, processes and objectives; 4. Extent of algorithm transparency; 5. Degree of data sharing between public organisations; 6. Degree of sovereignty on software; 7. Degree of digital sovereignty at federal level; 8. Degree of interoperability of data; 9. Degree of data quality; 10. Extent of human decision-making in processes; 11. Extent of access for citizens / businesses to fraud analytics report; 12. Degree of diversity in teams developing technologies; 13. Degree of sovereignty on data; 14. Degree of sovereignty on hardware; 15. Degree of interoperability of technologies; 16. Extent of process explainability; 17. Extent of biases in algorithms; 18. Degree of development of EBSI; 19. Degree of AI technology development; 20. Degree of development of analytics as a service
Economic driving forces	1. Extent of R&D funding related to AI & BCT; 2. Degree of involvement of the private sector in technology development in PA; 3. Degree of budgetary constraints for public organisations; 4. Importance of competition in tech markets (monopoly, oligopoly, etc.); 5. Extent of international competition (with China & USA); 6. Degree of private sector involvement in public sector; 7. Degree of start-ups involvement / inclusion in public sector; 8. Number of tech profiles graduating per year; 9. Extent of brain drain in IT; 10. Distribution of wealth among society
Legal driving forces	1. Degree of regulation towards private actors providing new technologies; 2. Extent of legal development regarding data sharing among public and private organisations; 3. Degree of legal enforcement of current rules (e.g. GDPR, ECHR, administrative law principles); 4. Extent of policy-specific regulation regarding new technologies (e.g. fiscal policy, social policy); 5. Degree of change in public procurement rules; 6. Degree of integration between digital practices and legislation; 7. Degree of horizontal regulation regarding new technologies within public sector; 8. Degree of compliance with legislation within society; 9. Outcome of standards development; 10. Extent of new self-sovereign identity regulation; 11. Extent of ethical clause regulation for public procurement; 12. Extent of innovation process regulation for public regulation

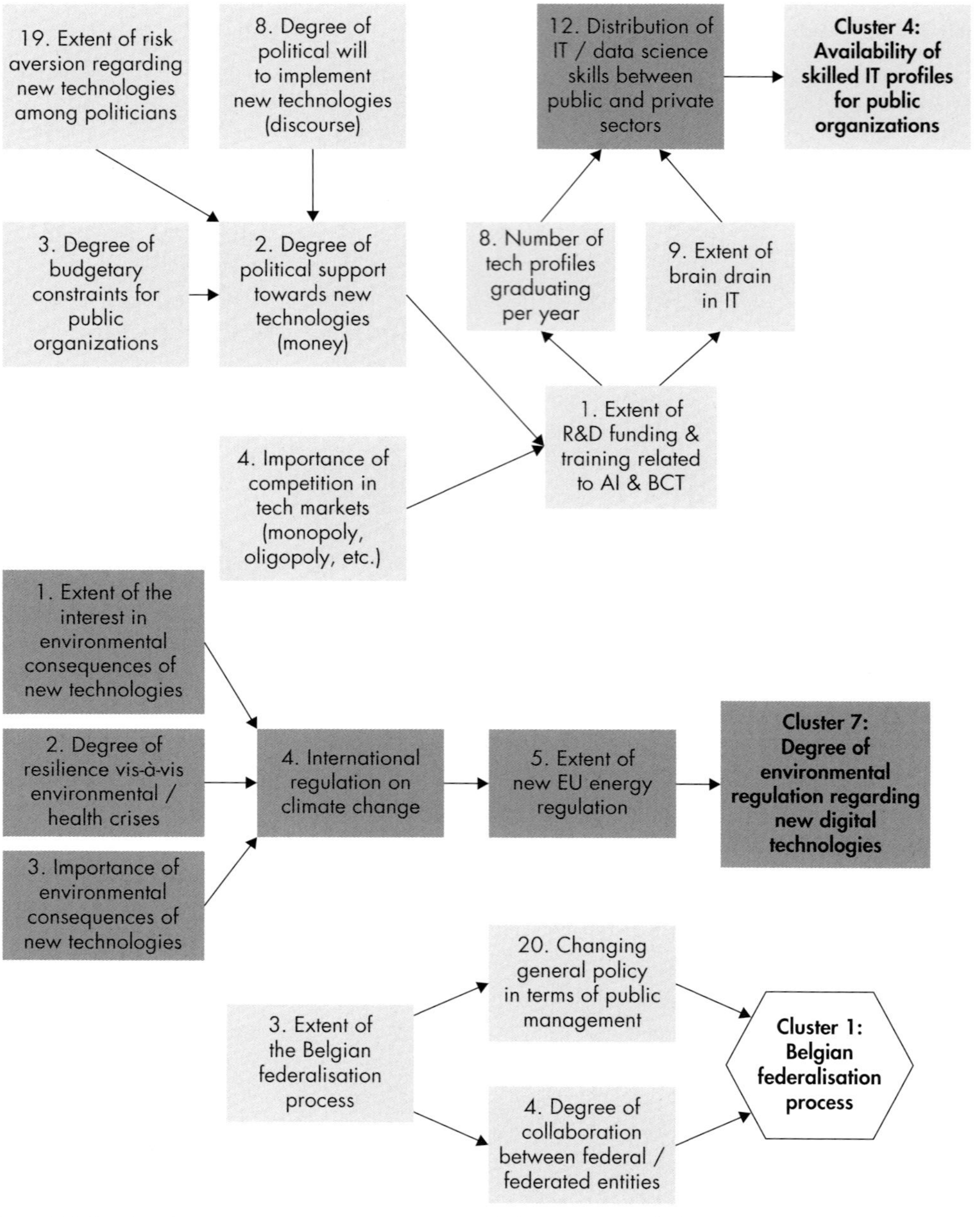

Figure 7.2. Results of the clustering process (examples of clusters 1, 4, and 7).

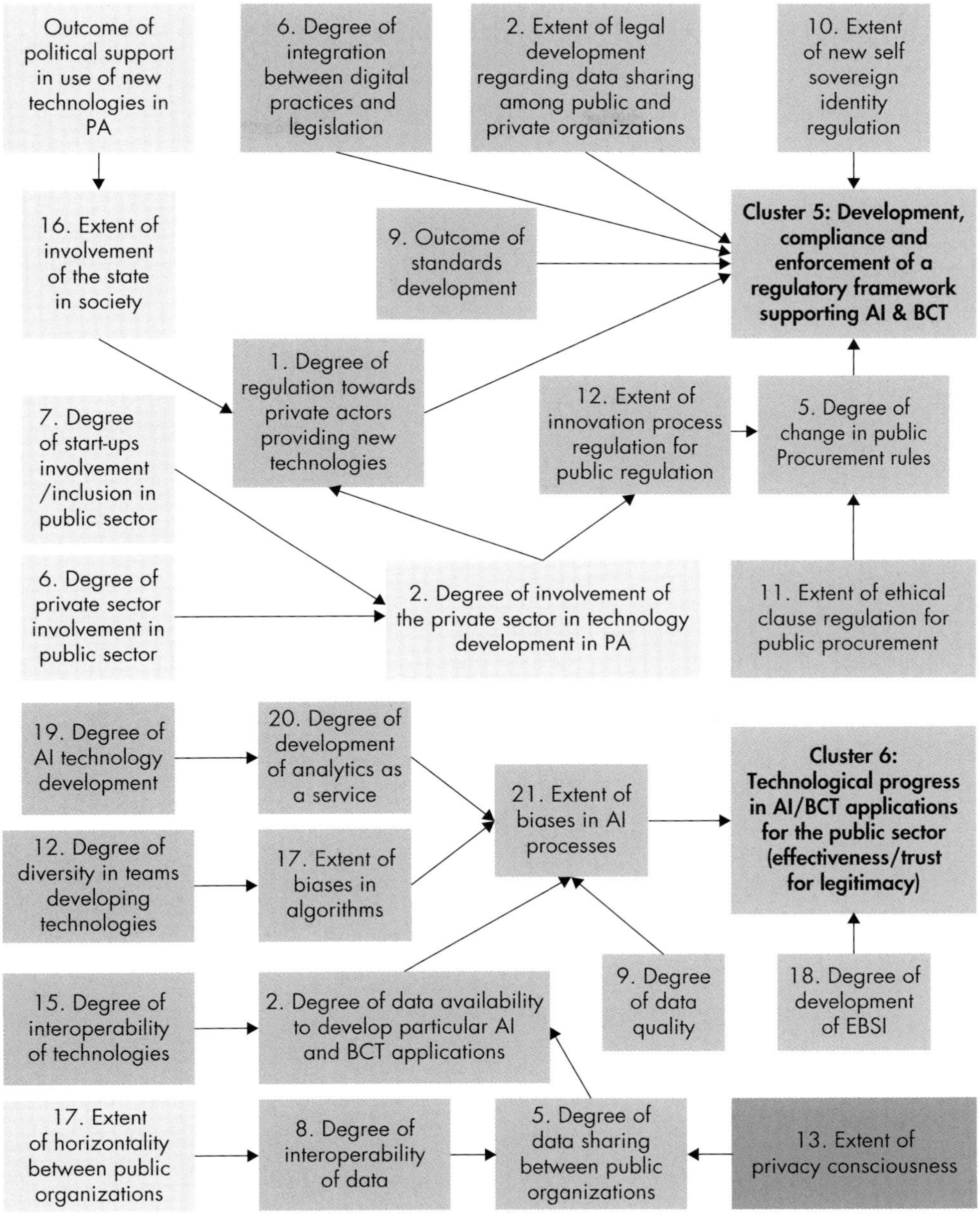

Figure 7.3. Results of the clustering process (examples of clusters 5 and 6).

degree of political financial support towards new technologies. The latter, and the importance of competition on tech markets (driving force 4) influence the extent of R&D funding and training related to AI and BCT.

In this regard, it is important to underline that establishing clusters was a collective endeavour during which researchers discussed the relevance of each relation between driving forces. Another team of researchers with a different background could have reached another clustering outcome. In scenario thinking, this is not an issue as long as the process is consistent and leads to distinctive scenarios.

In our process, while we discussed the possible clustering outcomes, we did not systematically assign two extreme outcomes to each of them (as phase 4 required). This decision was mostly guided by two considerations. First, the discussion did not lead us to identify one single axe of impact for each of these factors, which were highly multidimensional. Second, and more pragmatically, such an identification, although possible, was highly time-consuming. This is, however, not to say that this phase is not important, as it could be for other processes, but it was of lesser interest in our approach.

In a subsequent brainstorming session (phase 5), these clusters were then discussed in regard to the importance of their uncertainty and of their impact on the subject matter (i.e. the integration of BD, AI and BCT in federal public organisations). Concretely, an uncertainty-impact matrix was drawn, with two axes: one going from low to high degree of impact and another from low to high degree of uncertainty (Figure 7.4). The impact axis concerns the importance of the impact of the cluster on the subject matter, and disregards the direction of this impact. The uncertainty axis focuses on the uncertainty about the nature of the impact that was previously discussed. Each cluster was thus discussed along these lines to identify the most impactful and most uncertain clusters (placed in the upper right quadrant): political support in using AI and BCT in public organisations, and trust in public sector using new technologies. These were therefore chosen as the scenario macro-factors.

To exemplify this process, we consider the position of cluster 2 (political support in the use of new technologies in public administration). During the brainstorming session, we considered that this cluster has a high impact on the development of new digital technologies in government. Indeed, government and political actors are the main decision-makers regarding this topic. Concretely, by 2030, elected politicians could decide to support – or not – the development of such technologies. They could support them in providing financial support and in offering specific conditions for private companies to invest in them. Conversely, they could limit their development by adopting a conservative or prudent approach to new technologies or by supporting low-technologies principles. Similarly, we considered that the nature of the impact of such a cluster was uncertain by 2030 as no clear action had been taken at the time by political actors regarding this topic.

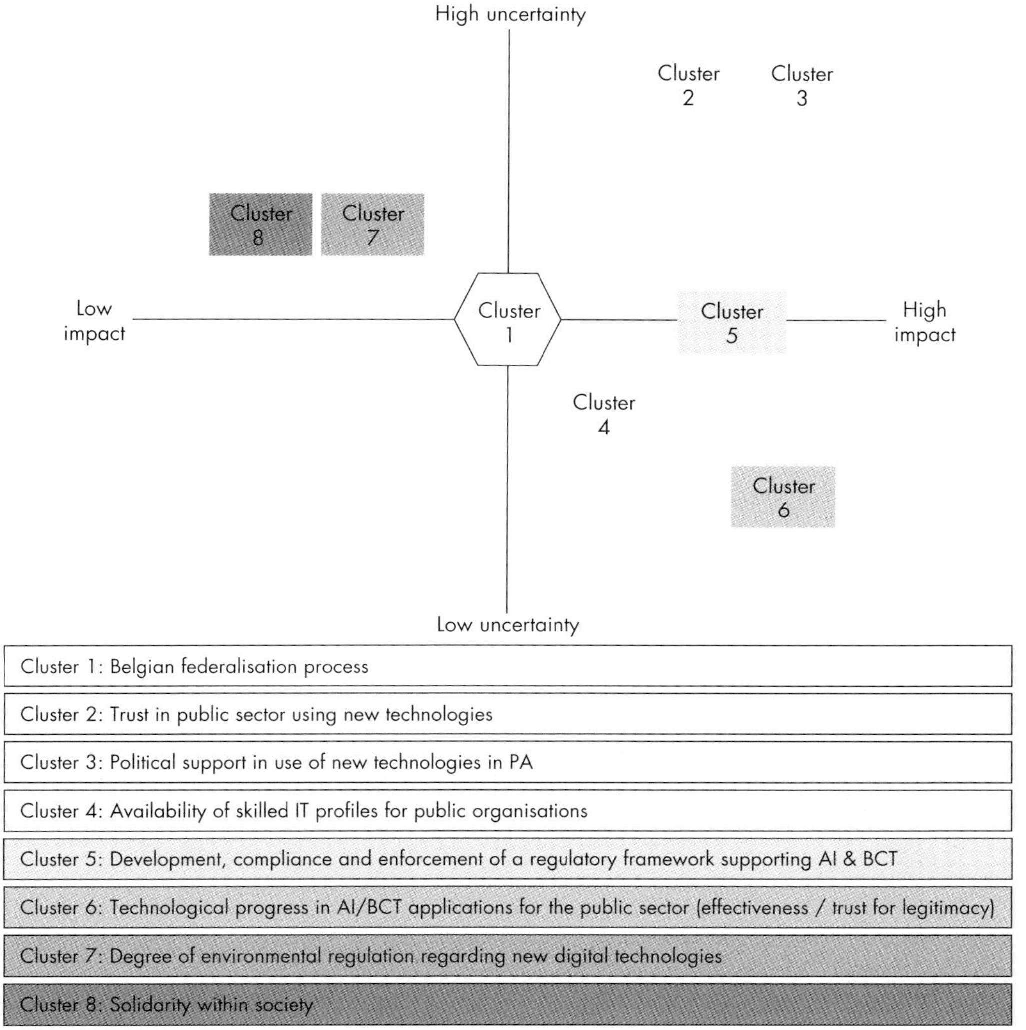

Figure 7.4. Uncertainty-impact matrix.

The stages 6 and 7 (framing and scoping the scenarios) were combined and conducted by a limited number of members of the research team, identifying the extreme outcomes for both macro-factors and building a set of descriptors for each scenario. The issues that were considered as descriptors in each scenario were identified from the previously collected data and from the multiple driving forces. Each scenario was also named and given a particular orientation. A tentative version was discussed and amended on multiple occasions, among the research team, in research seminars and in the follow-up committee of the project. It led to a final set of four scenarios (Figure 7.5).

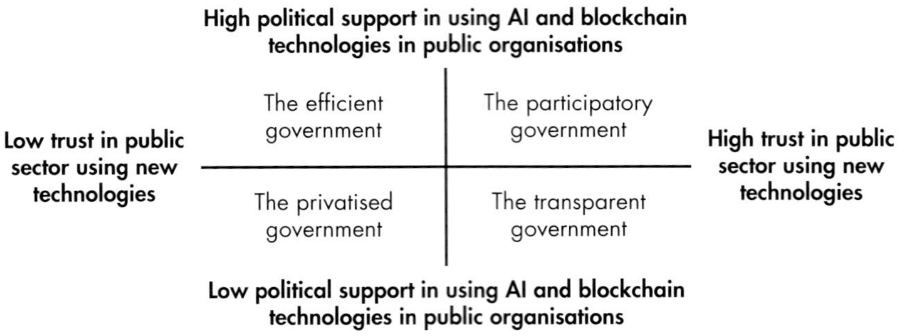

Figure 7.5. The four scenarios.

For each of these scenarios, a set of descriptors was agreed on. Table 7.2 illustrates some of the descriptors.

Table 7.2. Scenario descriptors.

	The efficient government	**The participatory government**	**The privatised government**	**The transparent government**
Political support in using AI and BCT in public organisations	High through major R&D funding towards private actors to develop these technologies	High with several public initiatives towards these technologies	Low, with the use of AI / BCT being low on the agenda	Low, with focus on open data and controllable technologies and rule-based algorithms
Trust in public sector using new technologies	Low among parts of the population	High through participatory deliberation	Low, as outsourcing has decreased technological expertise in the public sector	High, due to clear political orientations in using technologies in policy (security and transparency)
Public private relationships regarding technologies	Strong partnership with private actors as technology providers	Co-creation and participation with high public expertise	High presence of private actors in policy implementation	Public regulation vis-à-vis private actors to support open data and open source applications
Impact on policy	High efficiency and budget increases but opacity of algorithms and privacy issues	High legitimacy but time-consuming and costly processes	Outsourcing of implementation, with high financial returns but less control and socio-economic disparities	Continuity of current processes with a crossroad bank for policy

The first scenario was coined as the efficient government. It is characterised by a high political support for the use of AI and BCT in public organisations – which translates into generous funding allocated to private actors for the development of these technologies – as well as a strong partnership between private technology providers and the public sector. However, there is low trust in the public sector when using these technologies among citizens. Overall, the policy output of this scenario is characterised by a high efficiency due to budget increases in favour of these new technologies, but at a high cost for transparency and the privacy of data.

The second scenario was defined as the participatory government. This scenario is again characterised by a high political support for the use AI and BCT in public organisations – which translates into several public initiatives in favour of these technologies – as well as high levels of trust in the public sector using these technologies. The partnership between private and public actors is based on co-creation and participation with high public expertise. In this configuration, the policy impact is characterised by a high legitimacy of the technological choices and outcomes but these result from time-consuming and costly processes.

The third scenario was defined as the privatised government. It is characterised by a low political support for the use of AI and BCT in public organisations as well as low levels of trust in the public sector using new technologies because of its limited technical expertise. Private actors are highly involved in policy implementation due to a widespread use of outsourcing. Overall, the policy output of this scenario is characterised by low control over implementation (due to outsourcing), high financial returns, and important socio-economic disparities.

The fourth scenario was coined as the transparent government. This scenario is characterised by a low political support for the use of AI and BCT in public organisations and a strong regulation of private technology providers. This translates into a technological focus on open data, controllable technologies, and rule-based algorithms. This scenario is also associated with high levels of trust, due to a clear political commitment in favour of security and transparency regarding the use of technologies in policy. Overall, the policy output of this scenario is characterised by a continuation of current processes with a crossroad bank for policy.

Finally, based on the chosen descriptors for each of those scenarios, four narratives were developed, in the form of a very short story illustrating the lives of two characters per scenarios (8 in total). These narratives were storylines that included some specific key events and explained the context in 2030 and some of its implications to policies against tax fraud and social security infringements. They are internally consistent to enhance the capacity of participants to be immersed in them. They are also externally consistent, as each of them represents a specific and differentiated future that could happen. These scenarios are thus possibilities; they are neither determining the future, nor the way participants should see it (although one can see them as performative). The following script describes the narrative that was developed for the first scenario (the efficient government):

> Sophie is heading the R&D AI grant programme for the federal government. In the aftermath of the COVID crisis and building on the TECH4GOV coalition, a major public private partnership has been set up to develop data technologies. The federal government has massively funded tech providers and start-ups to build AI applications for policy. A couple of weeks ago, Sophie's programme was put forward in the media, as the newly developed fraud detection application has strongly increased federal tax revenues. The fight against tax fraud has never been higher on the agenda, and the massive use of public and private databases for policy has led to major efficiency gains and budget increase.
>
> However, such efficiency gains are not seen in a good light by everyone as they could come at the expense of transparency and privacy rights. Jan is highly critical of these policy orientations. His family, as many others, believes they were wrongfully declined social benefits through what they considered as opaque and discriminatory algorithms. That's why he recently signed the petition against the project of the government to implement data processing technologies from a major tech provider, as he fears a lack of transparency and is not confident his right to privacy will be respected.

A similar process was used for the second, third, and fourth scenarios. These four scenario narratives were translated into short, animated motion design videos whose goal was to support the first set of workshops.[2]

7.6 Co-creating through scenario workshops

As the second phase, the co-creation phase is mainly built around scenarios workshops. To make these workshops a reality, several aspects needed to be taken into account: designing the workshops protocol (1), selecting and inviting participants (2), choosing a location (3) and finally facilitating these workshops (4). These four sections mainly rely on our DIGI4FED experience.

7.6.1 Designing the workshop protocol

The co-creation phase mainly relied on two successive sets of workshops. These were supported by the four scenarios narratives built in the exploration phase (first set of workshops) as well as the results of an analysis of the first set of workshops (second set of workshops).

Designing the two scenario workshops

First, two workshops were designed in which scenarios served as the basis of discussions, to engage participants with a plurality of possible futures and to broaden their perspectives to include a wider set than currently existing ones (Cairns and Wright, 2018). We organised two workshops to distinguish between actors from the social fraud domain and those from the tax fraud domain. The

[2] The video can be accessed on the Cité research department website of Liege University: https://www.cite.uliege.be/cms/c_7331132/fr/des-scenarios-pour-penser-le-futur-des-nouvelles-technologies-dans-les-organisations-publiques.

objective of this first set of workshops was to elicit future expectations (including opportunities and risks) of participants regarding a model of governance aiming at integrating BD, AI and BCT in federal public organisations.

These two *scenario* workshops were designed to fit into a 4-hour period, in a physical setting (see below for the location). The protocol was designed to maximise speech time for participants, combining a plenary approach to subgroup discussions. As Belgian federal public organisations work in both Dutch and French, no constraints were imposed regarding the choice of a language of expression during the workshops. This choice was made to favour free speech from participants in a setting and a language in which they felt the most comfortable. Conceptualisations, complex thoughts, and impressions can be difficult to express and convey with nuance in a language that is less familiar. As such, we followed the advice of Rice and Ezzy (1999), who recommend allowing participants to express their thoughts in their own language.

The production of some visual support material was also arranged to facilitate the presentation of the workshop agenda and objectives to participants. PowerPoint presentations were devised in both French and Dutch versions to ensure participants' best possible comprehension. The animation video presenting the four scenarios was recorded in English with subtitles.

These first two workshops were audio-recorded; the recordings were transcribed and thematically analysed to highlight several issues around which to build the design of the third workshop. The idea was to identify the key issues, or *challenges*, that could possibly hinder the integration of BD, AI and BCT in federal public organisations.

Designing the solution-oriented workshop

The second set of workshops was based on these problems to be solved. In other words, the third workshop was designed to provide participants with feedback on the analysis of the scenario workshops. The analysis results (i.e. the identified challenges) served as a groundwork to elicit solutions to these challenges.

Contrary to the first two workshops, we organised a single workshop combining both policy domains. This change in design was decided following the first two workshops after seeing the enthusiasm of the actors to share their insights and experiences with colleagues they had never met before. Furthermore, the subsequent thematic analysis of the two scenario workshops revealed that the main issues voiced by participants regarding the integration of AI, BD, and BCT in the fight against tax fraud and social security infringements tended to converge across policy sectors. Similar macro-categories of themes were highlighted across workshops, such as 'Articulating the IT architecture with the legislation'; 'Data integration and coordination'; 'Balancing transparency, control and accountability'; or 'Resources (data expertise, funding, etc.)' of administrations. In that regard, grouping the stakeholders of two policy domains that are intertwined in some ways (both technically and in terms of content) made sense for the third workshop.

This third, solution-oriented, workshop was also designed to be held in a physical setting and for a duration of 4 hours. Again, we chose to allow participants to express themselves in their own respective languages (either Dutch or French). As for the first two workshops, we devised some visual support material to present the workshop agenda and objectives to participants. An additional presentation was prepared to share the main results of the scenario workshops (the ten challenges) to participants. The third workshop was also audio recorded with the consent of participants. These recordings were then transcribed to facilitate the implementation of a thematic analysis.

7.6.2 Selecting and inviting participants

The previously mentioned stakeholders mapping and interviews were key phases in the identification, selection, and invitation of participants. The different choices were made to generate heterogeneity within each policy domain group, fostering the expression of a plurality of points of view. The selection of participants was made based on several criteria: the institutional affiliation of the participant to one of the identified stakeholders, his/her expertise with some aspects of the policy domain, his/her interest in the topic and his/her level of seniority. For this latter aspect, we mostly targeted participants who were not in a senior management position. This choice was guided by our need to get specific insights from operational civil servants while reducing any dominance effect within the group.

Once selected, participants were invited by e-mails in their mother tongue, as they could speak either Dutch or French. Upon confirmation of their participation, they also received a second e-mail providing them with more information on the agenda of the workshops. When a participant declined to participate, he/she was asked to refer another member of his/her organisation when possible.

The first two workshops respectively gathered ten participants around the fight against social fraud and nine participants discussing tax fraud. Participants were civil servants from the policy domain or from a peripheral institution or service providers to public organisations.

The third workshop gathered 16 participants, most of whom had already participated in the two previous workshop instalments. Participants were evenly distributed between those involved in the fight against social fraud and those involved in the fight against social security infringements.

7.6.3 Choosing a location

A living lab is, often, a physical space, in which actors can meet and discuss, also informally (although lately, more and more living labs have taken place in a virtual way, also due to the COVID-19 pandemic). Also, it is important to choose a place in which participants feel free to express their opinions and share their experience. A rather neutral setting, especially with politically sensitive topics, is thus an element to consider. Pragmatically, choosing a central location for participants is also recommended, as well as considering logistical elements (drinks and food, seating, availability, geographical location).

In the context of the DIGI4FED project (which was mostly conducted during 2020 and 2021 in a pandemic context), we had delayed the organisation of the living lab to make sure it could take place in a physical setting, while strongly ensuring and enforcing all Covid-safe rules and recommendations. The workshops took place in the building of one of the federal public organisations, using two rooms with seats separated by Plexiglas and people were wearing masks.

The chosen location was also very close to public transport, in a central neighbourhood in Brussels, facilitating access for participants mostly working in the city.

7.6.4 Facilitating workshops

Facilitating the two scenario workshops

For the first set of workshops, a team of five people was set up to facilitate the debates and support the process; it was composed of two facilitators, two note-takers and one logistical focal point in charge of welcoming and guiding participants in the building, as well as organising the post-meeting meal.

The role of the two facilitators was to mediate the debates between the different actors, taking care that everyone received a relatively similar speech time and that no voice went unheard. The role of the facilitator is also to make links between previous assertions of one or several participants and to rephrase arguments so as to be sure that they are clearly understood by every participant around the table.

The role of the note-takers was to write down the main topics that were discussed and to summarise them for further use (see below).

A first short plenary moment was organised to welcome participants, to introduce the research and the day and to present them with the four scenarios via the 7-minute video clip. In that regard, many participants underlined that the video clip was dense, as we had foreseen. Participants were then divided into two subgroups. In each group, they watched the video clip a second time with a specific request: they were asked to identify the three most desirable and the three least desirable dimensions in any scenario, keeping in mind their own position and experience of the topic. This was aimed at focussing the debates on the content of the scenarios rather than on discussion of the way they were set up, thus keeping the use of scenarios as a tool to stimulate the debate. Therefore, it was very important in the facilitation process to keep the discussion focused on why, and due to which aspects, people were discarding one scenario or another.

From then, each subgroup started a discussion on the topics and issues participants wanted to raise. It appeared that each group had similar issues to highlight (policy efficiency, data privacy, type and maturity of technologies, operationalisation and issues related to the policy itself). However, participants reacted differently to the video clip. In one group, they referred a lot to scenarios, while they did not in the other, although they still referred to issues that were discussed in the video clip. We consider this can be related to the facilitation process that made more emphasis on scenarios in one group than in the other.

These subgroup discussions lasted for more than an hour before participants were asked to join back to the plenary session. During this plenary session, the two note-takers presented the main insights from each group and there was a larger debate about them. This part really helped participants to form a group to further discuss some of the issues that had been raised. It also highlighted some new issues, such as strategic and budgetary dimensions of integrating new digital technologies in public organisations.

The first two workshops were closed with a lunch, during which the research team and the participants informally discussed additional issues, notably in regard to specific institutions.

Facilitating the solution-oriented workshop

The third workshop was supervised by a team of 6 people in charge of facilitating the debates and contributing to the workshop process. The team consisted of three facilitators, one note taker, and two logistical assistants. As in the previous workshops, the three facilitators were entrusted with mediating the debates between stakeholders and making sure that every participant had the opportunity to intervene. Facilitators were also expected to present the workshop objectives and the material that would serve as the groundwork for discussions among stakeholders (i.e. the challenges that were identified via the analysis of the scenario workshops). The note-taker was to write down, summarise, and organise the main topics of discussion, thus helping facilitators in the process and producing material for future use.

The solution-oriented workshop activities were divided into three main successive steps. The first phase started with a brief general introduction presenting the DIGI4FED project, the workshop participants, and the day's agenda and objectives. Then, the facilitators invited participants to react to a 15-minute-long presentation on the key challenges that were highlighted by the analysis of the two previous scenario workshops. Participants were invited to formulate specific solutions to the challenges they deemed the most accessible and the easiest to solve. The discussions among participants regarding the details, the usefulness, and the relevance of the proposed solutions were moderated by facilitators. During this process, the most complex challenges were also identified by the facilitators and the note-taker. These challenges, for which no clear solution was reached, were left for further discussion later during the ulterior steps of the workshop. This first phase lasted approximatively one hour.

During the second phase, the facilitators invited participants to briefly discuss the challenges to which no obvious solutions were found during the previous phase. To that end, participants were split into three subgroups, each assisted by a facilitator. When arranging the subgroups, the facilitators made sure to get a proper equilibrium regarding subgroup participants' profiles (generalist or technical expertise) and sector affiliation (social security or taxation and public or private organisations). During the 30 minute long debates, the facilitators' role was to assist the discussions and ensure that every subgroup participant had the opportunity to express his or her views. These discussions were not audio recorded.

In a third phase, a one hour long plenary discussion was held to share the solutions reached by each subgroup. Participants were also invited to put solutions and challenges onto a timeline in order to elicit priorities for the development of a governance model as envisioned by the DIGI4FED project.

After a concluding note from one of the facilitators, the workshop ended for lunch during which participants interacted with the team of researchers.

7.7 Conclusions and recommendations for participatory innovation within the public sector

The implementation of the exploration and co-creation phases of the living lab process allowed us to reflect on some of the added value and the shortcomings associated with this approach. Hence, we conclude this chapter with some reflections about the value of participatory methods in public sector innovation and, in particular, with regard to the integration of new digital technologies in public administrations. We also formulate some practical and methodological recommendations for future projects and for researchers who would be interested in implementing participatory innovation methods within the public sector.

7.7.1 Using participatory methods in the public sector

Participatory methods and collaborative innovation are gaining increasing legitimacy in the public sector. Specifically, this chapter showed how participatory approaches and more precisely living labs can contribute to supporting the development of new digital technologies in the public sector. Reflecting on the insights provided here, we would like to highlight three key aspects showing how this approach to new digital technology in the public sector offers an added value compared to the more centralised approaches to technology institutionalisation.

First, the exploration phase of the living lab process allowed for the identification of multiple philosophical or ideological points of view regarding new digital technologies, going further than the literature review and document analysis to elicit stakeholders' views on the topic. The systematic analysis of these points of view offered a pluralistic landscape of positions, underlining questions regarding the role of the state, its governance of new technologies and the public-private interactions, its digital sovereignty or the (sometimes relative) importance of data protection.

Such a landscape of positions is rarely made available in a systematic fashion. This proved to be of great value in the process of scenario-building since this landscape of positions provided multiple contrasting visions regarding the future of the new digital technologies in the public sector.

Second, building on scenarios through scenario workshops, the co-creation phase served as a catalyst for debates among stakeholders. This allowed participants to build their knowledge and to use anticipatory knowledge about the futures of new digital technologies in the process. Concretely, the co-creation phase facilitated the identification of ten key challenges to the integration of new digital technologies in public organisations. These ten challenges were clustered into three main categories of challenges: (1) citizens' acceptability of data exploitation; (2) the articulation of data use with regulations; (3) the resources and data expertise within administrations. For instance, the challenges pertaining to the first category include citizens' acceptance of the new anti-fraud tools and the difficulty of preserving citizens' trust. Challenges pertaining to the second category include the interpretation and implementation issues associated with the GDPR, while the third category includes challenges such as the lack of internal expertise and the dependence on external (private) ICT contractors. The co-creation phase also contributed to the collective identification of eleven potential solutions (and their implementation timeline) to address the aforementioned challenges. Among these solutions we can, for example, mention the pooling of resources, expertise and capabilities of administrations; the privileging of EU (or national) ICT contractors; and the rethinking of regulations by actively integrating ICT tools and data use in their development.

Third, the analysis of the whole living lab process and of the interactions between stakeholders also helps to identify the key elements to support the decision-making process. These elements notably include the identification of institutions that are key to leading the digital transformation, the production of groundwork on which policy formulation can take place, the positioning of the issue on the agenda of stakeholders, including public ones, and the establishment of a network of stakeholders interested in new digital technologies in the public sector.

7.7.2 Recommendations for future research

Based on our observations, we propose some theoretical and practical recommendations for future research projects that integrate participatory innovation methods.

Our first observation is that scenario-building is a rather long and time-consuming process. More specifically, the first phase of the living lab process required us to mobilise a vast array of methods and techniques to support the construction of the scenarios. Indeed, we implemented an extensive literature review, a qualitative analysis of semi-structured interview data, and resorted to a matrix approach for scenario-building. Furthermore, this phase requires constant interaction with field actors to elicit and confirm their views regarding the issues at stake. As such, each of the steps leading to scenario-building required a close coordination among the six researchers involved in previous project-related data collection processes (literature review and interviews). Although the obtained scenarios proved fruitful and insightful in driving the co-creation phase workshops, we surmise that this process could have been greatly simplified. For instance, it would have been possible to considerably reduce the time spent at identifying the key drivers of change by engaging in another set of workshop(s) based on the insights provided by the literature review. This would have considerably shortened the qualitative data collection process as well as its analysis. However, in our case this solution proved impractical due to the COVID-19 restrictions at the time.

A second element is that the living lab process can necessitate a certain amount of flexibility and creativity from part of the involved researchers. For instance, staff changes within the stakeholder organisations led to unforeseen changes among the workshop participants. The team of researchers had to quickly react by finding new participants presenting an adequate profile. Similarly, our choice to combine the two originally planned solution-oriented workshops into one unique workshop was a pragmatic response in view of the results obtained via an analysis of the two previous workshop instalments. Although this choice was carefully debated to make sure that it would not jeopardise the overall project, we still had to opt for this solution in a timely manner. This is why our recommendation for future ventures would be to develop a comprehensive roadmap – or a decision tree – that would probably help researchers to choose between alternate courses of action when addressing contingencies during the multiple steps of the process. This would also be particularly useful in case of change of staff in the research team. However, some issues could not be planned ahead, and flexibility and open-mindedness should be key characteristics of the individual(s) leading the living lab process.

The third aspect to consider is the careful articulation of the various steps that are planned within each phase of the living lab process. From our experience, it appears that some little inconveniences could have been prevented during the co-creation phase. More precisely, the short timeframe that was envisioned between the two sets of workshops left little time for conducting the qualitative analysis that was expected to serve as the groundwork of the solution-oriented workshop. Thus, future participatory innovation projects which consider using a similar articulation of scenario and solution-oriented workshops should make sure to organise workshops in an adequate timeframe. Leaving too much time between workshops poses the risk of losing participants' interest in the process. Alternatively, leaving a shorter timeframe between both sets of workshops contributes to keeping participants invested. However, this shorter timeframe could prove challenging for researchers since they would have to perform a comprehensive analysis of the first set of workshops in a rather short period of time. It is thus necessary to find an equilibrium between the time researchers need to perform their analyses and the time necessary to keep participants invested in the participatory process.

A fourth observation is that designing and organising workshops is a collective endeavour that should preferably be undertaken by a team of researchers. The living lab process is highly demanding in terms of human resources, notably due to the logistics that are involved. For instance, workshop facilitation demands the active participation of three to four researchers. In addition to the facilitators, several other people may be involved for logistical aspects: welcoming and guiding participants; ensuring the adequate recording of debates; providing food and drinks and so on. In short, implementing such participatory processes remains difficult for researchers who are not involved in a collective research program.

Last but not least, the process described in this chapter did not consider the testing phase of a living lab approach. We however recommend envisaging this option. Such a testing phase could be envisioned through several participatory techniques. Two possible options are serious gaming and the Delphi method. First, serious gaming is used across all disciplines (from the humanities and social sciences to computer science). Serious games are distinguishable from classic games, in that their main purpose is not to entertain, although there is still an entertaining, experiential, and emotional sphere in a serious game (Laamarti *et al.*, 2014). They are a learning tool aiming at emphasising consequences of actions in reality. This helps participants to project themselves into a second-degree universe in which they can make decisions individually or collectively (Parotte *et al.*, in press). Second, the Delphi method is a structured communication tool that allows for dialogue among a panel of actors (similar to those who participated to the workshops). It facilitates the search for consensus, the documentation of points of dissensus and the co-construction of a shared evaluation of a complex situation while controlling for possible power struggle between participants (Brady, 2015; Landeta, 2006; Turoff, 1970). While the serious game requires face-to-face interactions, Delphi surveys can be organised online which is a positive alternative during, for example, a pandemic.

Supplementary material

Supplementary material can be found online at https://doi.org/10.3920/978-90-8686-930-5_7

Figure S1. Results of the clustering process.

References

Adam, B. and C. Groves. 2007. *Future matters: action, knowledge, ethics*. Brill, Leiden, the Netherlands, 218 pp. https://doi.org/10.1163/ej.9789004161771.i-218

Alavi, H.S., D. Lalanne, and T, Rogers. 2020. The five strands of living lab. *ACM Transactions on Computer-Human Interaction* 27: 1-26. https://doi.org/10.1145/3380958

Alter, N. 1993. Innovation et organisation: deux légitimités en concurrence. *Revue Française de Sociologie* 34: 175-197. https://doi.org/10.2307/3322487

Bakici, T., E. Almirall, and J. Wareham. 2013. The role of public open innovation intermediaries in local government and the public sector. *Technology Analysis & Strategic Management* 25: 311-327. https://doi.org/10.1080/09537325.2013.764983

Bason, C., H. Hollanders, C, Hidalgo Gomez, R. Kattel, G. Korella, L. Christine, B. le Masson, M. Mazzucato, A. Mungiu-Pippidi, M. Pröhl, and J. Oravec. 2013. *Powering European publc sector innovation: towards a new architecture*. Available at: https://ec.europa.eu/futurium/en/system/files/ged/42-public_sector_innovation_-_towards_a_new_architecture.pdf.

Bekkers, V., J. Edelenbos, and B. Steijn. 2011a. Linking innovation to the public sector: contexts, concepts and challenges. In V. Bekkers, J. Edelenbos, and B. Steijn (eds.) *Innovation in the public sector. Linking capacity and leadership*. Palgrave Macmillan, Basingstoke, UK, pp. 3-32.

Bekkers, V., J. Edelenbos, and B. Steijn, eds. 2011b. *Innovation in the public sector. Linking capacity and leadership*. Palgrave Macmillan, Basingstoke, UK, 252 pp. https://doi.org/10.1057/9780230307520

Bergvall-Kareborn, B. and A. Stahlbrost. 2009. Living Lab: an open and citizen-centric approach for innovation. *International Journal of Innovation and Regional Development* 1: 356-370.

Berryhill, J., K.K. Heang, R. Clogher, and K. McBride. 2019. *Hello, world: artificial intelligence and its use in the public sector*. OECD Publishing, Paris, France, 184 pp. https://doi.org/10.1787/19934351

Bishop, P., A. Hines, and T. Collins. 2007. The current state of scenario development: an overview of techniques. *Foresight* 9: 5-25. https://doi.org/10.1108/14636680710727516

Bootz, J.-P. 2010. Strategic foresight and organizational learning: a survey and critical analysis. *Technological Forecasting and Social Change* 77: 1588-1594. https://doi.org/10.1016/j.techfore.2010.06.015

Bradfield, R., G. Wright, G. Burt, G. Cairns, and K. Van Der Heijden. 2005. The origins and evolution of scenario techniques in long range business planning. *Futures* 37: 795-812. https://doi.org/10.1016/j.futures.2005.01.003

Brady, S.R. 2015. Utilizing and adapting the Delphi method for use in qualitative research. *International Journal of Qualitative Methods* 14: 1-6. https://doi.org/10.1177/1609406915621381

Brandsen, T., T. Steen, and B. Verschuere. 2018. Co-creation and co-production in public services: urgent issues in practice and research. In T. Brandsen, T. Steen, and B. Verschuere (eds.) *Co-production and co-creation: engaging citizens in public services*. Routledge, New York, NY, USA, pp. 3-8.

Brown, K. and S.P. Osborne. 2012. *Managing change and innovation in public service organizations*. Routledge, Oxon, UK, 262 pp.

Cairns, G. and G. Wright. 2018. *Scenario thinking. Preparing your organization for the future in an unpredictable world*. Palgrave Macmillan, Basingstoke, UK.

Dekker, R., K. Geuijen, and C. Oliver. 2021. Tensions of evaluating innovation in a living lab: moving beyond actionable knowledge production. *Evaluation* 23: 347-363. https://doi.org/10.1177/1356389021997848

Dell'Era, C. and P. Landoni. 2014. Living lab: a methodology between user-centred design and participatory design. *Creativity and Innovation Management* 23: 137-154.

De Vries, H., V. Bekkers, and L. Tummers. 2016. Innovation in the public sector: a systematic review and future research agenda. *Public Administration* 94: 146-166. https://doi.org/10.1111/padm.12209

European commission, 2009. *Mapping foresight. Revealing how Europe and other world regions navigate into the future*. Available at: https://op.europa.eu/en/publication-detail/-/publication/3fab185a-16fa-491e-bceb-41a8d9972f06

Fobé, E. and M. Brans, M., 2012. *Toekomstverkenningen in een beleidscontext. Theoretische beschouwingen over een geïnformeerde beleidspratijk en de rol van beleidsgerichte toekomstverkenning*. Available at: https://lirias.kuleuven.be/1866351?limo=0.

Fuglsang, L. and J.S. Pedersen. 2011. How common is public sector innovation and how similar is it to private sector innovation. In: V. Bekkers, J. Edelenbos, and B. Steijn (eds.) *Innovation in the public sector. Linking capacity and leadership*. Palgrave Macmillan, Basingstoke, UK, pp. 44-60.

Gascó, M. 2017. Living labs: implementing open innovation in the public sector. *Government Information Quarterly* 34: 90-89. https://doi.org/10.1016/j.giq.2016.09.003

Hartley, J., 2005. Innovation in governance and public services: past and present. *Public Money and Management* 25: 27-34. https://doi.org/10.1111/j.1467-9302.2005.00447.x

Hartley, J., E. Sørensen, and J. Torfing. 2013. Collaborative innovation: a viable alternative to market competition and organizational entrepreneurship. *Public Administration Review* 73: 821-830. https://doi.org/10.1111/puar.12136

Havas, A., D. Schartinger, and M. Weber. 2010. The impact of foresight on innovation policy-making: recent experiences and future perspectives. *Research Evaluation* 19: 91-104. https://doi.org/10.3152/095820210X510133

Hollanders, H., A. Arundel, B. Buligescu, V. Peter, L. Roman, P. Simmonds, and N. Es-Sadki. 2013. *European public sector innovation scoreboard 2013. A pilot exercise.* Available at: https://www.researchgate.net/profile/Bianca-Buligescu/publication/279507400_European_Public_Sector_Innovation_2013/links/5593c86b08ae5af2b0ebb35e/European-Public-Sector-Innovation-2013.pdfdecision-m.

Koch, P. and J. Hauknes. 2005. On innovation in the public sector. Available at: https://nifu.brage.unit.no/nifu-xmlui/bitstream/handle/11250/226573/d20-innovation.pdf?sequence=1.

Koschatzky, K., 2005. Foresight as a governance concept at the interface between global challenges and regional innovation. *European Planning Studies* 13: 619-639. https://doi.org/10.1080/0965431050010736

Laamarti, F., M. Eid, and A. El Saddik. 2014. An overview of serious games. *International Journal of Computer Games Technology* 2014: 358152. https://doi.org/10.1155/2014/358152

Landeta, J. 2006. Current validity of the Delphi method in social sciences. *Technological Forecasting and Social Change* 73: 467-482. https://doi.org/10.1016/j.techfore.2005.09.002

Lascoumes, P. and P. Le Galès. 2007. Introduction: understanding public policy through the instruments – from the nature of instruments to the sociology of public policy instrumentation. *Governance* 20: 1-21. https://doi.org/10.1111/j.1468-0491.2007.00342.x

Linstone H.A. and M. Turoff. 1975. *The Delphi method. Techniques and applications.* Addison-Wesley, Reading, MA, USA, 616 pp. https://doi.org/10.2307/3150755

Moore, M. and J. Hartley. 2008. Innovations in governance. *Public Management Review* 10: 3-20. https://doi.org/10.1080/14719030701763161

Nesti, G. 2017. Co-production for innovation: the urban living lab experience. *Policy and Society* 37: 310-325. https://doi.org/10.1080/14494035.2017.1374692

Organisation for Economic Co-operation and Development (OECD). 2011. *The call for innovative and open government.* OECD Publishing, Paris, France, 312 pp. https://doi.org/10.1787/9789264107052-en

Organisation for Economic Co-operation and Development (OECD). 2014. *Innovation the public sector: from ideas to impact.* Available at: https://www.ospi.es/export/sites/ospi/documents/documentos/Innovating-the-Public-Sector-conference-conclusions.pdf

Organisation for Economic Co-operation and Development (OECD). 2015. *The innovation imperative in the public sector. Setting an agenda for action.* OECD Publishing, Paris, France, 72 pp. https://doi.org/10.1787/9789264236561-en

Parotte, C., F. Ciraolo, N. Flore, H. Faniel, and R. Charlier, in press. Expérimenter le pathway evaluation process (Pep) Serious Game à l'ULiège. Evaluation de la méthode prospective. Rapport d'expertise. ULiège, Liège, Belgium.

Pollitt, C. 2011. Innovation in the public sector: an introductory overview. In V. Bekkers, J. Edelenbos, and B. Steijn. *Innovation in the public sector. Linking capacity and leadership.* Palgrave Macmillan, Basingstoke, UK, pp. 35-43.

Rice, P.L. and D. Ezzy. 1999. *Qualitative research methods: a health focus.* Oxford University Press, Oxford, UK.

Rogers, E. 2003. *Diffusion of innovations.* Free Press, New York, NY, USA.

Ruijer, E. and A. Meijer. 2019. Open government data as an innovation process: Lessons from a living lab experiment. *Public Performance & Management Review* 43: 613-635. https://doi.org/10.1080/15309576.2019.1568884

Schaffers, H. and P. Turkama. 2012. Living labs for cross-border systemic innovation. *Technology Innovation Management Review* 2: 25-30.

Schumpeter, J. 1942. *Capitalism, socialism and democracy*. Harper and Brothers, New York, NY, USA.

Schuurman, D., L. De Marez, and P. Ballon. 2016. The impact of living lab methodology on open innovation contributions and outcomes. *Technology Innovation Management Review* 6: 7-16. http://doi.org/10.22215/timreview/956

Sleeswijk Visser, F., P.J. Stappers, R. Van der Lugt, and E.B. Sanders. 2005. Contextmapping: experiences from practice. *CoDesign* 1: 119-149. https://doi.org/10.1080/15710880500135987

Spaniol, M.J. and N.J. Rowland. 2019. Defining scenario. *Futures & Foresight Science* 1: e3. https://doi.org/10.1002/ffo2.3

Sveiby, K.-E., P. Gripenberg, and B. Segercrantz, eds. 2012. *Challenging the innovation paradigm*. Routledge, New York, NY, USA, 272 pp.

Turoff, M. 1970. The design of a policy Delphi. *Technological Forecasting and Social Change* 2: 149-171. https://doi.org/10.1016/0040-1625(70)90161-7

Van der Heijden, K., R. Bradfield, G. Burt, G. Cairns, and G. Wright. 2006. *The sixth sense. Accelerating organizational learning with scenarios*. John Wiley & Sons, Chichester, UK.

Van der Steen, M. and M. van Twist, M. 2012. Beyond use: evaluating foresight that fits. *Futures* 44: 475-486. https://doi.org/10.1016/j.futures.2012.03.009

Yuan, Q. and M. Gasco-Hernandez. 2021. Open innovation in the public sector: creating public value through civic hackathons. *Public Management Review* 23: 523-544. https://doi.org/10.1080/14719037.2019.1695884

Zwetkoff, C., C. Elsen, L. Vigneron, and J. Pardo. 2018. *L'utilisateur au coeur du processus d'innovation*. Livre blanc méthodologique. WeLL Living Lab, Liege, Belgium, 86 pp.

Chapter 8

Conclusion – strategies and policy decisions in the new digital-era governance

E. Tan* and J. Crompvoets

KU Leuven, Public Governance Institute, Parkstraat 45 – bus 3609, 3000 Leuven, Belgium; evrim.tan@kuleuven.be

Abstract

This concluding chapter discusses how to create viable technology adoption strategies and policies in the new digital-era governance. Specifically, the chapter elucidates the policy processes and decisions that need to be taken to implement AI, blockchain, and big data solutions in public governance processes. Building on the research insight about the use of digital technologies in the fight against fraud, the chapter expounds on the drivers for the implementation of the AI, big data, and blockchain technologies in administrative decision-making processes. The chapter extrapolates new challenges and issues practitioners need to face in the new digital-era governance, and concludes with a new research agenda for the role of data-processing and new digital technologies in digital public governance.

Keywords: technology adoption, social security, taxation, policy strategies, future of public governance

8.1 Introduction

In consecutive chapters focussing on the technological dimensions (Chapter 2), management and operational dimensions (Chapter 3), legislative and regulative dimensions (Chapter 4-5), trust dimensions (Chapter 6), and the policy innovation dimensions (Chapter 7), we have elaborated on various challenges and drivers that influence the adoption of the new digital technologies such as AI, big data, and blockchain in the public sector domain. What remains untold is that how policymakers can use these research insights to develop viable strategies for the adoption of new digital technologies in public governance processes. Addressing this question is the aim of this concluding chapter.

This concluding chapter is structured in two parts. In the first part, we will focus on elucidating the interrelationships among various drivers influencing the adoption of new digital technologies in public services. Understanding the interrelationships between these drivers is a crucial starting point to assess the reliability of adoption strategies in digital public governance. We will first go through the theoretical models in the literature about the key drivers in the technology adoption.

DOI 10.3920/978-90-8686-930-5_8, © Wageningen Academic Publishers 2022

Next, building on the research conducted among the taxation and social security organisations in Belgium, we will construct an interpretive structural model (ISM) to identify the drivers associated with the new digital technologies and later to illustrate the way different drivers influence each other in the use of new digital technologies to fight against fraud.

Building upon this insight, in the second part, we will elaborate on the main issues and potential solutions emerging in the new digital era governance. Focussing on effective policy design choices for practitioners, we will substantiate the arguments set out in the Introduction. We will conclude the chapter with an assessment of what lies ahead, which particular questions are waiting to be addressed, and a look at what public governance and digital governance scholars can do to address these questions.

8.2 Theoretical assumptions about technology adoption in public governance

The theoretical assumptions about how new digital technologies can be adopted in the public sector take their origins in the information sciences. The so-called adoption models theorise and identify key drivers influencing users' intended behaviour and their actual behaviour to use the new technology or not. Most of these theoretical models in the literature stem from Technology Acceptance Model (TAM) and Unified Theory of Acceptance and Use of Technology (UTAUT). Despite the widespread testing of TAM and UTAUT models in different sectoral domains, the specificities of a government context have led some researchers to develop e-government specific adoption models later. Below we briefly summarise the key assumptions of these models.

8.2.1 Technology acceptance model

TAM is designed to predict the relationship between information technology acceptance and the actual use on the job. 'Perceived usefulness' and the 'perceived ease of use of the system' are two key factors used to predict the users' acceptance of digital technology (Figure 8.1; Davis 1989). An extended version of TAM, referred to as TAM2, incorporated 'social influence' and 'cognitive instrumental processes' (Venkatesh and David 2000). Social influence includes factors such as subjective norms, voluntariness, and image. Cognitive instrumental processes refer to job relevance, output quality, result demonstrability. TAM has been tested heavily in different empirical settings, including the public sector domain such as healthcare and higher education, and is estimated to predict the technology acceptance in 40% of the situations (Taylor and Todd 1995, Venkatesh and Davis 2000). Over the years, the empirical testing of the model contributed to its extension by exploring the impact of external constructs on users' attitudes, behavioural intention, and actual use of technology. Several factors, such as facilitating conditions, and systems quality have been included in the extended versions of TAM incorporating external constructs (Fathema *et al.*, 2015).

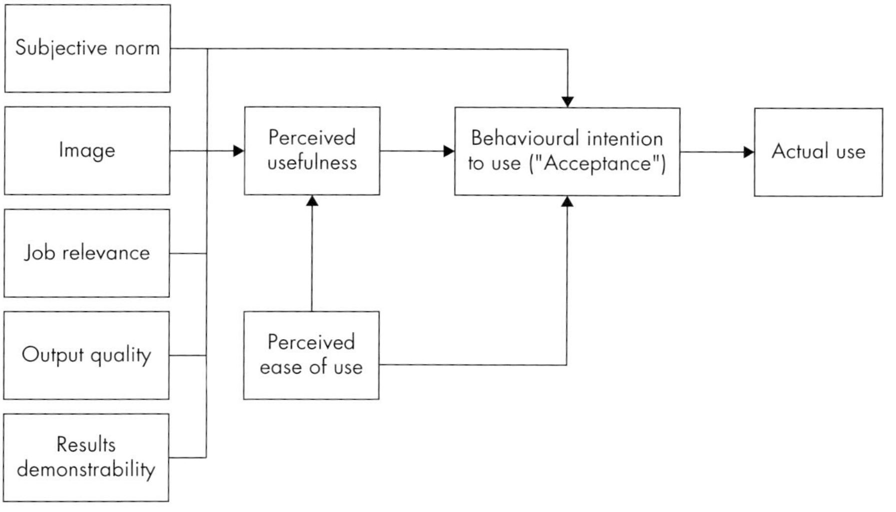

Figure 8.1. The extended version of the TAM-TAM2 model (Venkatesh and Davis, 2000).

8.2.2 Unified theory of acceptance and use of technology model

UTAUT has been developed by Venkatesh *et al.* (2003) to create a unified model that can bring the factors emphasised by eight different models including TAM and other prominent approaches, such as Theory of Planned Behaviour (TPB; Ajzen, 1991), Theory of Reasoned Action (TRA; Fishbein and Ajzen, 1975), Motivational Model (MM; Davis *et al.*, 1992), Model of PC Utilisation (MPCU; Thompson *et al.*, 1991), Innovation Diffusion Theory (IDT; Moore and Benbasat 1996), and Social Cognitive Theory (SCT; Compeau and Higgins, 1995). The UTAUT model postulates that there are four key constructs that act as determinants of user behaviour in technology adoption: 'performance expectancy', 'effort expectancy', 'social influence', and 'facilitating conditions' (Figure 8.2). In addition, UTAUT also highlights the role of gender, experience, age, and voluntariness of use as moderator variables. During its empirical validation, Venkatesh *et al.* (2003) excluded the variables such as anxiety, self-efficacy, and attitude toward using technology as not determinants of behavioural intention and removed from the UTAUT model). In 2012, a modified version of UTAUT called UTAUT2 has been modelled by Venkatesh, Thong, and Xu, and extended the original model by including new constructs such as habit, hedonic motivation, and price value (Venkatesh *et al.*, 2012).

8.2.3 A unified model of electronic government adoption

Despite the widespread testing of TAM and UTAUT models, the performance of TAM and UTAUT models in the public governance context was a mixed bag. Several scholars criticised the inadequacy of the models to fit into the e-government complexities, and the need for a theoretical development in the context of e-government adoption research (Dwivedi *et al.*, 2017; Rana *et al.*, 2017). In 2017, Dwivedi *et al.* proposed a specific model for e-government adoption

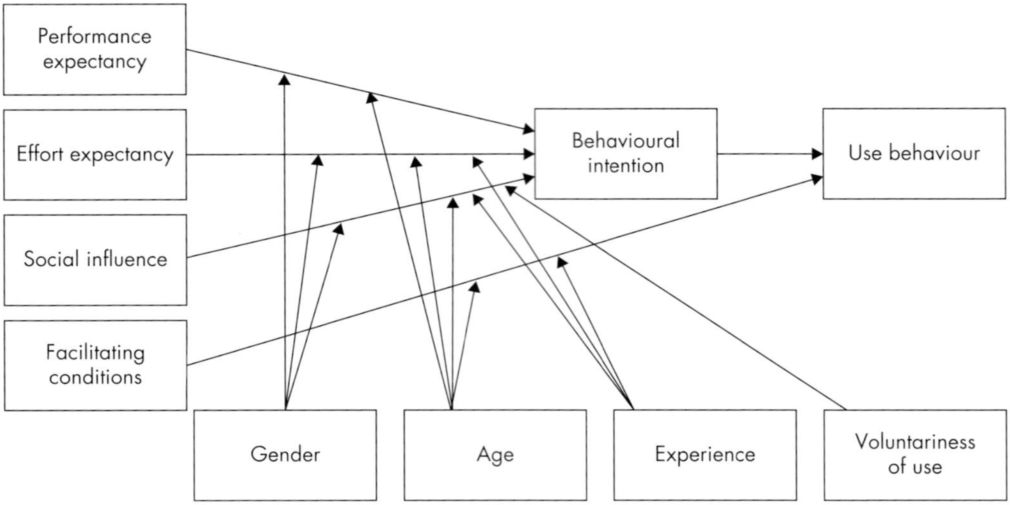

Figure 8.2. The UTAUT model (Venkatesh *et al.*, 2003).

called a unified model of e-government adoption (UMEGA). With the aim of developing a parsimonious model fitting to every e-government situation, Dwivedi and his colleagues dropped moderators from their proposed model and selected five key drivers, i.e. performance expectancy, effort expectancy, social influence, facilitating conditions, and perceived risk, to explain the user's attitude and behavioural intention to adopt e-government technologies (Figure 8.3). They reported that their model provided a significantly higher explanatory power (80% of variances in behavioural intention of non-adopters) in comparison to other technology adoption models. Later, an extended version of UMEGA was developed and tested by other scholars introducing new constructs, such as computer self-efficacy, trust in the Internet, and trust in government (Mensah *et al.*, 2020; Verkijika and De Wet, 2018).

These theoretical models are illuminative in the way different drivers influence user behaviours to adopt new digital technologies. However, the empirical testing of these theoretical assumptions in specific sectoral and organisational contexts can reveal different underlying relationships among identified drivers. Furthermore, the idiosyncrasies of different digital technologies may differently affect the perception of users about their disruptive and transformative nature. To have a better understanding of how these drivers may affect technology adoption strategies in specific public sector domains, we will refer to our research about the use of new digital technologies to fight taxation and social security fraud in Belgium.

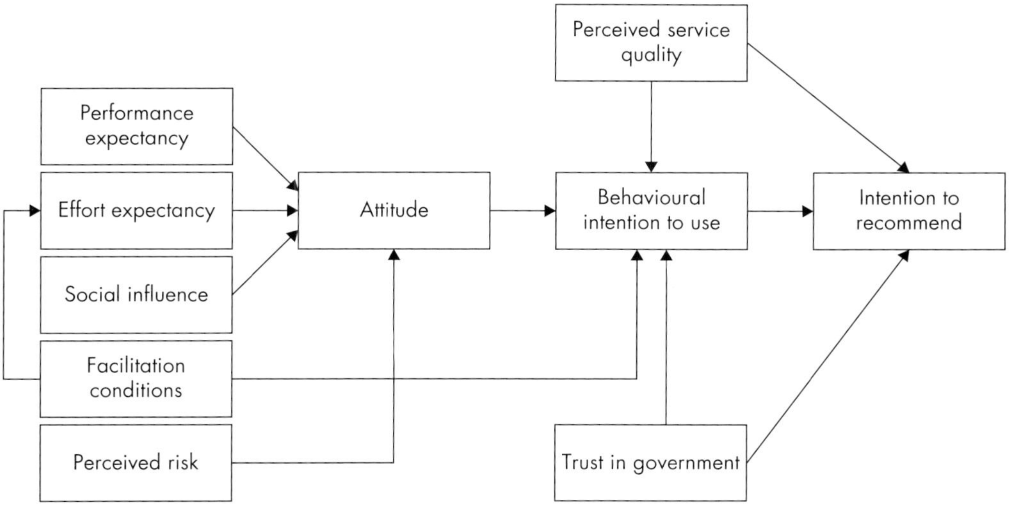

Figure 8.3. The extended UMEGA (Mensah *et al.*, 2020).

8.3 The use new digital technologies in the taxation and social security domain

Before going into the details of the research that took place in Belgium, we will first take a more specific look at the use of new digital technologies in the taxation and social security domain and at the institutional landscape in Belgium. The annual report of OECD Tax series show that the importance of new digital technologies is growing among tax authorities where data-driven insight is used against tax fraud as in smart compliance with risk management and compliance by design (OECD, 2021). Nowadays, new digital technologies such as AI, big data, and robotics are used in automating repetitive processes, extracting key data, scanning tax reports, identifying tax evasion, identifying tax deductions and credits, forecasting the burden of tax, improving transparency, and fighting against corruption (Naveen, 2020). Through supervised or unsupervised machine learning it is possible to find patterns of anomality in large amounts of data for predictive analysis and smart auditing that can overperform traditional data-mining techniques (De Roux *et al.*, 2018, Van Vlasselaer *et al.*, 2017). For instance, a good example of big data analysis to detect tax fraud is the British Connect System (Maciejewski, 2017). Furthermore, blockchain technology can improve the compliance by design, and even create new means of taxation and social security distributions through the use of smart contracts and token technologies. For the moment, there are various projects at different stages of development where governments leverage blockchain technology to fight against VAT fraud (e.g. EU), improving security and authenticity of custom data share (e.g. Mercosur countries), tracking taxes on real estate transactions and non-resident income tax (e.g. Finland, Sweden), and creating single platforms to track the transfer of sensitive data on banking, health, and business records within government organisations (e.g. Estonia).

8.4 The institutional landscape of Belgium

Belgium is a federal state including a federal government, three regional governments, three community governments and numerous local governments. The competence of tax and social security policy is mostly located at the federal level while regional and local governments also have the competence for some tax areas; however, the fight against tax and social security fraud is a federal matter. In addition, the European level also plays a role, notably providing guidelines on ethics in artificial intelligence.

Tax policy is mostly organised in a centralised way around the Federal Public Service (FPS) Finance. Specifically, for the fight against tax fraud, the FPS Finance is the main actor: several directorates-general play a role, including tax collection, fraud detection, customs, and excise. A second layer of actors includes the public sector organisations that can provide information about potential infractions to the FPS Finance. These actors notably include the financial intelligence processing unit (CTIF/CFI), several services of the police forces in charge of corruption (OCRC/CDBC) and of major crimes (DJSOC), the national bank (that has a point of contact – PCC/CAP – to alert it about frauds), the financial services and markets authority (FSMA), the college of attorneys-general and the gaming commission. A third layer of actors is composed of people and public and private organisations that have a mandate to provide information about fraud they encounter to the FPS Finance: lawyers, accountants, notaries, banks, major auditing firms and judges. In addition, there are transversal actors that are present in this field: actors at the EU levels that enforce specific tax policy provisions and actors at the national level that act as external stakeholders (technology providers, unions, business federations and NGOs in charge of fair taxation and/or privacy rights).

Moving to the second policy area, the fight against social security fraud is part of the federal social security policy. Federal social security policy provides several social services to citizens about welfare benefits (e.g. unemployment, retirement, children, maternity leave) and healthcare reimbursement. This policy domain is rather fragmented. The FPS Social Security oversees part of the policy design, but most of the policy implementation, including the fight against social security infringements, is led by a multitude of public bodies. These are called social security public institutions (SSPI). Each of these SSPI has its own specific responsibility: providing employment benefits, providing healthcare reimbursements, collecting social security contributions from contractual workers or from self-employed workers, providing retirement benefits, providing holidays benefits, and facilitating data sharing between SSPI. Among these topics, there are five areas in which specific actions are undertaken to combat what is broadly defined as social security infringements. These are led by five distinct organisations for, namely: employment (ONEM/RVA), healthcare insurance (INAMI/RIZIV), social security for contractual workers (ONSS/RSZ), social security for self-employed workers (INASTI/RSVZ), and control of social legislation (FPS Employment). These five organisations are working together under the coordination of the SIRS/SIOD in the fight against social security infringements. In addition, peripheral actors can be identified in this policy landscape: SMALS, a non-profit organisation that acts as the federal IT support service, and the Crossroads Bank for Social Security (CBSS) that organises data sharing among SSPI. There are also non-governmental actors such as social partners, social secretariats, and mutual health funds, that play a role in the organisation and management of social security. Eventually, the EU institutions also play a role in providing specific regulations in terms of social fraud (especially in the matter of social dumping).

Lastly, there are some overarching political and administrative actors that play a role in both policy domains, such as the federal government, the federal parliament, the data protection authority (DPA/APD) and the Court of Audit.

8.5 Drivers affecting the adoption of new digital technologies in the fight against fraud

Having set the institutional landscape and the technological developments in the taxation and social security domains to fight fraud, the DIGI4FED research team conducted a series of semi-structured interviews in 2020 and 2021 with 66 public officials and technical, business and policy experts from the public sector and stakeholder organisations in the taxation and social security domains in Belgium. The purpose of the interviews was to understand the perceptions of the key players and the issues that may affect the adoption of technologies such as big data, AI, and blockchain, in the fight against fraud. The collected data has been thematically analysed to construct the key drivers in technology adoption and later, through interpretive structural modelling, to explore the underlying the relationships among identified drivers. The adoption models have provided the conceptual framework to categorise the identified themes and subthemes into key drivers. Table 8.1 gives the categorisation of constructs, their definition and categorisation according to the adoption models, and the themes identified through thematic analysis. The subthemes that are used for the categorisation of themes are included in Appendix 8.A.

Following the analysis of the interview data, 13 variables are identified as the drivers in the adoption of new technologies in the fight against fraud. These variables are categorised according to the drivers identified by the technology adoption models to illustrate the conceptual relations among them. Taking a closer look at each variable:

1. Technological maturity. This variable captures the perception of interviewees about the fitness and technological readiness of new digital technologies to replace existing tools and technologies in the fight against fraud. Technological maturity is related with several underlying elements. For instance, the biases and noises associated with AI algorithms such as systemic biases in the training data, misinterpretation of data, possible errors in probabilistic AI, or using too many algorithms are related with the technological readiness of AI-based solutions. However, the lack of mature use cases that rely on blockchain or predictive AI that are as effective as current data matching operations indicate technological maturity problems. Through these underlying themes, technological maturity influences performance expectancy of users for the new digital technologies.

Table 8.1. Key drivers in technology adoption.

Drivers	Codes	Constructs	Elements	Definitions
Performance expectancy	1	Technological maturity	Bias and noise Technology convergence Blockchain/DLT AI/machine learning Fraud detection technologies	This variable captures the maturity of new digital technologies that are used in the fight against fraud.
	2	Perceived usefulness	Automation Improved social security and taxation Better data collection & analysis Past experiences Indirect added value of new digital technologies	This variable captures the perception of stakeholders about the usefulness of new digital technologies in improving the fight against fraud.
Self-efficacy	3	Capacities, skills and competencies	Resources Digital skills Training	This variable captures the resources, digital skills, and training of the administrations concerning the use of new digital technologies in the fight against fraud.
	4	Management/ operational systems	Guidelines Rules & standards Principles Processes Strategies	This variable captures management systems and means in the administrations concerning the use of new digital technologies in the fight against fraud.
Perceived risk	5	Perceived risk	Legal challenge Control of data Democratic challenge Administrative challenges Societal challenges	This variable captures the perception of stakeholders about the risk of using new digital technologies in the fight against fraud.
Effort expectancy	6	Governance system	Data governance Open governance Multi-level governance Network governance	This variable captures the modes of governance in relation to new digital technologies that influence the fight against fraud.
	7	Technical infrastructure	Security Quality of database Data collection & analysis Software Computer maturity Reliance/dependence on external actors	This variable captures the technical capacity of the system infrastructure that influence the use of new digital technologies in the fight against fraud.

Social influence	8	Public values	Appropriateness of technology Respecting privacy Tax fairness	This variable captures the public values in relation to the use of new digital technologies in the fight against fraud.
	9	Trust	Trust in administration Trust in society Trust in technology Trust in system Trust in tech providers/private sector	This variable captures the trust dimensions in relation to the use of new digital technologies in the fight against fraud.
	10	Socio-cultural elements	Digital culture Digital divide Willingness to share data	This variable captures the socio-cultural conditions in relation to the use of new digital technologies in the fight against fraud.
Facilitating conditions	11	Interoperability	Technical interoperability Semantic interoperability Organisational interoperability Regulative interoperability	This variable captures the interoperability conditions in relation to the use of new digital technologies in the fight against fraud.
	12	Policy priorities	EU-level policy priorities Fight against fraud Political support Geo-political aspects	This variable captures the national and international policy priorities in relation to the use of new digital technologies in the fight against fraud.
	13	Regulations	Data Taxation/social security Transcending laws Justification of decisions	This variable captures the national and supranational regulations in relation to the use of new digital technologies in the fight against fraud.

2. Perceived usefulness. Perceived usefulness captures the perception of interviewees about the added value associated with the new digital technologies in the fight against fraud. Automation of processes and better data collection and analysis techniques to improve the performance in fraud detection are directly associated with the perceived usefulness of new digital technologies. Furthermore, past experiences with digital solutions (e.g. crossroad banks), and indirect influences of new digital technologies in related sectoral areas (e.g. the impact of better predictive medicine on social security), are associated with the perceived usefulness of the new digital technologies. Perceived usefulness along with technological maturity influence the performance expectancy of technology adopters.

3. Capacities, skills, and competencies. This variable contains the themes about digital skills, resources, and training concerning the inclusion of new digital technologies in the fight against fraud. This variable not only captures the competencies in the public sector organisations but also includes relational capacities in the stakeholder organisations such as lack of digital skills in companies, unions, and society, as well as the availability of training programmes for citizens and the public at large. In the technology adoption literature, this variable is associated with the self-efficacy concept signifying 'the belief in one's ability to successfully perform a technologically sophisticated new task' (McDonald and Siegall, 1992).
4. Management/operational systems. Existing guidelines, rules, principles, processes and standards in relation to the adoption of new digital technologies in fraud detection are comprised in this variable. This variable specifically captures the managerial inputs from macro-level strategies (e.g. data strategies, mid-to-long term digital investment plans) to micro-level operational processes and principles (e.g. only-once principle, data confidentiality) showing that public sector organisations generate public value through data-driven technologies in the fight against fraud. This variable is also fitting to the self-efficacy concept as it reflects the perceptions on the ability of existing systems to adopt new digital technologies.
5. Perceived risk. Perceived risk is a multifaceted variable and captures different risks associated with legal, systemic, administrative, and societal dimensions. For instance, how to ensure the protection of citizen rights, or follow rules on data confidentiality, these capture perceived risks about legal dimensions. Or the control of data and technologies by the private sector (e.g. GAFAM). or the public sector, which fall into the systemic and democratic challenges. Perceived risk is one of the drivers in technology adoption identified in the UMEGA model.
6. Governance system. This variable contains different governance mechanisms that influence the use of new digital technologies in the social security and taxation policies. Concepts and processes related to data governance, open governance, multi-level governance, and network governance fall into the scope of this variable. The presence and effectiveness of governance mechanisms is part of the effort expectancy in the adoption of new digital technologies in the fight against fraud.
7. Technical infrastructure. Another variable that influences effort expectancy of adopters is the technical infrastructure. Technical infrastructure refers to the perception of respondents on the adequacy of existing information systems to adopt new digital technologies. This variable contains elements concerning the security of infrastructure (e.g. data infrastructure, network infrastructure), quality of databases, means and tools used in data collection and analysis, type of software (e.g. open vs proprietary solutions), and the maturity of computer infrastructure.
8. Public values. This variable is about the public values associated with the use of new digital technologies in the fight against fraud. Specifically, appropriateness of technologies such as the added value of AI or asking moral and ethical questions before adopting a technology, respecting the privacy of individuals (e.g. while profiling fraudsters), and ensuring tax fairness are related value components. This variable is one of the social influence drivers.

9. Trust. The trust variable contains various trust components concerning administration, technology, system, technology providers, and society. This variable captures the meaning about the trust and distrust of interviewees in various actors in the taxation and social security domain, and about their capabilities and intentions in bringing added value in the fight against fraud. This variable is also considered as one of the social influence drivers.
10. Socio-cultural elements. The last variable related to social influence drivers concerns the socio-cultural elements. Digital culture in administration and society, digital cultural differences between Wallonian and Flemish authorities, the digital divide in the society in terms of the access and use of digital technologies, and the willingness to share data are the elements influencing the changes in this variable.
11. Interoperability. The last three variables are categorised under facilitating conditions. The first such a variable is interoperability. Interoperability captures the ease of data transfer across different organisations: regulative, semantic, organisational, and technical interoperability are the underlying themes of this variable. Higher interoperability facilitates the extracting and use of data to supplement data-driven technologies in the fight against fraud.
12. Policy priorities. This variable captures the national and international policy priorities in relation to the use of new digital technologies in the fight against fraud. Policy priorities set by the European Union (e.g. Green Deal, digital innovation), policy priorities on the fight against fraud (e.g. targeting small business vs large business), political support on digital transformation, and relevant geo-political considerations (e.g. impact of technology on environment, the use of rare materials in technological solutions) are topics influencing the variances in policy priorities.
13. Regulations. The last variable identified from the interview transcripts is regulations. National and supernational legislations on data (e.g. GDPR), on taxation and social security, and other transcending laws (e.g. environmental protection) that may affect technology adoption are covered by the variable. Furthermore, regulations on the justification of decisions in the fight against fraud are important for the way regulations affect the technology adoption in the taxation and social security domains.

8.6 Interrelationships among drivers in the adoption of new digital technologies

A thematic analysis across interview data has revealed 13 variables to understand the dynamics in the adoption of new digital technologies in the fight against fraud. These variables correspond largely to the theoretical assumptions of technology adoption models. Nonetheless, technology adoption models provide a limited insight into the interrelationships among these drivers. Understanding how changes in certain variables influence changes in other variables is important for developing effective technology adoption strategies. To address this issue, an analysis-based interpretive structural modelling (ISM) and MICMAC have been conducted.

ISM is a mathematically-derived, methodical, and cooperative method that allows researchers to examine contextual relationships among factors identified through expert opinion, and to establish hierarchical levels of challenges (Warfield 1974). ISM analysis involves the development of a directed graph through a hierarchical configuration of the relationships as interpreted by the researchers. MICMAC analysis complements ISM in the development of a graph that classifies factors based on driving power and dependence power (Ahmad *et al.*, 2019). ISM has been used in the e-government literature to explore critical success factors in e-service delivery (Lal and Haleem 2009), citizen's perceptions of e-government's trustworthiness (Janssen *et al.*, 2017), challenges for implementing the Internet of Things (IoT) in smart cities (Janssen *et al.*, 2019). We use the ISM method to explore and describe the dependencies between the variables (i.e. constructs) identified by the thematic analysis.

The development of the model goes through three levels of analysis. The first level of analysis is about the development of a structural self-interaction matric (SSIM). In developing SSIM, there are four possible ways of relating variables to each other, that are represented by 'V', 'A', 'X', and 'O' symbols (Hughes *et al.*, 2020). The symbols are to be interpreted as: V = Variable i influences variable j; A = Variable j influences variable i; X = Both variables i and j are influenced by each other; O = Variables i and j are not related to each other or do not influence each other. The relationships among these variables are interpreted by the research team based on the thematic relationships revealed through the thematic analysis. More specifically, if an underlying concept for a construct is related with another driver, this relationship is used to interpret the nature of the relationship.

Following SSIM, first an initial reachability matrix (IRM), and later the final reachability matrix (FRM) have been developed. IRM illustrates the relationships described by SSIM in a binary way, where '1' shows an existence of the relationship and '0' shows a non-relationship. In the next step, the IRM is converted into an FRM, where the transitive relations are included. Transitive relations correspond to: if a variable X influences variable Y, and variable Y influences Z, then variable X should influence variable Z even if there is no mutual relationship interpreted between variables X and Z. When such a relationship is found, an initial 'no relationship' – i.e. '0' – has been recoded as '1'. FRM also shows the driving and dependence power of each variable. The driving power for each variable is the total number of variables, including itself, which it may help to achieve. On the other hand, dependence power is the total number of variables, including itself, which may help in achieving it. These driving powers and dependence powers are used later in the classification of variables as part of the MICMAC analysis.

The third step in the development of the ISM relies on the level partitioning. For that, the FRM is used to develop reachability and antecedent sets for each of the variables in the matrix. The reachability set, R(Pi), for a particular variable includes the variable itself and other variables that might help to achieve it (i.e. the corresponding value is 1). Similarly, the antecedent set, A(Pi) consists of the variable itself and other elements that might help in achieving it. The variables for which the interaction of these sets, R(Pi) $\cap$ (A(Pi) = R(Pi), and the reachability set match, are considered as the top-level variables of the ISM hierarchy. Each iteration of the level partition matrix identifies the hierarchy of variables in achieving other variables. The top-level variables would not help to achieve other variables above their own level of hierarchy. Once the top-levels are identified, they are separated, and the same process is repeated. The iteration process continues until all levels of partitions are established. Accordingly, five levels of partitions are identified in the ISM diagraph.

- Level 1 is the highest level in the hierarchy of the ISM diagraph and contains the following variables: technological maturity, capacities, skills and competencies, perceived risk, public value, policy priorities, regulation.
- Level 2 is the second level in the hierarchy of the ISM diagraph and contains the following variables: management/operational systems, governance, technical infrastructure, socio-cultural elements
- Level 3 is the third level in the hierarchy of the ISM diagraph and contains only perceived usefulness.
- Level 4 is the fourth level in the hierarchy of the ISM diagraph and contains only interoperability.
- Level 5 is the lowest level in the hierarchy of the ISM diagraph and contains only trust.

Before completing the ISM analysis, we visually categorised the driving and dependency powers among the variables through a MICMAC diagram. The MICMAC diagram contains four categories, namely autonomous, driving, dependent, and linkage. Autonomous shows variables with weak dependency and driving powers, showing variables that are mostly disconnected from the system. Driving shows variables with higher driving power and weak dependency power, mostly signifying variables that determine the changes in other variables without necessarily being dependent on the changes in other variables. Dependency shows variables with higher dependency power and with weak driving power, mostly containing variables that vary with changes in other variables without necessarily affecting the changes in other variables. Linkage shows the variables that have both higher dependency and driving power. The variables in the 'linkage' quadrant are the most influential variables in the system. The cut-off point for each quadrant is arbitrarily designated by the number of variables. Since there are 13 variables, the cut-off point is designated at 6.5 in both axes.

The MICMAC diagram in Figure 8.4 shows that all variables fall into the linkage category, which means that all identified constructs are highly influential in the adoption of new digital technologies in the administrative systems focussing on the fight against fraud in taxation and social security. Most variables show higher driving and dependency power suggesting that changes in any of them can lead to drastic changes in the adoption strategies of the new digital technologies in the fight against fraud.

Lastly, based on the level partitioning, and the direction of relationships identified in SSIM, we have developed the ISM model for the adoption of new digital technologies in the fight against fraud in taxation and social security systems. The ISM model in Figure 8.5 shows the direct and indirect relationships among variables. The direction of relationships is shown with an arrow. The ISM model suggests that trust (variable 9) plays a very significant role in driving the rest of the variables and is positioned at the base of the ISM hierarchy. The level of trust influences interoperability (variable 11) of information systems. Interoperability conditions in return directly influence the perceived usefulness (variable 2) of new digital technologies in

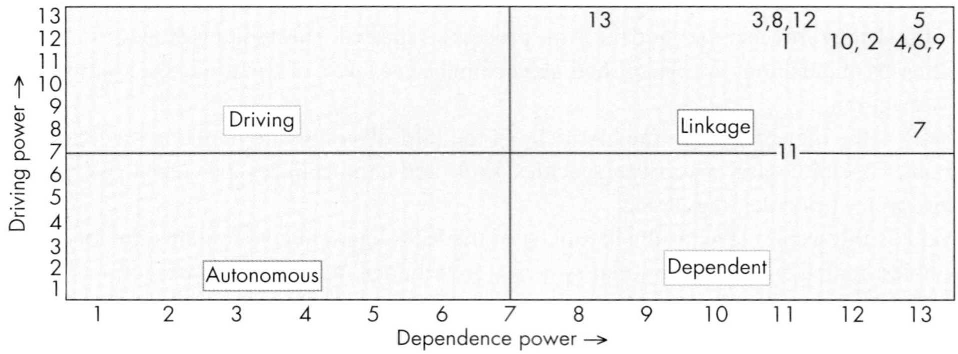

Figure 8.4. MICMAC diagram.

the fight against fraud. Perceived usefulness influences the interrelationships between socio-cultural elements (variable 10), governance conditions (variable 6), technical infrastructure (variable 7), and management/operational systems (variable 4) at place. In this level, we observe that governance conditions, technical infrastructure, and management systems influence each other and as such should be assessed together. The model also shows that socio-cultural elements are closely interlinked with governance conditions. The changes and conditions in level II affect the interrelationships in the highest level of hierarchy. In the highest level, we observe direct mutual influences between policy priorities (variable 12) and regulation (variable 13), policy priorities and perceived risk from the use of new digital technologies in the fight against fraud (variable 5), and policy priorities and public values (variable 8). Thence, policy priorities are not only directly influenced by these variables but also perceived risk, public values, and regulations are directly influenced by the priorities set by the policymakers. Furthermore, policy priorities and the maturity of technologies (variable 1) are influential on the capacity conditions (variable 3) which in return influence public values concerning the appropriateness of technology, privacy, and tax fairness. Capacity conditions are also directly influential in perceived risks associated with new digital technologies along with regulations, technological maturity, and policy priorities. Finally, technological maturity is influenced directly by the perceived risk and capacity conditions.

8.7 Policy insights for practitioners – the use of new digital technologies in the fight against fraud

So, how can we interpret these findings for the adoption of new digital technologies in the fight against tax and social security fraud? The empirical data for our research comes from the Belgian case and specifically the use of new digital technologies for the fight against fraud in taxation and social security schemes. Therefore, these relationships and the revealed interrelationship among key drivers should be interpreted within a specific context. Nevertheless, the following conclusions can be drawn from the analysis:

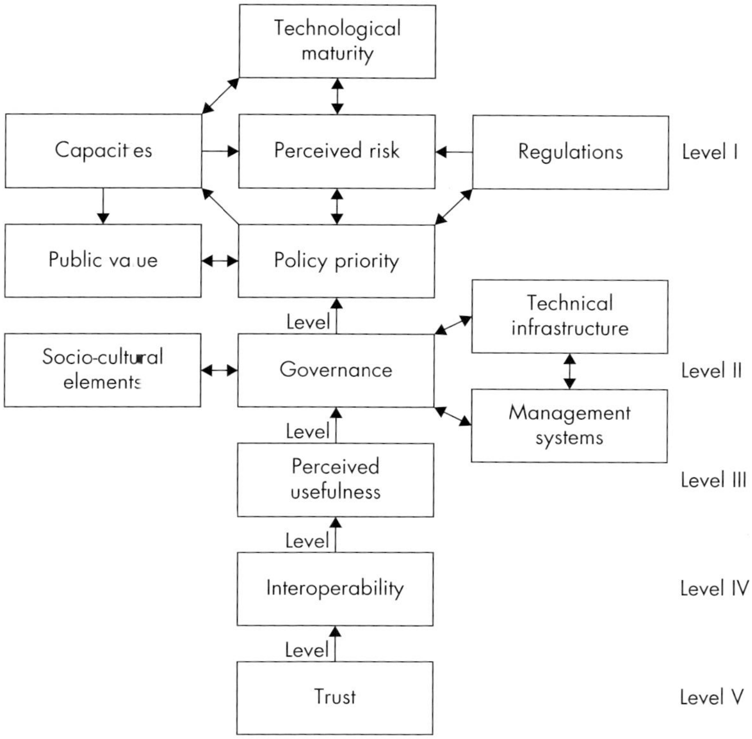

Figure 8.5. ISM model.

1. The thematic analysis of interview data has led to the creation of 13 variables that are influential on the adoption of new digital technologies in the fight against fraud. The MICMAC analysis showed that all these variables have strong dependence and driving powers, suggesting that changes in one of them may trigger changes in other variables. As such, any policy intervention should keep in mind this unstable nature of these variables and prioritise periodic reassessment of changes in the observed variables.
2. The ISM suggests that 'trust' factors constitute the bottom of the hierarchy in the adoption of new digital technologies in the fight against fraud. The trust variable consists of elements such as trust in administration, technology, technology providers, system, and society itself. A study by Janssen *et al.* (2017) shows that trustworthiness of e-government services is determined by the cognitive factors, personal qualities (e.g. responsiveness, competence), prior experiences in the service area, and perceived prior knowledge on technology. A takeaway from this finding is that any policy design for the implementation of new digital technologies in the fight against fraud should integrate the key stakeholders in the sectoral area and a variety of individuals (both service producers and receivers) to map out the key trust challenges.
3. Following the trust factors, the ISM hierarchy suggests that interoperability is the second most important driver that determines the changes in the rest of the drivers. Interoperability refers to the technical, organisational, regulatory, and semantic factors that determine the ease of

data transfer across data providers and consumers. Our analysis suggests that assessing the interoperability conditions in the existing data ecosystem is the second decision that needs to be taken by the policy designers.

4. The third driver in the ISM hierarchy is the perceived usefulness of the new digital technologies in the fight against fraud. In our research, this variable captures the perceived usefulness of automation of services, digitisation of social security and taxation services, the use of advanced data collection and analysis methods in social security and taxation services, past experiences with the use of digital technologies/infrastructures in the fight against fraud, and the indirect added value of digitisation in other related sectoral areas (e.g. development of better predictive medicines) on taxation and social security systems. Our analysis suggests that the perceived usefulness of new digital technologies concerning these dimensions is closely related with the prior trust and interoperability conditions.
5. The next variables in the ISM hierarchy are closely related to the digital governance. Variables of governance, technical infrastructure, management/operational systems, and socio-cultural factors create a cluster of drivers that are closely interrelated to each other, and this stage in the policy design mostly captures the decisions concerning the integration of new digital technologies in the existing digital governance mechanisms. The model suggests that in this system of relations, modes of governance concerning new digital technologies (e.g. multi-level governance, data governance, network governance, open governance) is the central factor that is directly influenced by socio-cultural conditions concerning digitalisation (e.g. digital divide, digital culture, willingness to share data), and by technical infrastructure conditions regarding new digital technologies (e.g. security, quality of database, data collection and analysis methods, software, computer maturity, and reliance on external actors), and management/operational systems concerning new digital technologies (e.g. guidelines, rules and standards, principles, processes, and strategies).
6. The top level of the ISM hierarchy is composed of a cluster of six variables, namely technological maturity of new digital technologies in the fight against fraud; existing capacities, skills and competencies concerning new digital technologies in the administration; regulations concerning new digital technologies; public values concerning new digital technologies; policy priorities in the use of new digital technologies in the fight against fraud; and the perceived risks associated with the use of new digital technologies in the fight against fraud. According to the ISM, these variables are the most influential variables in the adoption of new digital technologies in the fight against fraud. Any policy intervention should assess these variables jointly, as there are various direct and indirect relations among them. Yet, the density of relationships can be used to identify subclusters of relationships among these factors. Such a subcluster can be identified among regulations, policy priorities, and perceived risks. In this set of relationships, laws and regulations concerning data, social security and taxation, justification of decisions in the fight against fraud and transcending laws affecting the use of new digital technologies directly influence the policy priorities and perceived risks in the use of new digital technologies in the fight against fraud. In the meantime, perceived risks concerning new digital technologies about legal, societal, administrative and democratic challenges and the control of data directly influence and are influenced by the policy priorities at the EU level such as policy priorities in the fight against fraud, the level of political support in the use of

new digital technologies in the fight against fraud, and other geo-political aspects that can affect the policy decisions concerning the adaptation of new digital technologies (e.g. impact of high-energy consuming technologies on environment). Such policy priorities in return also influence the regulatory conditions that can affect the way new digital technologies can be used in the policy areas.

7. Another subcluster at the top-level is among the variables of capacities, public value, and policy priorities. Here, the capacity conditions in the administrations concerning resources, digital skills, and training on the new digital technologies are expected to influence public values about the appropriateness of new digital technologies in the fight against fraud, respecting privacy of individuals and ensuring tax fairness. In the meantime, these public value conditions are influential on the policy priorities, and policy priorities in return influence the capacity conditions.
8. A third subcluster is between the technological maturity of new digital technologies, capacity conditions, and perceived risks. Here, the maturity of technologies such as AI and machine learning systems, blockchain and distributed ledger technologies, and other fraud detection technologies leveraging advanced data analysis methods, plus the convergence of these technologies, and the way the maturity of these technologies influences the bias and noise considerations, all this directly influences and is influenced by the capacity conditions and associated perceived risks.
9. Finally, perceived risks associated with the use of new digital technologies in the fight against fraud appear to be at the centre of the interrelationships identified at the top level. Perceived risks are influenced directly by the capacity conditions, technological maturity, regulations, and policy priorities. This empirical finding complements the multifarious risks expressed in previous chapters concerning the legal, democratic, administrative, societal, and data governance aspects. The next section will focus on highlighting particular risks associated with the new digital era governance, and what public managers can do to mitigate them.

There are some caveats for the interpretation of the model to other areas of application. The drivers and their identified interrelationships are dependent on the contextual and sector-specific conditions of the study. Constructs concerning the institutional and regulatory landscape of the Belgian social security and taxation system, and the sector-specific conditions of tax and social security fraud affect the way the drivers influence the technology adoption. Similar research in other policy domains and country cases can provide deeper insight on how certain drivers (e.g. trust, public values) affect technology adoption. Furthermore, different digital technologies such as AI and blockchain have different affinities and fortes in influencing public governance processes. For example, blockchain can create trust in the way that data is handled and managed, whereas AI can lead to faster and deeper data insight to use in public governance processes. The way these technologies are employed in management and governance systems can have an enhancing or delimiting impact on the way drivers influence each other. The challenge is here three-fold. First, these technologies may invoke different value trade-offs concerning transparency, efficiency, effectiveness, and accountability in public services processes. Second, the effective implementation of these technologies necessitates strategic planning that is both internal and external to organisational processes. For instance, deployment of AI-based systems raises questions about the type of data to use in feeding into the system (e.g. training data, input data, feedback data), the way that data will be retrieved from public or private sources, cybersecurity and data protection protocols, and the precautions against data bias and behavioural limitations on the use of data insight in policy-making processes. Third, the integration of these technologies into organisational infrastructure is different. While AI-based solutions may demand more in-house solutions for effective integration of the technology, blockchain-based solutions necessitate effective coordination of stakeholder groups and outside actors in digital solutions.

Keeping in mind the context-dependent factors, our findings suggest that challenges concerning trust, interoperability, configuration of digital governance systems, and complex relationships between socio-technological conditions are the key considerations for practitioners in the new digital era governance. In the following sections, we will take a closer look at some of these challenges and postulate policy actions to address these challenges.

8.8 New challenges and issues for practitioners in the new digital era of governance

In the Introduction, we have portrayed it as a risky venture for public administration to adjust to the challenges and risks associated with the new digital technologies. Throughout the book, technological, organisational, legal, societal, and policy-level challenges are scrutinised with empirical insights, shedding a light on how new digital technologies can disrupt the existing institutions in the public sector domain. Our analysis highlights many novel challenges for public sector organisations, which call for innovative approaches in strategic thinking. In the second part of this chapter, we will bring the interdisciplinary research insights together to discuss how decision-makers can address these challenges of the new digital era of governance. Following the levels of analysis construed from the ISM model, we will focus on the dynamics of trust, interoperability issues, mechanisms of digital governance, and policy designs for a digital future.

8.9 How to create trust in digital governance?

Building trust is a core challenge in the new digital era of governance. It is not only the starting point for developing a technology adoption strategy for government, but it is also arguably the most obscure challenge for governments to tackle. Trust challenges faced by governments are manifold. On the one hand, political and administrative institutions struggle to maintain citizen trust by showing their intention and capability in using new digital technologies. On the other hand, user trust in data sharing and the ethical application of AI and advanced data analytics have become politically sensitive issues. More algorithmic transparency has been called for to address trust challenges as sophisticated algorithms are often 'black boxes'. Furthermore, political trust challenges are also associated with the predictive and manipulative power of machine learning and deep learning programs over individual users and behaviours. The topic has attracted mass media reporting, as seen in the US Facebook-Cambridge Analytica scandal or the Dutch childcare benefits scandal. Both incidents featured major society-wide trust breaches that stemmed in part from the use of large-scale databases and AI. As Bjorn Kleizen and his colleagues showed in Chapter 6, our current empirical and theoretical knowledge base about how society perceives the trustworthiness of AI applications used in public governance, and what policies and factors are influential in engendering trust in the AI-based decisions made by the government, is vastly limited.

There are several limitations to engendering trust in AI projects employed by government. The first issue is about how trust challenges are addressed by algorithms and explainable AI (XAI) research. Trust in relation to AI research is currently mostly understood as being able to interpret the input-output relations of an AI system, meaning that one can explain, given a certain input, how the output was generated.[1] Trust, in this context, refers to a technical property more like transparency (e.g. explainability, expected levels of performance). Yet, ongoing debates around AI hint at transparency and interpretability as being insufficient for trust. The European Commission's High-level Expert Group on AI proposes that there are three main characteristics of trusting AI: (1) the AI technology itself; (2) designers and organisations behind the development, and those involved in the deployment and use of AI; and (3) the socio-technical systems involved in the AI life cycle. A wide array of techniques has been proposed to achieve either truly transparent models, or post-hoc explanation techniques to build trust in algorithms (Barredo Arrieta *et al.*, 2020, Ribeiro *et al.*, 2016, Došilović *et al.*, 2018). A trade-off is inevitable as inherently transparent models cannot approximate the intricacies of high dimensional insights that are produced by deep architectures, and explanations can only be verified by experts who can verify causality (Murdoch *et al.*, 2019). From a user standpoint, these approaches see trust as a by-product rather than the ultimate goal, which is hard to quantify (Barredo Arrieta *et al.*, 2020, Lipton 2018). Quantification is typically user-driven and based on subjective measures (Yu *et al.*, 2017; Zhou *et al.*, 2016, 2019) and results are obtained from explanation mechanisms such as correlation or explanations under guarantees of accuracy, which remain anecdotal due to a lack of causal proof (Yu *et al.*, 2017). These trends indicate that there is a lack of a quantitative and objective evaluation methodology or topology of relevant components within XAI to trust. The concept of trust is linked to confidence, transparency, fairness, and so on, but never measured as either a component, mediator, or result of these concepts which have been quantified with reasonable success (Barredo Arrieta *et al.*, 2020). For digital system acceptance, Shin (2020) argues that trust is rather established by fairness, accountability, transparency, and explainability, reversing the equation for a user study to grasp the impact of these components on trust in personalised AI systems. This offers a first insight into what users might deem important, but the connection with AI solutions and XAI calculations have not been established yet. User evidence is crucial for the evaluation of XAI (Islam *et al.*, 2020) and an inside-out approach asking users what they would consider valuable information regarding a model has not been investigated as such.

The limitations of XAI research and methodologies to engender algorithmic trust can be theoretically substituted by other technological solutions leveraging blockchain and blockchain-linked technologies such as self-sovereign identity frameworks and zero-knowledge protocols. The

[1] For an example, see the White Paper on Artificial Intelligence (COM(2020)65) published by the European Commission (2020).

combination of these technologies to supplement consent-based, pseudonymous personal data in data analytics solutions can create deterministic, technology-led solutions to trust breaches. Yet, the existing technology solutions are either enterprise-led or lack technological maturity for wider adoption. Eventually, decision-makers need to trust in the reliability of blockchain-linked technologies in ensuring security, scalability, performance, and stability goals to test with such solutions to remedy AI-led trust challenges.

Trust is also an important intervening variable when investigating how people interact, use, and rely on technology. Parasuraman *et al.* (2000) distinguish between different forms of trust in technology (i.e. relying too much, too little, or not at all on technology), and their impact on trusting the role of technology in decision-making processes. Current developments push this debate even further with the increasing adoption of AI by political and public institutions, while citizen-government trust relationships are frequently featured in both social science and academic legal debates. The social and technological dimensions of these dynamics make it extremely difficult to concisely define the term trust and to develop strategies for increasing trust in political/public governance processes, technological systems, and especially the growing intersection thereof. There are many semantic complexities concerning trust. First, trust can be defined in many ways depending on the field or context (Madsen and Gregor, 2000). It also entails different themes like competence, benevolence, and reliance (Benbasat and Wang, 2005; Chopra and Wallace, 2003; Cramer *et al.*, 2008; Grandison and Sloman, 2000). Second, trust has been also recognised as changing over time (Holliday *et al.*, 2016; Nourani *et al.*, 2020). Third, measuring trust is challenging and can be done either explicitly or implicitly. Explicit measurements apply questionnaires or interviews to ask people about their trust perceptions. One-dimensional approaches measure trust with a single Likert-type question (Millecamp *et al.*, 2019). Although this method is quick and easy, it is susceptible to people interpreting 'trust' differently. Therefore, multidimensional approaches use Likert scales to measure trust as an ensemble of multiple constructs. For example, McKnight *et al.* (2002) introduced the concept of trusting beliefs, consisting of the constructs of competence, benevolence, and integrity. Later research added more constructs, including perceived transparency and intention to return (Berkovsky *et al.*, 2017; Pu and Chen, 2007) Overall, while a multidimensional approach is more nuanced than its one-dimensional counterpart, it requires longer questionnaires and is, therefore, more time-consuming. Implicit measurements try to avoid the potential bias in self-reported explicit measurements by measuring trust through an intermediary. Examples are loyalty-measured by the number of logins after sign-up (McNee *et al.*, 2003; Tintarev and Masthoff, 2011); accepting recommendations (Cramer *et al.*, 2008), and time spent on a page, click-through rate, or page-exiting manner (Fox *et al.*, 2005). In the context of explaining recommender systems, implicit measurements for trust have not yet been widely adopted, possibly because intermediaries like loyalty require long(er)-term studies.

Algorithmic trust and semantic determinants of trust need to be complemented with design considerations. Previous research has shown that providing explanations for AI systems can increase acceptance of AI outcomes (Cramer *et al.*, 2008; Herlocker *et al.*, 2000) and increase users' trust (Berkovsky *et al.*, 2017; Pu and Chen, 2006). A wide range of design dimensions has been proposed to support model-agnostic explanations to increase user trust. For example, Adadi and Berrada (2018) categorised explanation methods into four groups: visualisations, knowledge extraction, influence methods and example-based explanation.

Despite the rich set of explanation methods that have been proposed, there are important challenges. First, little work has been done to evaluate whether the proposed methods indeed enhance human interpretability (Tjoa and Guan, 2021). Researchers have criticised the lack of human-subject tests to verify whether the methods work for actual users (Adadi and Berrada, 2018; Gedikli *et al.,* 2014). Previous research in recommender systems has shown that simpler justifications are often preferred over more complex explanations (Herlocker *et al.*, 2000), but at the same time, there is a trade-off between interpretability and completeness of explanations (Gilpin *et al.,* 2018). Second, many explanation methods are static. Several researchers have voiced the need for interaction with explanations as a core requirement to support trust and understanding (Abdul *et al.,* 2018; Adadi and Berrada, 2018). User interaction with explanations is also key to enabling users to steer the process with input and feedback (Jin *et al.*, 2019). Although recent visual analytics techniques provide sophisticated visualisation and interaction capabilities, a key challenge is to tailor visual analytics techniques to the level of expertise of end-users, as most techniques require significant expertise (Chen *et al.*, 2018). Third, the results of several user studies have indicated that the effectiveness of explanations depends on both personal and situational characteristics (Bunt *et al.*, 2012). Although several researchers have started to work on identifying personal and situational characteristics as well as ways to automatically detect, such characteristics (Berkovsky *et al.*, 2019; Conati *et al.*, 2017; Hoppe *et al.*, 2018) adapting trustworthy AI interfaces to these characteristics is a largely unexplored research area.

Beyond the semantics, technical and design issues, we may also discern many organisational and managerial challenges giving rise to potential trust problems. Actors in a position to provide oversight (e.g. senior management, political superiors and/or the judiciary) may lack the required expertise to accurately assess potential issues with data analytics projects (Busuioc, 2021). Lack of capacities and expertise inside the organisations may cause citizens and civil society organisations to lose their trust in the use of advanced analytics and AI in public governance. Low levels of trust, in turn, may impact the legitimacy of AI algorithms and thus hinder governments from utilising advances in AI technologies to their fullest benefit. Low level trust in governance can be caused by endogenous reasons such as lack of proficiency in data and in AI to evaluate information or lack of organisational culture and management processes to cultivate innovation. Or exogenous reasons such as digital culture or digital divide can lead to citizen distrust and societal discomfort and further the information asymmetry between citizens and service providers (Van de Walle and Bouckaert, 2003).

Moreover, essential components of trust relationships, including transparency and the provision of information, as well as debates on the justification of algorithms are often confounded by technical complexities (Ahonen and Erkkilä, 2020). As mentioned in Chapter 6, cultural associations of algorithms may also affect social judgements. Terms such as big data and AI sometimes seem to invoke cultural associations with a police state or big brother in which all actions are monitored, or with 'general AI' capable of sentience and intelligent action. Given such social interpretation processes, there may be a substantial disconnect between the policies designed to secure AI processes and the trustworthiness that government and non-government actors attribute to such processes externally. Agile management techniques, open government data, and collaboration with relevant stakeholders may address certain administrative capacity problems and socio-cultural barriers in fostering trust. How actors participate in these processes, and what other administrative and regulative constraints exist for decision-makers are important to understanding the success of these methods in creating trust. Management tools and techniques such as an ethical code of conduct and leadership styles may matter in creating trust among employees (Brendel *et al.*, 2021). Yet, it seems like managerial approaches to AI solutions remain inherently value-based choices and fractured in most public sectors.

Given this constellation, we pose that creating trust in new digital technologies cannot be achieved through purely technical consideration of the information system itself – trust is a property emerging from the embedding of the information systems in a specific context. This context is defined by (a) the technical properties of the employed model (i.e. data-driven, rule-driven); (b) the role of humans in relation to all aspects of model selection, data selection, model training, and model employment; (c) considerations on the role of the institution applying the data-driven system and the way digital solutions are applied in governance processes (i.e. objectives of the institution, how does it mediate between state and individuals) and (d) the ability to challenge the outcomes of the system by the people subjected to its decisions. Public sector organisations need to experiment with different policies and solutions and combine technological solutions with communication efforts to smooth the transition.

8.10 How to build interoperability into digital governance?

Digitisation of public services has raised the importance of interoperability as a key consideration for governments. In short, interoperability refers to the ability (or capability) of two or more systems and/or networks, devices, applications, components, to exchange information and to use the information that has been exchanged (Andročec, 2017; Brutti *et al.*, 2019; Casiano Flores *et al.*, 2021). However, this working definition seems to be limited to addressing the interoperability challenges as the increasing complexity among data sharing and using systems has expanded the scope of interoperability beyond the technical considerations. Currently, one of the most conceptually encompassing definitions of interoperability is used by the European Interoperability Framework (EIF). The changing scope of interoperability can also be seen in the evaluation of the EIF. The first version of the framework in 2004 focussed solely on ICT considerations, whereas the 2010 and 2017 revisions have expanded its scope into regulative, organisational, and semantic dimensions (Chantillon *et al.*, 2021a).

The current conceptual model of EIF adopts a layered and principle-driven approach towards interoperability (European Commission, 2017). The model identifies 12 interoperability principles, namely, subsidiarity and proportionality, openness, transparency, reusability, technological neutrality and data portability, user-centricity, inclusion and accessibility, security and privacy, multilingualism, administrative simplification, preservation of information, and assessment of effectiveness and efficiency. These principles are fundamental behavioural aspects for driving actors in establishing interoperable European public services (European Commission, 2017).

Besides the twelve underlying principles, the model also builds on four interoperability layers and the transversal 'integrated public service governance':

1. Legal interoperability is about 'ensuring that organisations operating under different legal frameworks, policies and strategies are able to work together' (European Commission, 2017). In the European Union context, that national legislations of Member States do not block the European public services within and between Member States, and clear agreements are set about how to deal with differences in legislation across borders.
2. Organisational interoperability refers 'to the way in which public administrations align their business processes, responsibilities, and expectations to achieve commonly agreed and mutually beneficial goals' (European Commission, 2017). More specifically, organisational interoperability focusses on the clarification and formalisation of business processes in an agreed way with commonly accepted modelling techniques that ensure associated information is exchanged across European service providers, stakeholder organisations, and user communities in an uninterrupted way.
3. Semantic interoperability is 'the precise format and meaning of exchanged data and information ... preserved and understood throughout exchanges between parties' (European Commission, 2017). Both semantic aspects about the meaning of data elements and the relationship between them, and syntactic aspects about the exact format of the information to be exchanged in terms of grammar and format fall within the scope of semantic interoperability.
4. Technical interoperability 'covers the applications and infrastructures linking systems and services'. Aspects of technical interoperability include interface specifications, interconnection services, data integration services, data presentation and exchange, and secure communication protocols. (European Commission, 2017). Especially, a plethora of legacy systems and fragmented ICT services are considered as the key challenges for technical interoperability.

The final element of EIF is an 'integrated public service governance' that integrates the four interoperability layers in the definition of organisational structures, roles, and responsibilities for the delivery and operation of public services, service level agreements, establishment and management of interoperability agreements, change management procedures, and plans for business continuity and data quality (European Commission, 2017).

Some recent works endorsed by the European Commission ISA^2 programme[2] have developed further tools and conceptual frameworks for the implementation of the EIF principles and models across European public services. For example, the 'European Framework for Interoperability Skills and Competencies in the Public Sector (EFISC)' identifies the skills and competencies related to interoperability in the public sector (Casiano Flores *et al.*, 2021). Another framework called

[2] Interoperability solutions for public administrations, businesses and citizens (ISA^2) programme has been adopted in 2015 and implemented until 2020. In 2021, the programme has been rebranded as 'Interoperable Europe'.

'European interoperability framework for Smart Cities and Communities (EIF4SCC)', provides 'local administration leaders with definitions, principles, recommendations, practical use cases, and a common model to facilitate service delivery to the public across domains, cities, regions and borders' (Chantillon *et al.*, 2021b).

EIF serves as a useful tool to conceptualise the interoperability challenges in the new digital era governance. But the framework is not without its shortcomings. One shortcoming of the framework is about the absence of cultural interoperability (Chantillon *et al.*, 2021b). Chantillon and his colleagues define cultural interoperability as 'the approach taken by individuals and organisations to take into consideration their social and cultural differences and, if applicable, organisational cultural differences'. In this definition, cultural interoperability appears to be including both digital culture inside the organisation and the socio-cultural elements affecting the choices in digital governance configurations. Another overlooked challenge is that interoperability principles might create conflicting data governance goals and thereby broaden interoperability challenges in data governance. For instance, centralised data governance models are for the moment more effective in implementing advanced data analytics techniques and AI solutions, but decentralised systems supported by distributed ledger and blockchain technologies provide more open and transparent data governance systems.

Furthermore, technological neutrality may fail to be ensured due to the choices with information infrastructure or technological solutions (Tan *et al.*, 2022). As an example, for the moment, there are two competing interoperability solutions at the European scale for cross-border digital public services. One solution is based on the Single Digital Gateway (SDG) approach. Following the adoption of the gateway regulation in 2018, the European Commission and national administrations have started to connect national portals through Your Europe portal to provide information and services for citizens and businesses engaging in cross-border services. The SDG adopts a centralised system and, once completed, will provide a single point of entry to European citizens and businesses for public administration services such as online income tax declarations. The idea is that, through the portal, public administrations will exchange information and have access to user data. The system is expected to save companies more than €11 billion per year and boost cross-border activity.[3] The downside is that the system may contradict the increasing data sensitivities among citizens and services about the use and reuse of personal data.

Alternatively, decentralised digital ecosystems are developed by the GAIA-X and EBSI initiatives. For instance, the EBSI promotes a decentralised data governance system, where business and citizen data are stored in decentralised nodes and protected through encryption technologies that foster pseudonymity of identities. It seems as if both types of solutions are endorsed by the European Commission for the moment, and it is not clear how these ontologically different approaches to data governance at the European level will create technical interoperability for the European public services. A similar concern can be raised concerning legal interoperability.

[3] See https://ec.europa.eu/growth/single-market/single-digital-gateway_en.

Two goals, albeit not necessarily complementary, drive the legislative discussions at the European regulative space. On the one side, protection of personal data is articulated in every legislative piece concerning the use of data in public services. One the other side, how not to curb business innovation or miss the new digital revolution about AI and blockchain are issues that shape the legislative discussions.

In this context, regulative sandboxes might provide solution in the short term. For instance, The European Commission's White Paper on Artificial Intelligence (COM(2020)65) foresees such regulatory tools (European Commission, 2020). Similarly, in November 2021, the European Commission published a draft regulation regarding a pilot regime for market structures based on distributed ledger technology. The pilot regime is designed as a sandbox to enable eligible market participants to operate DLT-based multilateral trading facilities and securities settlement systems while being exempted from certain requirements under the regulatory framework that have hindered the application of DLT in the financial instruments markets. This regulatory sandbox mechanism could be a solution to develop and test different regulatory mechanisms for the use of new digital technologies and automated tools. Nevertheless, there are still many unknowns regarding the design and methodology of experimental regulations in the world of EU experimental law-making.

8.11 How to create a digital governance system?

> The main requirement in the design of organisational communication systems is not to reduce scarcity of information but to combat the glut of information, so that we may find time to attend to that information which is most relevant to our tasks-something that is possible only if we can find our way expeditiously through the morass of irrelevancies that our information systems contain. (Administrative Behaviour, 4th ed., H.A. Simone, 1997: 23)

The polymath and a giant of public administration, AI, and economics disciplines, Herbert Simone, has been already emphasising in the early ages of the Internet revolution that the main purpose of organisation communication systems should not be to attract more information but to gain insight from the abundance of information. We have already mentioned in the Introduction that data-driven technologies have expanded the capacities of public sector organisations to gain insights from various information sources. Thence, the question, how can decision-makers build a digital governance system that optimises the data collection and analysis to provide better policy insights?

A digital governance system sets the framework for a public sector organisation that clarifies the roles, accountability and decision-making mechanisms between citizens, government, and other public and private sector actors in organising their information exchange in public governance. New digital technologies present new ways and tools in extracting, analysing, and using the data to optimise this information exchange. We have already covered in depth the novel attributes of new digital technologies in detecting and effecting government functions (Chapter 1) and how different technologies can impact the processes in turning data insights into policies (Chapter 2).

However, the formerly presented ISM-model confirms that the decision to configure a digital governance system is dependent on the interrelationships between the relevant socio-cultural conditions (e.g. digital divide, digital culture, willingness to share data), existing managerial systems in digital governance (e.g. guidelines, rules, processes, strategies), technical infrastructure (e.g. security and quality of database, software, computer maturity, digital tools in data collection, storing, processing) and relevant governance mechanisms (e.g. data governance, open governance, multi-level governance, network governance). Innovation in the digital governance systems is also influenced by the perceived usefulness of new technologies in bringing an added value in comparison to existing solutions and policy objectives set by the decision-makers. A comprehensive digital governance system should consider all these dimensions, along with the underlying trust and interoperability contingencies.

Assessing the socio-cultural elements is critical to setting the appropriate digital governance mechanisms. Digital divide among social groups, digital culture inside and outside the administration, and willingness of society to share data are important aspects for digital governance systems. The e-government literature shows us socio-economic conditions, education level, age groups, gender, and prior experiences are some intervening variables on the acceptability of digital governance solutions and means. Furthermore, trust factors also affect the willingness to share personal or interpersonal data with public sector organisations. Trusting the intermediary institutions in data collection and storing, trust in the administration and system infrastructure to assure the security and fair use of data, and trust in the political actors not to exploit personal data for political gains affect the willingness of public to share data.

The socio-cultural conditions affect the means and methods in data governance. Data governance contains the rules, roles, policies, standards, and mechanisms in the collection and use of data in public governance processes. Data governance guidelines need to specify the role attributions in data collection, data storing, data curating, data protection, data control, data processing, and their integration in business processes. Furthermore, managerial processes concerning change management, risk management, and innovation management need to be specified. Administrative simplification and automation of processes are other key considerations in data governance. Elimination of unnecessary tasks, data sorting, data matching and cross-checking across institutions, purpose delineation, and streamlining are important for process automation. Automation decisions are also related to the accessibility and acceptability of automated decision-making instruments in data-driven decisions. Resistances in administration or public sentiments can affect the way the data-driven technologies are introduced in data governance systems. Automated decision-making entails further questions about the accountability mechanisms. How to ensure explainability and fairness of AI-led decisions? What are the roles of human agents in the automated processes? Do the human agents participate in decision-making processes, or do they monitor the automated system? How about bias and noise factors embedded in the data infrastructure and algorithmic decisions? We have already shared several cases and explanations about data bias throughout the book. As such, algorithmic decisions need to be carefully assessed and integrated into decision-making systems. Alternative transparency mechanisms in the use of algorithms in the decision-making processes might be necessary to increase public acceptance on algorithmic decisions. Depending on the technical infrastructure and the socio-cultural conditions, a combination of technology-driven (e.g. self-sovereign identity system, blockchains, XAI), design-driven (e.g. government as a platform), and policy-driven (e.g. open government, data watchdogs, only-once principle) solutions can be developed for digital governance systems. Furthermore, co-production, living labs, scenario testing, sandboxes, and communication campaigns are tools public sector organisations can use to assess and improve the reliability of digital governance systems.[4]

A final remark needs to be made about the policy implications of the design choices within the information infrastructure. Integration of new digital technologies can improve the security, efficiency, and effectiveness of data governance processes, but design choices often call for the assessment of the accompanying trade-off conditions. The use of strong AI vs human-centred input, digitisation vs reduction in human resources, efficiency vs transparency and explainability in algorithmic decisions, decentralisation of data infrastructure vs more complexity in data analysis, privacy enhancement technologies vs the cost of more advanced cryptographic solutions, federated vs autonomous data governance systems, the cost of investment vs the return of investment, openness vs not sharing too much information about business secrets, all are some trade-off conditions influencing the design choices with the information architecture. Due to these trade-off conditions inherent in the system infrastructure, technology neutrality or technology-agnostic approaches in digital governance might be difficult to attain. The political nature of the design choices calls for the participatory and integrative mechanisms in the design phases. Business units, data scientists, technology providers, policy makers, and data prosumers including profit and not-for-profit organisations are the natural stakeholders in the co-creation of the information infrastructures. Understanding and testing the implications of the design choices are necessary for the scalable deployment of the data-driven technologies in public governance.

8.12 How to predict the future of digital governance?

We are not only living in an era of data abundance but also with a plethora of new digital technologies and systems, seemingly up and coming every other day. Advancements in sensing, communication, processing, and actuation technologies are taking place simultaneously and concurrently (Lember *et al.*, 2019). Rampant innovation with government technologies presents alternative pathways for public managers to choose from for the digital transformation of their organisation. But the history of digital innovation suggests to us that the success is not only dependent on the innovativeness but also on the timing. For instance, the first commercialised smart phone with a touchscreen, IBM Simone, had to be discontinued within a year following its debut (Aamoth, 2014). Until the big outbreak of the iPhone in 2007, many people were not even aware of the term 'smartphone' which was already coined in 1995 (Sager, 2012). Similarly, many blockchain technologies such as zero knowledge protocols, digital tokens, and decentralised databases date back as early as the 1980s, but not until the emergence of Bitcoin in 2008 did these technological innovations make an appearance in the public debate.

---

[4] See Chapter 7 for a detailed structure.

It is also fair to ask, how 'disruptive' are the new digital technologies? In 1970, Marvin Minsky, one of the forefathers of AI research, presumed that 'from three to eight years we will have a machine with the general intelligence of an average human being.' (Live Magazine, 1970). We know that this prediction, as with many other predictions of a digital future, did not age well. The over-inflated promises of AI capabilities have led to the cancellation of research funds and shelving of AI projects in the late 1970s and 1980s (Crevier, 1993). AI research was only to make a comeback in the 2000s following the achievements of deep learning systems, the advancements in the computing powers, and the big data to support them. A similar predicament seems to be looming over the blockchain technologies. The blockchain hype of the mid-2010s has been upturned following the collapse of the DAO project and the cryptocurrency crash in 2018, both leading to the abandonment of many blockchain projects in the public sector. So how can decision-makers in public administration assess which technological innovation holds a future in public governance and is worthy of investment? In this section, we will take a closer look at how decision-makers can prepare themselves for the uncertainty surrounding the future of digital governance.

Our former analysis through the ISM-model provides some answers to this question: (1) Trust and interoperability conditions are important contingent factors to assess the plausibility, desirability, and feasibility of changes in the information systems; (2) The perceived benefits of digital technologies in the administrative systems and policy objectives drive the willingness of decisionmakers to invest in new technologies; and (3) The match of socio-cultural conditions with the characteristics of the digital technologies and their ease of adoption in the existing digital governance systems may help decision-makers on their investment decisions. Yet, all the investment decisions with new digital technologies eventually boil down to the interplay between the policy priorities of the sectoral area of application, technological maturity of the proposed digital solutions, associated resources and digital skills in the digital governance domain, comprehensiveness of the regulative framework to reduce the risks associated with new digital technologies, public values influencing the appropriateness of digital solutions, and the perceived risks associated with the implications of digital technologies for public administration and society. A careful analysis of these dynamics is embedded in reliable and pertinent digital transformation strategies.

We know so far that design choices with information management systems and the use of new digital technologies in and of themselves are politically sensitive issues. Along with increasing capabilities of data-driven technologies to manipulate, predict, and shape human behaviour, the political sensitivity around digital technologies has only been amplified. This political sensitivity is seemingly creating new political priorities and vibrancy about the role of public and private sector organisations in identifying the rights of people to control and use digital technologies. The projection of such political sensitivities can be traced back to the emergence of pirate parties in the European political spectrum. The original 'Piratpartiet' was formed in Sweden in 2006 with three main campaign points: (1) reformation of copyright law, (2) abolition of patents, and (3) respect of an individual's right to privacy from government surveillance (Khetani, 2012). In 2009, these points were formalised through the Uppsala declaration during the convention of the Pirate Parties International[5] (PPI) before the upcoming European Parliament elections. Digital rights were associated with all three policy objectives: (1) banning of digital rights management technologies and objection to media and hard levies to compensate for digital copying, (2) removal of software patents, and (3) the right to anonymity in communication that encompasses all digital communication. In the following national and European elections, pirate parties have gained notable electoral support to the surprise of many. As of the 2020s, the electoral success of pirate parties has remained marginal at local, national, and European elections, but digital policies such as net neutrality, data privacy rights, the right to be forgotten, and open content spearheaded by pirate parties have gained wider public visibility and saliency in political debates.

Political scandals about mass surveillance, destabilisation of democratic processes through data manipulation, and massive data hubs controlled by a handful of technology companies have only contributed to the rise of data politics. Yet, politics of digital governance are not only affected by data politics. Geopolitical dimensions of the control of digital technologies, both in terms of software and hardware, also play a role for their wide scale adoption by government. The control of the rare materials to produce smart phones and other appliances, the competition among states to control or intimidate technology giants (e.g. Huawei scandal), antitrust concerns (e.g. probes against Apple, Google, Facebook), orchestrated cyberwarfare attacks to information systems (e.g. 2007 cyberattacks to Estonia) are some geo-political issues concerning digital governance policies. Last but not least, energy consumption and environmental impact of digital technologies stir political sensitivities. For instance, the high energy consumption of the proof-of-work blockchains such as Bitcoin was influential on the policy decision of the European Commission to abandon them in the design choice of the EBSI blockchain (Tan *et al.*, 2022).

In another case, a mega data centre investment by Meta, the parent company of Facebook, in the Dutch Flevopolder, is expected to slurp the electricity equivalent of two times of Amsterdam's annual energy consumption and potentially derail carbon emission goals set by the European Green Deal (Stikker, 2021). The secret acquisition of lands from the Flevopolder municipality by Meta under false names, the active interventions of the Minister of Economic Affairs and Climate, and the Dutch Data Protection Association on behalf of Meta during the talks with local authorities, and the putting of pressure on the grid operator to give Meta priority for connection to the scarce energy network all appear to be just the tip of political scandals associated with the investment process. This case adds just another example to the political characteristics of investment decisions with new digital technologies.

Existing capacity conditions inside and outside of administration, and the available resources and skills for digital governance are other sets of factors shaping investment decisions. Organisations need the necessary human resources, both quantitatively and qualitatively, and the financial resources to acquire, develop, and manage data-driven technologies and systems. As mentioned

[5] The international association of pirate parties, Pirate Parties International (PPI), includes as of 2021, members from 39 countries.

in the Introduction, public sector organisations are usually disadvantaged in attracting digital talents. In the absence of digital skills, public sector organisations need to be able to acquire the external knowledge and capabilities from profit and non-for-profit organisations. This requires, however, managerial skills, awareness and foresight about the digital needs and means for digital transformation. The political sensitivity of digital technologies also calls for more soft skills involving critical thinking alongside with data literacy and the technical skills necessary to understand the risks associated with new digital technologies. A key challenge is that occupational trainings and traditional public administration training programmes rarely include training modules to supply the needs and demands of the public sector organisations about the new digital technologies. We will come to this issue later in the concluding section, but for now, the challenge for public sector organisations is not only to attract the right people but also to have access to the right IT training and education programmes. The rapid changes with technologies and associated technological skills only add to the challenge of designing specialised programmes to suit the public needs. In time, the wider accessibility of training programmes and a wider pool of experts with specialised training can facilitate wider experimentations with the new digital technologies in the public sector.

The ISM model shows that capacities of public sector organisations also shape the public values concerning digital governance. Ensuring fairness in public services delivered through the means of digital technologies, appropriate use of technologies in improving the lives of citizens, prioritising the human dignity and benefits, ensuring the privacy of data subjects, and avoiding discrimination towards service receivers are some public values subjected to organisational and institutional capacities in the use and integration of digital technologies. Human biases, lack of skills to oversee algorithmic processes, lack of understanding and foresight in the way new digital technologies improve the existing business processes, and not being able to control the data sources or technologies supplemented by tech providers are direct results of capacity gaps inside public administration. Lack of capacities increases the perceived risks associated with data-driven technologies and reduces the willingness of decisionmakers to invest in new digital technologies to improve public governance processes. Alternatively, administrative and social capacities that cultivate a culture of digitalisation are highly likely to encourage the political and administrative decision-makers to experiment with novel technological solutions.

Certainly, technological maturity is another variable in guesstimating the prospect of digital technologies in public governance. Risks associated with data biases, stability and security of technological solutions, throughput and latency of data networks, scalability of solutions, reliability of datasets, the maturity of use cases may all affect the feasibility and desirability of proposed solutions. System failures and data breaches stirred by digital solutions contain political risks that decision-makers cannot easily afford. The childcare benefit scandal in the Netherlands has been suggestive of the way the shortcomings of new digital technologies can spiral out toward government collapse. The problem is, without testing and experimenting, it is difficult to make progress on the adoption of technological solutions. Especially, disruptive technologies such as blockchain and AI require insight from use cases in the public sector. One solution to this conundrum could be implementing sandboxes and user group experiments to assess the risks associated with digital solutions. A gradual introduction of solutions without replacing the older systems can give enough data and policy insight to test the reliability of new solutions. Furthermore, benchmarks, performance indicators, and citizen surveys can provide the leverage for policymakers to initiate and upscale innovative projects with new digital technologies.

Regulations are other key factors in investment decisions. Data legislations, sector-specific regulations, and transcending legislations such as environmental laws can increase or decrease the likelihood of certain technological solutions in public governance. Public sector organisations require legislative clarity before investing time and money in new digital solutions. The Covid-19 pandemic and rapid advancements in data-driven technologies have accelerated the regulative efforts concerning data privacy, secondary use of data, AI, cryptocurrencies, and data governance in almost every regulative space. However, not all regulative approaches follow similar paths. We witness the emergence of regulative spaces with tighter and looser approaches for the use and reuse of data sources. Some countries have higher sensitivities on the human aspects of digital technologies. Other countries prioritise market dynamics as the engine of innovativeness, and are hesitant on implementing prescriptive, hard laws that may undermine technological progress. More authoritarian governments try to control data sources through state-sanctioned digital tools, state-controlled companies or outright banning of alternative digital solutions. Nonetheless, technological progress is fickle, and we do not know how effective regulative tools can be in shaping the future of digital governance. Especially, in Western countries, competing goals between international competition, democratic sensitivities, and massive gains in public welfare through digital innovation seem to lean on less prescriptive regulations in data governance. We also observe the emergence of regulative sandboxes as some safer means to test and anticipate the impact of regulations on technological progress. There are still many unknowns regarding the design and methodology of experimental regulations, but the evolution of these instruments can be a catalyst for wider testing with new digital technologies.

We expect the mixture of these elements to direct the investment decisions in digital governance. Yet, the leading institutions in the new digital era governance can vary in different politico-administrative systems. In the Introduction, we portrayed three possible digital futures where state, market, and communal actors can all hold a leading position in public governance.

In a digital new public management system (dNPM), private technology providers can delimit the pathways in digital transformation by creating vendor lock-ins in exchange for more optimised and better solutions. For instance, a memorandum of understanding signed between Apple and Georgia, Arizona, Kentucky, and Oklahoma governments in November 2021 gives a high degree of control to Apple over the government agencies responsible for issuing identification cards to use state-issued identification cards in the iPhone's Wallet app (Son and Leswing 2021). Apple requires the states to maintain the systems needed to issue and service credentials, hire project managers to respond to Apple inquiries, market the new features and push for its adoption with other government agencies. The apparent reason why states are willing to give the reins of their e-identity systems to Apple is the advantage of Apple's technology to provide more secure and convenient ways for agencies to issue credentials and cut down on fraud. The risk with vendor lock-in systems is not only associated with the cost for taxpayers or the ethical implications of the control of private sector organisations over citizenry. Another risk is that technology providers are not only providing hardware or software solutions, but they generate value by controlling vast sources of data. Even in the case of restricting the privacy of data accumulated by private service providers, the exploitation of metadata can lead to the uncompetitive market advantages over other service providers. It is not surprising to see that the digital market appears to be creating monopolies and oligopolies despite the vibrant start-up and R&D constellations.

The digital neo-Weberian state (dNWS) systems may prefer state-led digital solutions over the private sector. EU countries, Singapore, and China provide some cases where the state takes the lead in creating data crossroads, blockchain infrastructures, or single data gateways. In such setups, supranational and cross-regional infrastructure solutions may create de-facto choices in technology adoption. For instance, the choice of EBSI infrastructure for cross-border public services may condition national public administrations in Europe to transform their systems according to the governance rules and technological solutions adopted by other member states or at the supranational level.

Unlike the dNPM and dNWS, the open-source solutions and citizen-controlled systems in digital communitarianism may create a more dispersed outlook in digital public governance. In such a system, citizens, and citizen groups, may share certain responsibilities with government and private service providers as data controllers, ethical hackers, or as platform managers. The risk is that in dispersed systems, governance and coordination are always a challenge. New digital technologies may create certain remedies for coordination problems, but we need further testing and experimenting to ensure that citizen-led digital governance solutions lead to improved governance outcomes.

8.13 A new research agenda for the new era of digital governance

Our central claim in this book has been that the new digital era governance requires a new way of thinking about our approaches, theoretical models, and the practices surrounding the role of digital technologies in public governance. The new digital era governance is centred around data exploitation strategies and prioritises decisions about governance of and by data-driven technologies. The changes with digital technologies call for rethinking the roles, capacities, and skills within the public administration. Not only administrations, but the spheres of politics, market, and society in public governance are shifting at the helm of the digital transformation. The future of public governance seems to be heading towards new social, market, and political spaces defined by technological capabilities and imaginations of system architects. Witnessing the rapid digital transformation during a global pandemic in the way we work, meet, and socialise with each other, when we turn our gaze to the future, we see metaverses, decentralised autonomous organisations, and autonomous agents shaping the public governance domain.

At this juncture, what can the scholars and professionals of public administration and governance do to prepare themselves and public sector organisations for such a digital prospect? We array below some topics and questions that we expect to gain more prominence in the research agenda of digital governance.

- What are the conditions under which algorithms, AI systems, and big data-driven solutions are considered as trustworthy by citizens?
- Under what conditions are automated decision-making and implementation in public governance acceptable?
- Which practices are effective for avoiding bias, discrimination, and unfairness in AI-induced public services?
- What types of incentives should be established to data sharing by citizens and developing open-source digital solutions by technology providers? How can these incentives be integrated in public procurement processes?
- What are the leading practices in digital governance that provide improved public service outcomes that comply with ethical and trust standards?
- What type of skillsets and training programmes are needed to build capacities for AI and blockchain solutions?
- How do the regulatory landscapes affect the implementation of new digital technologies in public administrations?
- What do accountability and legitimacy mean in a code-bound governance system?
- What should be the roles of automated agents in public policy and administrative processes?
- How do we accommodate old and new digital solutions in public governance without leaving behind some parts of society? How do we address digital divide and variances in digital cultures across countries and regions?
- To what extent are the existing modes of governance (e.g. hierarchy, network, market) transferable in the newly emerging cyber and techno-physical spaces to coordinate public actions? Can new digital technologies create new modes of governance?
- How can public administration address emerging digital identities and ethical concerns about personal data?
- What centralised and decentralised solutions are acceptable in different sectoral domains and contextual circumstances?
- Which factors affect the design choices of decision-makers in digital public governance systems?

Addressing these questions require cross-disciplinary insights and development of lexicons that are interpretable across multiple disciplines such as computer sciences, information sciences, design sciences, and social sciences. We need crossbreeding of concepts, research methods, and analytical frameworks across these disciplines. We need interdisciplinary collaborations, testing and experimenting of technological solutions within different user groups, developing scaling up strategies for different governance contexts, and integrate these research insights into new interdisciplinary modules in public administration curricula to train a new generation of researchers and practitioners in digital governance.

References

Aamoth, D. 2014. *First smartphone turns 20: fun facts about Simon*. TIMES. August, 15. Available at: https://time.com/3137005/first-smartphone-ibm-simon/.

Abdul, A., J. Vermeulen, D. Wang, B.Y. Lim, and M. Kankanhalli. 2018. Trends and trajectories for explainable, accountable and intelligible systems: an HCI research agenda. In *Proceedings of the 2018 CHI conference*, pp. 1-18.

Adadi, A. and M. Berrada. 2018. Peeking inside the black-box: a survey on explainable artificial intelligence (xai). *IEEE Access* 6: 52138-52160, https://doi.org/10.1109/ACCESS.2018.2870052

Ahmad, M., X.-W. Tang, J.-N. Qiu, and F. Ahmad. 2019. Interpretive structural modeling and MICMAC analysis for identifying and benchmarking significant factors of seismic soil liquefaction. *Applied Sciences* 9 (2): 233. https://doi.org/10.3390/app9020233

Ahonen, P., and T. Erkkilä. 2020. Transparency in algorithmic decision-making: ideational tensions and conceptual shifts in Finland. *Information Polity* 25: 419-432. https://doi.org/10.3233/IP-200259.

Ajzen, I. 1991. The theory of planned behaviour. *Organizational Behaviour and Human Decision Processes* 50(2): 179-211.

Andročec, D. 2017. Overcoming service-level interoperability challenges of the IoT. In Z. Mahmood (ed.) *Connected environments for the internet of things*. Computer Communications and Networks. Springer, Cham, Switzerland. https://doi.org/10.1007/978-3-319-70102-8_5

Barredo Arrieta, A., N. Díaz-Rodríguez, J. Del Ser, A. Bennetot, S. Tabik, A. Barbado, S. Garcia, S. Gil-Lopez, D. Molina, R. Benjamins, R. Chatila, and F. Herrera. 2020. Explainable artificial intelligence (XAI): concepts, taxonomies, opportunities and challenges toward responsible AI. *Information Fusion* 58: 82-115. https://doi.org/10.1016/j.inffus.2019.12.012.

Benbasat, I. and W. Wang. 2005. Trust in and adoption of online recommendation agents. *Journal of the Association for Information Systems* 6(3):00065 https://doi.org/10.17705/1jais.00065

Berkovsky, S., R. Taib, and D. Conway. 2017. How to recommend? User trust factors in movie recommender systems. In *Proceedings of the 22nd International Conference on Intelligent User Interfaces (IUI '17)*. 287-300. Association for Computing Machinery, New York, NY, USA. https://doi.org/10.1145/3025171.3025209

Berkovsky, S., R. Taib, I. Koprinska, E. Wang, Y. Zeng, J. Li, and S. Kleitman. 2019. Detecting personality traits using eye-tracking data. *Proceedings of the 2019 CHI Conference on Human Factors in Computing Systems*. Paper 221, 1-12. Association for Computing Machinery, New York, NY, USA. https://doi.org/10.1145/3290605.3300451

Brendel A.B., M. Mirbabaie, T.-B. Lembcke, and L. Hofeditz. 2021. Ethical management of artificial intelligence. *Sustainability* 13(4): 1974. https://doi.org/10.3390/su13041974

Brutti, A., P. De Sabbata, A. Frascella, N. Gessa, R. Ianniello, C. Novelli, S. Pizzuti, and G. Ponti. 2019. Smart city platform specification: a modular approach to achieve interoperability in smart cities. In F. Cicirelli, A. Guerrieri, C. Mastroianni, G. Spezzano, and A. Vinci (eds.) *The internet of things for smart urban ecosystems*. Internet of Things (Technology, Communications and Computing). Springer, Cham, Switzerland. https://doi.org/10.1007/978-3-319-96550-5_2

Bunt, A., M. Lount, and C. Lauzon. 2012. Are explanations always important? a study of deployed, low-cost intelligent interactive systems. In *Proceedings of the 2012 ACM IUI conference*, pp. 169-178.

Busuioc, M. 2021. Accountable artificial intelligence: holding algorithms to account. *Public Administration Review* 81: 825-836. https://doi.org/10.1111/puar.13293

Casiano Flores, C., M. Chantillon, A.P. Rodriguez Müller, and J. Crompvoets. 2021. A vision to the future recommendations for the 2017 European Interoperability Framework. *Revista Cubana De Transformación Digital* 2(3): 31-45. https://doi.org/10.5281/zenodo.5545955

Chantillon, M., C. Casiano Flores, J. Crompvoets, M. Sallamo, M.E. Antunes, M.G. Barron, J. Barroca, P. Vicente, A.V. Raposo, and G. Sidique. 2021a. *Final study report – proposal for a European interoperability framework for smart cities and communities* (EIF4SCC). Publications office of the European Union, Luxembourg, Luxembourg. https://doi.org/10.2799/085469

Chantillon, M., C. Casiano Flores, J. Crompvoets, M. Sallamo, M.E. Antunes, M.G. Barron, J. Barroca, P. Vicente, A.V. Raposo, and G. Sidique. 2021b. *Proposal for a European interoperability framework for smart cities and communities* (EIF4SCC). Publications office of the European Union, Luxembourg, https://doi.org/10.2799/816559

Chen, S,J. Li, G. Andrienko, N. Andrienko, Y. Wang, P.H Nguyen, and C. Turkay. 2018. Supporting story synthesis: bridging the gap between visual analytics and storytelling. *IEEE Transactions on Visualization and Computer Graphics* 26(7): 2499-2516. https://doi.org/10.1109/TVCG.2018.2889054

Chopra K., and W.A. Wallace. 2003. Trust in electronic environments. *Proceedings of the 36th Annual Hawaii International Conference on System Sciences*, pp. 10, https://doi.org/10.1109/HICSS.2003.1174902

Compeau, D., and C. Higgins, 1995. Computer self-efficacy: development of a measure and initial test. *MIS Quarterly* 19: 189-211. https://doi.org/10.2307/249688

Conati, C., S. Lallé, M.A. Rahman, and D. Toker. 2017. Further results on predicting cognitive abilities for adaptive visualizations. In *Proceedings of the 26th International Joint Conference on Artificial Intelligence* (IJCAI'17). AAAI Press, 1568-1574.

Cramer, H., V. Evers, S. Ramlal, M. van Someren, L. Rutledge, N. Stash, L. Aroyo, and B. Wielinga. 2008. The effects of transparency on trust in and acceptance of a content-based art recommender. *User Modeling and User-Adapted Interaction* 18: 455. https://doi.org/10.1007/s11257-008-9051-3

Crevier, D. 1993. *AI: the tumultuous search for artificial intelligence*. BasicBooks, New York, NY, USA.

Davis, F.D., R.P. Bagozzi, and P.R. Warshaw. 1992. Extrinsic and intrinsic motivation to use computers in the workspace. *Journal of Applied Social Psychology* 22(14): 1111-1132.

Davis, F.D. 1989. Perceived usefulness, perceived ease of use, and user acceptance of information technology. *MIS Quarterly* 13 (3): 319-340. https://doi.org/10.2307/249008

De Roux, D., B. Perez, A. Moreno, M. del Pilar Villamil, and C. Figueroa. 2018. Tax fraud detection for under-reporting declarations using an unsupervised machine learning approach. *Proceedings of the 24th ACM SIGKDD International Conference on Knowledge Discovery & Data Mining* (KDD '18), pp. 215-222. https://doi.org/10.1145/3219819.3219878

Došilović, F.K., M. Brčić, and N. Hlupić. 2018. Explainable artificial intelligence: a survey. In *41st International convention on information and communication technology, electronics and microelectronics* (MIPRO). IEEE Xplore. https://doi.org/10.23919/MIPRO.2018.8400040

Dwivedi, Y.K., N.P. Rana, M. Janssen, B. Lal, M.D. Williams, and M. Clement. 2017. An empirical validation of a unified model of electronic government adoption (UMEGA). *Government Information Quarterly* 34(2): 211-230.

European Commission. 2017. The new European interoperability framework. Promoting seamless services and data flows for European public administrations. Publications Office of the European Union, Luxembourg, Luxembourg. https://doi.org/10.2799/78681

European Commission. 2020. WHITE PAPER On Artificial Intelligence – A European approach to excellence and trust. COM/2020/65 final. Available at: https://eur-lex.europa.eu/legal-content/EN/TXT/?uri=COM:2020:65:FIN

Fathema, N., D. Shannon, and M. Ross. 2015. Expanding the technology acceptance model (TAM) to examine faculty use of learning management systems (LMS). *Journal of Online Learning and Teaching* 11 (2): 210-233.

Fishbein, M., and I. Ajzen. 1975. *Belief, attitude, intention and behaviour: an introduction to theory and research*. Addison-Wesley, Reading, MA, USA.

Fox, S., K. Karnawat, M. Mydland, S. Dumais, and T. White. 2005. Evaluating implicit measures to improve web search. *ACM Transactions on Information Systems* 23(2): 147-168. https://doi.org/10.1145/1059981.1059982

Gedikli, F., D. Jannach, and M. Ge. 2014. How should I explain? A comparison of different explanation types for recommender systems. *International Journal of Human-Computer Studies* 72(4): 367-382.

Gilpin, L.H., D. Bau, B.Z. Yuan, A. Bajwa, M. Specter, and L. Kagal. 2018. Explaining explanations: an overview of interpretability of machine learning. In *2018 IEEE 5th International Conference on Data Science and Advanced Analytics (DSAA)*. IEEE Xplore 80-89. https://doi.org/10.1109/DSAA.2018.00018

Grandison T., and M. Sloman. 2000. A survey of trust in internet applications. *IEEE Communications Surveys & Tutorials* 3(4): 2-16. https://doi.org/10.1109/COMST.2000.5340804

Herlocker, J.L., J.A. Konstan, and J. Riedl. 2000. Explaining collaborative filtering recommendations. *Proceedings of the ACM Conference on Computer Supported Cooperative Work*. 241-250. https://doi.org/10.1145/358916.358995

Holliday, D.S. Wilson, and S. Stumpf. 2016. User trust in intelligent systems: a journey over time. In *Proceedings of the 21st International Conference on Intelligent User Interfaces* (IUI '16). Association for Computing Machinery, New York, NY, USA, pp. 164-168. https://doi.org/10.1145/2856767.2856811

Hoppe, S., T. Loetscher, S.A. Morey, and A. Bulling. 2018. Eye movements during everyday behavior predict personality traits. *Frontiers in Human Neuroscience* 12: 105. https://doi.org/10.3389/fnhum.2018.00105

Hughes, D.L., N.P. Rana, and Y.K. Dwivedi. 2020. Elucidation of IS project success factors: an interpretive structural modelling approach. *Annals of Operations Research* 285: 35-66. https://doi.org/10.1007/s10479-019-03146-w

Islam, S.R, W. Eberle, and S.K. Ghafoor. 2020. Towards quantification of explainability in explainable artificial intelligence methods. *The Thirty-Third International FLAIRS Conference* (FLAIRS-33). https://aaai.org/ocs/index.php/FLAIRS/FLAIRS20/paper/view/18410/17523

Janssen, M., S. Luthra, S. Mangla, N.P. Rana, and Y.K. Dwivedi. 2019. Challenges for adopting and implementing IoT in smart cities. *Internet Research* 29(6): 1589-1616.

Janssen, M., N.P. Rana, E.L. Slade, and Y.K. Dwivedi. 2017. Trustworthiness of digital government services: deriving a comprehensive theory through interpretive structural modelling, *Public Management Review* 20(5): 647-671. https://doi.org/10.1080/14719037.2017.1305689

Jin, Y., N. Tintarev, N.N. Htun, and K. Verbert. 2019. Effects of personal characteristics in control-oriented user interfaces for music recommender systems. *User Modeling and User-Adapted Interaction* 30: 199-249. https://doi.org/10.1007/s11257-019-09247-2

Khetani, S. 2012. A short history of the pirate party, who just clobbered Merkel's allies in German elections. Business Insider, April, 5th. Available at: https://www.businessinsider.com/pirate-party-germany-elections-2012-4?r=US&IR=T#the-original-pirate-party-was-formed-in-sweden-in-2006-1

Lal, R., and A. Haleem. 2009. A structural modelling for e-governance service delivery in rural India. *International Journal of Electronic Governance* 2(1): 3-21. https://doi.org/10.1504/IJEG.2009.024962

Lember, V., T. Brandsen, and P. Tõnurist. 2019. The potential impacts of digital technologies on co-production and co-creation. *Public Management Review* 21(11): 1665-1686. https://doi.org/10.1080/14719037.2019.1619807

Lipton, Z.C. 2018. The mythos of model interpretability: in machine learning, the concept of interpretability is both important and slippery. *Queue* 16(3): 31-57. https://doi.org/10.1145/3236386.3241340

Life Magazine. 1970. Interview with Marvin Minsky. *Life Magazine*, Nov 20th: 68.

McDonald, T., and M. Siegall. 1992. The effects of technological self-efficacy and job focus on job performance, attitudes, and withdrawal behaviours. *Journal of Psychology* 126: 465-475. https://doi.org/10.1080/00223980.1992.10543380

McKnight, D.H., V. Choudhury, and C. Kacmar. 2002. Developing and validating trust measures for e-commerce: an integrative typology. *Information Systems Research* 13(3): 334-359. https://doi.org/10.1287/isre.13.3.334.81

McNee, S.M., S.K. Lam, J.A. Konstan, and J. Riedl. 2003. Interfaces for eliciting new user preferences in recommender systems. In *Lecture Notes in Artificial Intelligence* (Subseries of Lecture Notes in Computer Science), 2702, 178-187. https://doi.org/10.1007/3-540-44963-9_24

Maciejewski, M. 2017. To do more, better, faster and more cheaply: using big data in public administration. *International Review of Administrative Sciences* 83(1): 120-135. https://doi.org/10.1177/0020852316640058.

Madsen, M., and S. Gregor. 2000. Measuring human-computer trust. *Proceedings of the 11th Australasian Conference on Information Systems*. Available at: http://citeseerx.ist.psu.edu/viewdoc/summary?doi=10.1.1.93.3874

Mensah, I.K., G. Zeng, and C. Luo. 2020. E-government services adoption: an extension of the unified model of electronic government adoption. *SAGE Open* 10: 2158244020933593. https://doi.org/10.1177/2158244020933593

Millecamp, M., S. Naveed, K. Verbeert, and J. Ziegler. 2019. To explain or not to explain: the effects of personal characteristics when explaining feature-based recommendations in different domains. In *IntRS '19: Joint Workshop on Interfaces and Human Decision Making for Recommender Systems*, 19 Sept 2019, Copenhagen, DK. 2450, 10-18. Available at: http://ceur-ws.org/Vol-2450/paper2.pdf.

Moore, G.C., and I. Benbasat. 1996. Integrating diffusion of innovations and theory of reasoned action models to predict utilization of information technology by end-users. In J. Kautz, and K. Pries-Hege (eds.) *Diffusion and adoption of information technology.* Chapman and Hall, London, pp. 132-146.

Murdoch, W.J., C. Singh, K. Kumbier, R. Abbasi-Asl, and B. Yu, 2019. Interpretable machine learning: definitions, methods, and applications. arXiv preprint. *arXiv*: 1901.04592.

Naveen, J. 2020. How AI and robotics can change taxation. *Forbes*, 9 January 2020. Available at: https://www.forbes.com/sites/cognitiveworld/2020/01/09/how-ai-and-robotics-can-change-taxation/

Nourani, M., J.T. King, and E.D. Ragan. 2020. The role of domain expertise in user trust and the impact of first impressions with intelligent systems. *Proceedings of the AAAI Conference on Human Computation and Crowdsourcing* (HCOMP) 2020. Available at: https://ojs.aaai.org/index.php/HCOMP/article/download/7469/7276/10871

Organisation for Economic Co-operation and Development (OECD). 2021. *Tax administration 2021: comparative information on OECD and other advanced and emerging economies.* OECD Publishing. Paris, France. https://doi.org/10.1787/cef472b9-en

Parasuraman, R., T.B. Sheridan, and C.D. Wickens. 2000. A model for types and levels of human interaction with automation. *IEEE Transactions on Systems, Man, and Cybernetics – Part A: Systems and Humans* 30(3): 286-297. https://doi.org/10.1109/3468.844354

Pu, P. and L. Chen. 2006. Trust building with explanation interfaces. In *Proceedings of the 11th international conference on Intelligent user interfaces* (IUI '06). Association for Computing Machinery, New York, NY, USA. 93-100. https://doi.org/10.1145/1111449.1111475

Rana, N.P., Y.K. Dwivedi, B. Lal, M.D. Williams, and M. Clement. 2017. Citizens' adoption of an electronic government system: towards a unified view. *Information Systems Frontiers* 19(3): 549-568.

Ribeiro, M.T., S. Singh, and C. Guestrin. 2016. Why should I trust you? Explaining the predictions of any classifier. *Proceedings of the 22nd ACM SIGKDD international conference on knowledge discovery and data mining*. 1135-1144. https://doi.org/10.1145/2939672.2939778

Sager, I. 2012. Before IPhone and Android came Simon, the first smartphone. *Bloomberg Businessweek*. June 29. Available at: http://www.businessweek.com/articles/2012-06-29/before-iphone-and-android-came-simon-the-first-smartphone.

Shin, D. 2020. User perceptions of algorithmic decisions in the personalized AI system:perceptual evaluation of fairness, accountability, transparency, and explainability. *Journal of Broadcasting & Electronic Media* 64 (4): 541-565. https://doi.org/10.1080/08838151.2020.1843357

Simone, H.A. 1997. *Administrative behaviour: a study of decision making processes in administrative organization*. 4th ed. The Free Press, New York, NY, USA.

Stikker, M. 2021. Komst datacenters verloopt ondoorzichtig en ondemocratisch. *Het Financiële Dagblad*, December, 10th. Available at: https://fd.nl/opinie/1422605/komst-datacenters-verloopt-ondoorzichtig-en-ondemocratisch-ijl1aaTZWxJL.

Son, H., and K. Leswing. 2021. Apple is sticking taxpayers with part of the bill for rollout of tech giant's digital ID card. *CNBC*, November, 14th. Available at: https://www.cnbc.com/2021/11/14/apple-sticking-taxpayers-with-part-of-the-bill-for-digital-id-rollout.html.

Tan, E., S. Mahula, and J. Crompvoets. 2022. Blockchain governance in the public sector: a conceptual model for public management. *Government Information Quarterly* 39(1): 101625. https://doi.org/10.1016/j.giq.2021.101625

Taylor, S., and P. Todd. 1995. Understanding information technology usage: a test of competing models. *Information Systems Research* 6(4): 144-176.

Thompson, R.L., C.A. Higgins, and J.M. Howell. 1991. Personal computing: toward a conceptual model of utilization. *MIS Quarterly* 15(1): 124-143.

Tintarev N., and J. Masthoff. 2011. Designing and evaluating explanations for recommender systems. In F. Ricci, L. Rokach, B. Shapira, and P. Kantor (eds.) *Recommender systems handbook*. Springer, Boston, MA, USA. https://doi.org/10.1007/978-0-387-85820-3_15

Tjoa, E. and C. Guan. 2021. A survey on explainable artificial intelligence (XAI): toward medical XAI. *IEEE Transactions on Neural Networks and Learning Systems* 32(11): 4793-4813. https://doi.org/10.1109/TNNLS.2020.3027314.

Van de Walle, S. and G. Bouckaert. 2003. Public service performance and trust in government: the problem of causality. *International Journal of Public Administration* 26(8-9): 891-913. https://doi.org/10.1081/PAD-120019352

Van Vlasselaer, V., T. Eliassi-Rad, L. Akoglu, M. Snoeck, and B. Baesens. 2017. Gotcha! network-based fraud detection for social security fraud. *Management Science* 63(9): 3090-3110.

Venkatesh, V., J. Thong, and X. Xu. 2012. Consumer acceptance and user of information technology: extending the unified theory of acceptance and use of technology. *MIS Quarterly* 36(1): 157-178.

Venkatesh, V., M.G. Morris, G.B. Davis, and F.D. Davis. 2003. User acceptance of information technology: toward a unified view. *MIS Quarterly* 27(3): 425-478. https://doi.org/10.2307/30036540

Venkatesh, V., and F.D. Davis. 2000. A theoretical extension of the technology acceptance model: four longitudinal field studies. *Management Science* 46(2): 186-204. https://doi.org/10.1287/mnsc.46.2.186.11926

Verkijika, S.F., and L. De Wet. 2018. E-government adoption in sub-Saharan Africa. *Electronic Commerce Research and Applications* 30: 83-93.

Warfield, J.N. 1974. Developing interconnection matrices in structural modelling. *IEEE Transactions on Systems, Man, & Cybernetics* (1): 81-87.

Yu, K., S. Berkovsky, R. Taib, D. Conway, J. Zhou, and F. Chen. 2017. User trust dynamics: an investigation driven by differences in system performance. In *Proceedings of the 22nd International Conference on Intelligent User Interfaces* (IUI '17). Association for Computing Machinery, New York, NY, USA. 307-317. https://doi.org/10.1145/3025171.3025219

Zhou, J., S.Z. Arshad, K. Yu, and F. Chen. 2016. Correlation for user confidence in predictive decision making. *OzCHI '16: Proceedings of the 28th Australian Conference on Computer-Human Interaction*. 252-256. https://doi.org/10.1145/3010915.3011004

Zhou, J., Z. Li, H. Hu, K. Yu, F. Chen, Z. Li, and Y. Wang. 2019. Effects of influence on user trust in predictive decision making. *CHI EA '19: Extended Abstracts of the 2019 CHI Conference on Human Factors in Computing Systems, May 2019*. LBW2812, 1-6. https://doi.org/10.1145/3290607.3312962

Appendix 8.A. Codification list

Constructs	Elements	Topics[1]
Regulations	Data	GDPR, national laws, data combination, data anonymisation (difficulty of doing, and viable use of it)
	Taxation/social security	national laws, EU laws, sanctions
	Transcending laws	environmental laws
	Justification of decisions	the challenge of explainability with deep learning
Level of trust	Trust in administration	tax authorities, inability to develop in-house solutions, regulative agencies, trust in centralised entity vs the necessity of blockchain
	Trust in society	no adherence to tax, easy acceptance of new technologies (not critical thinking), lack of skills regarding new technologies
	Trust in technology	Cloud systems, social acceptability, algorithms, blockchain, political distrust, digitalisation improves trusts, social distrust in new digital technologies (related to fake news, conspiracy theories, anti-science speeches)
	Trust in system	Surveillance capitalism, undeclared purposes (e.g. commercial, political)
	Trust in tech providers/ private sector	implementation of ethical standards, GAFAM (big tech), higher efficiency and expertise than public sector, not adapted to local challenges (better in-house), distrust of private sector in civil servants, distrust of banks to share data
Management/ operational systems	Guidelines	Legal guidelines, a precise framework to declare fraud suspicion (for banks, casinos, etc.)
	Rules & standards	data confidentiality, ethical rules and standards, lack of standards to assess overall quality of taxation system (e.g. ISO)
	Principles	only-once, trade-offs (weak/strong AI vs ethics, digitalisation vs reduction in human resources, efficiency vs transparency, cost reduction vs quality of services, decentralisation of data vs difficulty of analysis, privacy vs the cost of more advanced cryptographic solutions, cost of investment vs return of investment, openness vs not sharing too much information about business), lack of flexibility in project requirements

	Processes	risk assessment, human in/on the loop, safeguards against AI bias, AI-led data management, streamlining, process automation, reducing administrative burden, administrative simplification (e.g. public procurement), automation of rights, elimination of unnecessary tasks, reporting obligations for companies, treating big companies same as small companies, follow-up after the detection of tax fraud, data sorting, data protection, very long specific investigations vs agile way of working (thanks to data insights), data matching and cross-checking across institutions, purpose delineation, making proportionate cross-checks, administrative burden on authorisation of projects (delays timely detection of new types of fraud), public registry for used algorithms (e.g. Amsterdam algorithm register)
	Strategies	data strategy, 5-10 year digital investment plans, a central data management political vision, a common data collection and exploitation strategy, horizontal control, use of AI and data on most problematic cases, big fish vs small fish, incentives to entrepreneurs, pre-filling of tax returns, specificities of small and medium-sized enterprises, improving targeting of controls (focus on fraudsters leave others alone), early detection of frauds, creation of technology watch groups in administrations (to improve anticipation), subscription-based IT systems, new business models for media to inform society, operational strategy on data literacy, starting small with a few tenants on big data and AI solutions and incremental increase in time with trusted tenants
Perceived risk	Legal challenge	protection of citizen rights, implications to tax lawyers, rules on data confidentiality, evolution of identity
	Control of data	tech vs government
	Democratic challenge	authoritarian drift, development of a public GAFAM to preserve democracy and equality, avoiding Chinese system
	Administrative challenges	disrupting administrative practices and culture, the need of reinventing itself, difficult to keep public services relevant, budget cuts and personal reduction, will human inspectors have time to verify AI's decisions

	Societal challenges	job loss, decrease in working time, disruption in labour market (rapidly changing tech skills), environmental impact, increasing digital divide
Governance system	Data governance	sensitive data, reliability of databases (e.g. updating, info about foreign workers), secure networks to exchange data with mutuality, collaboration between data centres, exchanges between databases (e.g. social security data and financial data), siloed organisations, automatic collection of transaction data at transaction level, data sharing with banks, access and control of data, inclusivity, frequency of checks on data quality, internal audit processes (e.g. data anonymisation)
	Open governance	open data, open sources, better control for experienced individuals and groups, giving citizens access, reduced control and access to software, access to data by the private sector, use of open data to better understand the tool, user friendliness of the interface
	Multi-level governance	global data platform for tax authorities, international coordination at the EU level, social dumping, collaboration between Belgian and foreign supervisory bodies, administrative fragmentation (local-regional-federal), institutional complexity, EU as a normative actor (e.g. Green Deal, free movement of workers), EU as a coordinator
	Network governance	institutional rivalry, coordination between services, collaboration with companies and citizens, expertise centres, collaboration with other public actors (police, social security, justice), understanding the challenges of entrepreneurs, need for listening to many actors (including private actors and NGOs), control of compliance, finding innovation champions
Technical infrastructure	Security	data security, network security, data privacy
	Quality of database	social security data about foreign workers, up-to-dateness of data
	Data collection & analysis	Use of AI with other tools (e.g. IoT), blockchain/DLT, data hubs (across sector, across countries), automation, predictive analysis, setting algorithms properly (parameter choices, big vs small fish), data mining, sensitivity of data, classification of data, probabilistic vs deterministic models, open data, big data platform

	Software	subscription-based IT systems, supply model of software, open vs proprietary software, entanglement with obsolete legacy software
	Computer maturity	
	Reliance/dependence on external actors	GAFAM, SAS, in-house solution vs outsourcing
Public values	Appropriateness of technology	why we use AI, avoiding discrimination, asking ethical/moral questions, avoid nervousness of an association with 'Big Brother'.
	Respecting privacy	profiling fraudsters
	Tax fairness	
Technological maturity	Bias and noise	risk of bias with AI algorithm, misinterpretation, possible errors in probabilistic AI, systemic biases due to algorithm, bias in training data, using too many algorithms
	Technology convergence	blockchain and AI, AI and IoT, machine learning for risk assessment of machine learning, BCT as a means of transparency and trust
	Blockchain/DLT	not reached to maturity, disruptive potential, difficult to find good use cases, GDPR is an obstacle
	AI/machine learning	absence of AI use cases that are effective as current data-matching operations
	Fraud detection technologies	fraud analytics, predictive models, proactive/anticipatory use of machine learning, tools for better targeting and sorting data, nowcasting tools
Interoperability	Technical interoperability	lack of harmonisation of IT systems in administration, compatibility of systems and means of data exchange between administrations
	Semantic interoperability	homogenisation of data, lack of standardisation to ensure data quality, cross-border differences in the use of data and metadata (e.g. easier to do with some countries (the Netherlands) than others)
	Organisational interoperability	coordination between services, a harmonised social security database at the European level, aligning data strategies of multiple actors involved in social security, different applications used by insurance companies (a need for a common application for coordination with the RIZIV (Belgian National institute for health and disability insurance)
	Regulative interoperability	data exchange between EU countries about workers, EU as a coordinator (e.g. European Labour Authority)

Capacities, skills and competencies	Resources	human resources (not enough manpower to process data, growing need for IT skills), financial resources, lack of resources at SMEs for digital solutions/sharing faster and better data with tax authorities
	Digital skills	knowledge of algorithms and BCT, data management, critical thinking, data literacy, data storage and sharing, lack of digital skills in society, lack of digital skills in unions, lack of digital skills for new technologies in companies/SMEs
	Training	problem of initial training of civil servants, lack of competence and poor training in tax administration, technological developments, foresight on technological changes, access to appropriate IT training, rapid change of technological skills, development of civic and professional skills in public, inclusivity of training (including migrants), training citizens about data protection, information campaigns for citizens and public agents
Policy priorities	EU-level policy priorities	Competition with US, and Asian tech providers (EU wants to lead in producing data standards because it missed out digital transition), Green Deal, digital innovation
	Fight against fraud	Higher priority of fighting tax fraud against social fraud, targeting small business vs big business, too much time and money in fighting social fraud less effectiveness, ineffectiveness of tax policies against large players, finding a balance between fight against fraud and protection of individual liberties, more targeted and proportional measures against tax fraud, use of data and AI to select more problematic cases, government wants total transparency on assets for tax justice, but going other way, the need for a clear vision and consistency in targeting big fraudsters in specific sectors, impression of hunt for money than fraud
	Political support	Need for political support for data sharing, budget allocation, short-termism
	Geo-political aspects	Rare metals and potential conflicts, impact of technology on environment
Perceived usefulness	Automation	to free up some working time for other tasks, improves efficiency, elimination of unnecessary tasks, reduces administrative costs and pushes for simplification, it saves money and brain power that can be used to better look at data and make better decision

	Social security and taxation	for the public sector, for enterprises, for the funding of social security, citizens do not need to go to administration, gain speed and efficiency to target and control problematic companies; better target controls, data matching helps to thwart the creativity of fraudsters and allows to detect low-level social fraud that is generally difficult to identify, not only detecting undue payments but ensuring people receive everything they need on time
	Data collection & analysis	the opportunity to automate and reduce the administrative burden while improving services by anticipating problems (e.g. in health insurance companies), solving problems not able to be solved before; anticipating future problems, understanding better the reasons behind successful practices; better results with fewer and better qualified people, making more effective policies; new IT tools accelerate data analysis while reducing the risk of committing mistakes
	Past experiences	CBSS improved results on fight against social fraud, pre-filling of tax returns, usefulness of relational business data (SQL) rather than big data for fraud detection (ONSS), PoCs of predictive analysis with anonymised data were useful for inspectors, and that there was an enormous gain in efficiency and accuracy in terms of case selection and they wielded larger results
	Indirect added value of new digital technologies	Better predictive medicine impacts social security, fighting against unfair competition, making public authorities legitimate and relevant
Socio-cultural elements	Digital culture	Lack of digital and data culture in administration, lack of digital culture in society, different digital mindset between Wallonia and Flanders (data protection vs innovation with SMEs)
	Digital divide	Generational gap, socio-economic gap, gap between SMEs, digitalisation of society, access to Internet
	Willingness to share data	

[1] AI = artificial intelligence; BCT = block chain technology; CBSS = Crossroads Bank for Social Security; DLT = distributed ledger technology; EU = European Union; GAFAM = Google, Amazon, Meta (Facebook), Apple and Microsoft; GDPR = General Data Protection Regulation; IoT = Internet of Things; NGO = non-governmental organisation; ONSS = Office Nationale de Sécurité Sociale; PoC = proof of concept; SAS = SAS Institute; SME = small and medium sized enterprises.

Index

E

L

M

N

O

P